# Handbook of Environmental Pollution and Control

# Handbook of Environmental Pollution and Control

Edited by **Raven Brennan**

**SYRAWOOD**
PUBLISHING HOUSE

New York

Published by Syrawood Publishing House,
750 Third Avenue, 9th Floor,
New York, NY 10017, USA
www.syrawoodpublishinghouse.com

**Handbook of Environmental Pollution and Control**
Edited by Raven Brennan

International Standard Book Number: 978-1-68286-092-2 (Hardback)

Printed in the United States of America.

# Contents

# Preface

This book aims to highlight the current researches and provides a platform to further the scope of innovations in this area. This book is a product of the combined efforts of many researchers and scientists from different parts of the world. The objective of this book is to provide the readers with the latest information in the field.

Urban development, emerging consumption trends and reckless industrialization are some of the factors behind increasing environmental pollution. This book traces the progress of this field and highlights some of its key concepts such as greenhouse gas emissions, chemical toxics and pollutants, solid waste management and different control measures, etc. The various studies that are constantly contributing towards advancing technologies to work for the betterment of environmental health are examined in detail. It aims to equip students and experts with the advanced topics and upcoming concepts in this area.

I would like to express my sincere thanks to the authors for their dedicated efforts in the completion of this book. I acknowledge the efforts of the publisher for providing constant support. Lastly, I would like to thank my family for their support in all academic endeavors.

**Editor**

# Characterization of industrial waste and identification of potential micro-organism degrading tributyl phosphate

**Trupti D. Chaudhari[3], Susan Eapen[2] and M. H. Fulekar[1]***

[1]Department of Life Sciences, University of Mumbai, Santacruz (E), Mumbai-400 098, India.
[2]Nuclear Agriculture Biotechnology Division, Bhabha Atomic Research centre, Trombay-400 085 Mumbai, India.
[3]Research Scholar, Environmental Biotechnology Laboratory, Department of Life Sciences, University of Mumbai, Santacruz (E), Mumbai-400 098, India.

**The present research study has been carried out in the waste disposal site for characterization of physical, chemical and biological parameters to assess the microbial consortium present in the contaminated site and to isolate the potential micro-organism for biodegradation of Tributyl Phosphate. The ambient conditions present in the contaminated site shows the values: pH (6.61), Temperature (35.6), Moisture (50.72%), Nutrients; Nitrogen (0.41%), Phosphorus (27.87 mg/l), and Sulphur (993.5 mg/l) respectively. The biological parameters studied indicate Dissolved Oxygen (4.58 mg/l), Biological Oxygen Demand (4.62 mg/l), Chemical Oxygen Demand (146.1mg/l).The microbial consortium identified was found to survive and multiply in the present environment conditions. Microbial consortium was sequenced and compared using BLAST, ClustalW and PHYLIP. In order to identify potential microorganism, microbial consortium was exposed to increasing concentration of Tributyl Phosphate viz. 10, 25, 50, 75 and 100 mg/l in MSM, the potential microorganism was found to survive at higher concentration and utilized it as a sole source of carbon. This organism was identified as *Pseudomonas pseudoalcaligenes* strain DSM 50018T using 16S rRNA sequencing. This organism was found to have high potential for degradation of Tributyl Phosphate present in Low Level Nuclear Waste.**

**Key words:** Tributyl phosphate (TBP), low level nuclear waste, 16S rRNA sequencing, industrial effluent, biodegradation.

## INTRODUCTION

Rapid industrialization and urbanization have enhanced the levels of organic and inorganic contaminants in the environment. Nuclear wastes generated through chemical processing in nuclear industry or nuclear weapons program have also enhanced the level of organic contaminants. The waste generated from nuclear industry generally contains radio nuclides, heavy metals along with myriads of toxic organics. Several physico-chemical methods to decontaminate the nuclear waste have been established and employed.

However, in Low-Level Nuclear Waste, concentrations involved are low and volumes are large. Hence physical and chemical methods cannot be practiced to decon-taminate the Low-level nuclear waste. The organics as well as inorganic chemicals present in the nuclear waste find their ways in soil-water causing environmental pollution. Most of these compounds can be inactivated or degraded by microorganisms (Kumar et al., 1996).Among them, Tri butyl phosphate (TBP) has been poorly investigated because of its low toxicity in mammals. (Healy et. al., 1995 McDonald et al., 2002).Nevertheless, its wide utilization in defoamers, plasticizers, herbicides, hydraulic fluids and as a solvent for conventional nuclear fuel processing generates large amount of wastes. This compound is very stable in the natural environment and is hardly affected by natural photolysis and hydrolysis (Environment Protec-

*Corresponding author. E-mail: mhfulekar@yahoo.com.

tion Agency 1992). Recent studies have shown that nuclear waste contaminants can have both lethal and a sublethal effect on a variety of organisms. TBP is a known carcinogen and remain in the environment for a very long period of time. In spite of its low solubility in water (4 mM at 30°C), TBP presents an acute toxicity hazard to freshwater living organisms. The acute toxicity values for fish (96hr LC50) range from 4.2 to 18 mg/l. Toxicity values for six species of algae ranged from 1.1 mg/l (*Scenedesmus subspicatus*) to 5-10 mg/l (*Chlorella emersonii*).(SIDS Initial Assessment Report, 2001) These hazardous waste generated by nuclear industries have become a treatment and disposal problem causing environmental concern for organic contaminant such as TBP. The recent advancement in bioremediation will be beneficial to treat the organic contaminants in the Low-Level Nuclear Waste. Bioremediation refers to site restoration through the removal of organic contaminants by micro-organisms. In order to explore the identification of potential micro-organisms the physico-chemical and biological characterization of the disposal site is important (Fulekar, 2005b). In the present study the microbial consortium at the industrial waste disposal have been assessed which are surviving and growing in the presence of the organic/inorganic contaminant including chemical parameters like P, N, S which provide the nutrient along with contaminant as a carbon source. The microbial consortium found was exposed to increasing concentration of TBP to identify the potential microorganisms which are capable to degrade this compound. The 16SrRNA method have been employed to identify the potential organism (Fulekar and Jaya, 2008) (Fulekar, 2008) for bioremediation of organic compound with special reference to TBP which can be useful to treat Low Level Nuclear Waste containing TBP as contaminant present in Low Level Nuclear Waste (EPA, 1992; Raushel, 2002).

## MATERIALS AND METHODS

### Sampling site

The industrial effluent treated by the various groups of industry such as Fertilizer, Petrochemical, Power plants and other chemical industries are discharged at this site through unlined channels. The industrial effluent was collected from the waste disposal site located in an industrial belt at Chembur 60km away from Bhabha Atomic Research Centre (BARC), Chembur, Mumbai, India. The effluent/sediment samples were collected in pre-cleaned polythene bottles / bags for the characterization of physico-chemical and microbial assays.

### Physico-chemical analysis of sample

Soil was air dried ground and passed through a 2 mm pore size sieve and was stored in sealed containers at room temperature. The samples were analyzed for the various physico-chemical parameters like pH, Temperature, Electrical Conductivity (EC), Total Solids (TS), Total Suspended Solids(TSS), Total Alkalinity(TA), Phosphate (P), Total Hardness(TH), Total dissolved solids(TDS), Sodium (Na), Potassium (K), including biological characterization like

Biological Oxygen Demand (BOD), Chemical Oxygen Demand (COD) and Dissolved Oxygen (DO). The parameters were analyzed as per the "APHA, Standard Methods for Water and Waste Water Analysis" Volume 2, 1989 (APHA, 1989, 1979).

### Microbial Characterization

Microbial characterization of the samples was done within 24 h from the time of sample collection. The samples from the effluent and sediment were prepared for the identification of microbes. Total bacterial count was carried out using serial dilution method in duplicates under aseptic conditions.1ml of the saline sediment suspension was serially diluted to $10^{-1}$ to $10^{-4}$. 0.1 ml of serially diluted sample was spread plated on sterile nutrient media plates.(Kumar, 2004) Nutrient agar medium was used as it supports growth of all kinds of micro-organisms in a particular sample. For the effluent sample, 1 ml of the effluent was diluted to 10 ml followed by serial dilution and spread plating on sterile nutrient agar plates. The plates were incubated for 24 h at 37°C.Isolated colonies were further analyzed using 16S rRNA sequencing.

### Identification of potential micro-organism for bioremediation

**Compound used:** Commercial-grade Tri-n-butyl phosphate was obtained from Otto Kemi, India. The nutrient agar used for the isolation of bacteria was obtained from HiMedia Laboratories Pvt. Ltd., Mumbai, India and prepared according to manufacturer's instructions. All the solvents used for analysis were of HPLC grade while other chemicals were of AR grade.

**Spiking of the compound:** Erlenmeyer flask (250 ml) and nutrient culture media were autoclaved for 20 min at 121°C. Aliquots of 500 μl acetone containing the pesticide were aseptically added to the autoclaved and dried Erlenmeyer flasks allowing the acetone to evaporate. After complete evaporation of acetone, 100ml culture media was added. (Fulekarand Geetha, 2008)

### Media used

Mineral salts medium (MSM) enriched with TBP was used for isolation and characterization of TBP degrading bacteria. The MSM has the following composition in (g/l): $CaCl_2$, 0.025; $MgSO_4.7H_2O$, 0.2; NaCl, 0.1; $(NH_4)_2SO_4$, 5.0; $FeSO_4.7H_2O$, 0.015; $ZnSO_4.7H_2O$, 0.00171; $FeSO_4$, $7H_2O$, 0.0015; $CoCl2.6H2O$, 0.000483; $CuSO4.5H2O$, 0.000471; $NaMoO_4.2H_2O$, 0.000453.The carbon source in MSM was replaced with Tributyl Phosphate. (Thomas and Macaskie, 2006) The pH was adjusted to 7.0 by the addition of 3.0 ml of 1 M NaOH. Controls were TBP-free culture media. The MSM was autoclaved (121°C, 15 min). TBP was self sterile and was spiked in the flask.

### Isolation and taxonomic characterization of TBP degrading bacteria

The potential bacteria capable of degrading TBP were isolated from the effluent sediments. The sediment (1 gm) was suspended in 10 ml of nutrient broth. 1 ml of the nutrient broth containing 24 h old mixed culture was then inoculated in 250 ml Erlenmeyer flasks containing 100 ml of mineral salts medium supplemented with TBP (10 mg/l).(Singh D and Fulekar, 2007). The flasks were incubated on a rotary shaker at 150 rpm for 10 days at 30°C. (Fulekar and Geetha, 2008) At periodic intervals, a loop full of bacterial growth from the flasks was streaked onto mineral agar supplemented with TBP (10 mg/l) and the plates were incubated at 30°C for 5 - 6 days (Thomas and Macaskie, 1996). The number of bacteria that survived was as-

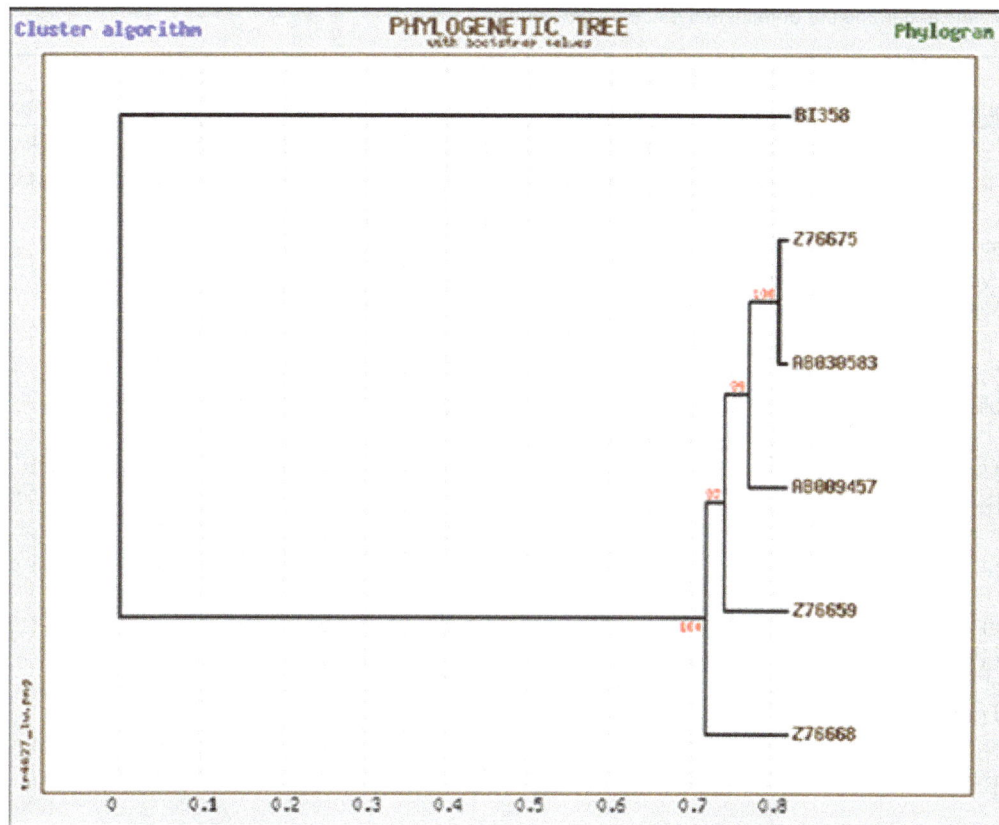

**Figure 1.** Phylogenetic tree of the most potential bacteria representing the evolutionary relationship with different bacteria.

sessed. After 10 days, 1 ml of this Nutrient culture media was transferred to MSM containing 25 mg/l of TBP. Similarly, 1 ml of the culture media was transferred to MSM containing 50, 75 and 100 mg/l. Bacterial isolate grown on TBP (100 mg/l) containing agar thereby depicting the ability to utilize TBP as the sole carbon source was isolated and was subjected to morphological, cultural and biochemical studies. The 16S rRNA gene of this bacterial strain that displayed maximum compound utilization potential was partially sequenced.

16SrRNA analysis was done using predetermined universal primers of 16S rRNA. The nucleotide sequences were used for BLAST analysis against the NCBI data base to obtain related sequences of related organisms. These sequences were aligned using CLUSTALW and a phylogenetic tree was constructed using the PHYLIP analysis programme.(Fulekar and Jaya 2008) Using the bootstrap values, the hit list for the 16S rRNA sequence homology identified the micro-organism as *Pseudomonas pseudoalcaligenes* strain DSM 50018T as shown in Figure 1.

## RESULTS AND DISCUSSIONS

The waste generated by various industries are being treated to comply with the standards prescribed under the EPA, 1986.The treated effluents are being discharged through the open channel at the disposal site. The present study has been carried out to study the physico-chemical and biological characteristics of the industrial waste disposal site. The microbial consortium which could survive in such contaminated site has also been assessed to identify the potential micro-organism degrading TBP.

The physical parameters (Table 1) studied shows that, the pH was in the range of 6.2 - 7.0 with an average of 6.61.The temperature was found to be 42 °C.Besides, the moisture content  and the alkalinity was found to be 50.72%  and 585.52 mg/l respectively. The chemical parameters (Table 1) highlighted DO (4.58 mg/l), BOD (4.62 mg/l), and COD (146.1 mg/l) which provided the ambient conditions at the contaminated site in which the adapted microorganisms survive and multiply. The chemical parameters - P (27.87 mg/l), N (0.41%), S (993.5 mg/l), Na (4695.5 mg/l), K (391.3 mg/l) also served as a nutrient to microorganism whereas the carbon source was obtained from the breakdown products of various organic contaminants The microbial consortium adapted at the contaminated site under these conditions was assessed. TBP was taken as one of the compound under study for the degradation by the microbial consortium /microorganism. The microbial consortium present at the site was analyzed by 16SrRNA sequencing and based on their sequencing they were identified as *Bacillus cereus* strain EK15, *Pseudomonas* sp. *N9-5, Pseudomonas pseudoalcaligenes* DSM 50018T, *Pseudomonas* sp. LM8, Shewanella NH21 (Table 2).

The microbial consortium assessed was further  expos-

**Table 1.** Physico-chemical parameters of industrial effluents/sediments.

| S/No. | Parameters | Effluent | | | Sediment | | | Average values |
|---|---|---|---|---|---|---|---|---|
| | | Sample I | Sample II | Sample III | Site I | Site II | Site III | |
| 1 | Color | Pale Yellow | Pale Yellow | Blackish | Black | Black | Black | Black |
| 2 | Odor | Pungent | Pungent | Pungent | Pungent | Pungent | Pungent | Pungent |
| 3 | pH | 6.8 | 6.3 | 7.0 | 6.5 | 6.2 | 6.9 | 6.61 |
| 4 | Temperature (ºC) | 40 | 42 | 39 | 30 | 28 | 35 | 35.6 |
| 5 | Conductivity (mS) | 16.8 | 17.2 | 16.8 | 17.0 | 16.1 | 15.9 | 16.63 |
| 6 | Total Solids (mg/l) | 26.13 | 29.5 | 35.13 | — | — | — | 30.2 |
| 7 | Total Dissolved Solids (mg/l) | 15.23 | 15.40 | 15.22 | — | — | — | 15.28 |
| 8 | Total Soluble Solids (mg/l) | 8.43 | 6.30 | 6.2 | — | — | — | 6.97 |
| 9 | Moisture Content (%) | NA | NA | NA | 62.66 | 52.31 | 37.19 | 50.72 |
| 10 | Bulk Density (Cm$^3$) | — | — | — | 0.68 | 0.63 | 0.67 | 1.98 |
| 9 | Alkalinity (mg/l) | 566.6 | 483.3 | 816.6 | 780 | 866.67 | 733.34 | 585.52 |
| 10 | Dissolved Oxygen (mg/l) | 3.16 | 4.83 | 2.4 | 5.36 | 6.13 | 5.63 | 4.58 |
| 11 | Biological Oxygen Demand (mg/l) | 3.0 | 4.80 | 2.36 | 6.83 | 5.44 | 5.32 | 4.62 |
| 12 | Chemical Oxygen Demand (mg/l) | 160 | 102.67 | 176 | 160 | 102 | 176 | 146.1 |
| 13 | Phosphate(mg/l) | 30.6 | 30.0 | 33.3 | 28.7 | 35.3 | 9.34 | 27.87 |
| 14 | Total Organic carbon (%) | 1.06 | 1.53 | 0.7 | 4.26 | 3.8 | 3.83 | 2.53 |
| 15 | Total Organic Matter (%) | 1.93 | 2.96 | 1.33 | 7.27 | 6.6 | 7.16 | 4.54 |
| 16 | Sulfate (mg/l) | 1100 | 1430 | 2560 | 380 | 420 | 71 | 993.5 |
| 17 | Nitrogen (%) | 0.035 | 0.0525 | 0.0175 | 1.32 | 0.42 | 0.65 | 0.41 |
| 18 | Sodium (mg/l) | 8820 | 8890 | 8522 | 752 | 650 | 539 | 4695.5 |
| 19 | Potassium (mg/l) | 550 | 542.0 | 535.0 | 96 | 102.0 | 94.0 | 319.3 |

exposed to TBP at increasing concentration varying from 10, 25, 50, 75 and 100 mg/l in MSM. The growth was observed as turbidity and optical density (OD) was measured daily at 600nm to assess the survival and growth of microorganisms at the respective TBP concentration. The microorganisms surviving at various concentration of TBP was assessed by spread plate method on miminal agar containing TBP (Table 3).

At 100 ppm only one bacterial colony was observed which was further assessed by biochemical tests, followed by its 16S rRNA sequencing including its phylogenetic tree (Table 4 and Figure 1). Based on the sequencing and phylogenetic analysis the microorganism was identified as *Pseudomonas pseudoalcaligenes* strain DSM 50018T. This microorganism, adapted in the contaminated site could degrade the higher concentration of TBP, utilizing it as a sole source of carbon. Earlier studies shows that *Pseudomonas pseudoalcaligenes* JS45 utilizes nitrobenzene as the sole source of nitrogen, carbon, and energy. (He and Chain, 1998).Another study by on Pseudomonas pseudoalcaligenes KF707 depicts its abi-

lity to utilize 2- and 4-fluorobiphenyl as sole carbon and energy sources.( Murphy et al., 2008).The organophosphorus hydrolyse gene was reported in *Pseudomonas pseudoalcaligenes* C2-1.This organo-phosphorus hydrolase gene was expressed in *Pichia pastoris plants*. The protein expressed by this gene is able to hydrolyze phosphoester bonds and reduce the toxicity of organophosphorus compounds. (Xiao-Yu Chu et al., 2008). In previous studies, TBP degrading mixed culture of Pseudomonas spp. were isolated that utilize TBP as a sole source of carbon and phosphorus. (Thomas and Macaskie, 1996) In the present study it was found that the isolated microorganism *Pseudomonas pseudoalcaligenes* DSM 50018T could utilize TBP as a sole source of carbon. The microorganism *Pseudomonas pseudoalcaligenes* DSM 50018T identified in the present research study could serve as a potential microorganism for the bioremediation of low level nuclear waste containing organic contaminants with special reference to TBP. This microorganism will be submitted to the gene bank for its beneficial use to biodegrade the hazardous compound s such as TBP.

**Table 2.** Microbial characteristics of industrial effluent and sediments.

| S/No. | Bacterial Identification Method | Description | 16S rRNA Gene Sequence |
|---|---|---|---|
| 1 | 16S rRNA Gene sequencing | Shewanella sp. NH21 | TAAGCGCACGCAGGGGCTTGTTAAGTTAGATGTGATATTTAACCT GGGAATTGCATTTAAGACTGGCTAAGGTGGAGAGGGGGGTGGAA TTTCCGGTGTAGCGGTGAAATGCGTAGAGATCGGAAGGAACATC AGTGGCGAAGGCGACCCCCTGGCCAAAGACTGACGCTCAGGTG CGAAAGCGTGGGGAGCAACAGGATTAGATACCCTGGTAGTCCAC GCCGTAAACGATGTCAACTTGGAGTTTGTGTTCTTGAAACGTGGA CTCCGGAGCTAACGCGTTAAGTTGACCGCCTGGGGAGTACGGCC GCAAGGTTAAAACTCAAATGAATTGACGGGGGCCCGCACAAGCG GTGGAGCATGTGGTTTAATTCGATGCAACGCGAAGAACCTTACCT ACTCTTGACATCCAGAGAATTTGCTAGAGATAGCTTASTGCCTTC GGGAACTCTAGACAGGTGCTGCATGGCTGTCGTCAGCTCGTGTT GTGAAATGTTGGGTTAAGTCCCGCAACGAGCGCAACCCTTATCCT TTGTTGCCAGCACGTAATGGTGGGAACTCAAAGGAGACTG |
| 2 | 16S rRNA gene | Bacillus cereus strain EK-15 | GTAAGCGCGCGCAGGTGGTTTCTTAAGTCTGATGTGCTAGTTCG GGCTCAACCGTGGAGGGTCATTGGAAACTGGGTTGAGGCAGAAG AGGAAAGTGGAATTCCATGTGTAGCGGTGAAATGCGTAGAGATAT GGAGGAACACCAGTGGCGAAGGCGACTTTCTGGTCTGTAACTGA CACTGAGGCGCGAAAGCGTGGGGAGCAAACAGGATTAGATACCC TGGTAGTCCACGCCGTAAACGATGAGTGCTAAGTGTTAGAGGGT TTCCGCCCTTTAGTGCTGAAGTTAACGCATTAAGCACTCCGCCTG GGGAGTACGGCCGCAAGGCTGAAACTCAAAGGAATTGACGGGG GCCCGCACAAGCGGTGGAGCATGTGGTTTAATTCGAAGCAACGC GAAGAACCTTACCAGGTCTTGACATCCTCTGAAAACCCTAGAGAT AGGGCTTCTCCTTCGGGAGCAGAGTGACAGGTGGTGCATGGTTG TCGTCAGCTCGTGTCGTGAGATGTTGGGTTAAGTCCCGCAACGA GCGCAACCCTTGATCTTAGTTGCCATCATTAAGTTGGGCACTCTA AGGTGACTGCGGTGACAAACCGGAGGAAGGTGGGGATGACGTC AAATCATCATGCCCCTTATGACCTGGGCTACACACGTGCTAC |
| 3 | 16S rRNA gene | Pseudomonas sp. N9-5 | TAAGCGCGCGTAGGTGGTTTGATAAGTTGGATGTGAAAGCCCCG GGCTCAACCTGGGAATTGCATCCAAAACTGTCTGACTAGAGTATG GCAGAGGGTGGTGGAATTTCCTGTGTAGCGGTGAAATGCGTAGA TATAGGAAGGAACACCAGTGGCGAAGGCGACCACCTGGGCTAAT ACTGACACTGAGGTGCGAAAGCGTGGGGAGCAAACAGGATTAGA TACCCTGGTAGTCCACGCCGTAAACGATGTCGACTAGCCGTTGG GATCCTTGAGATCTTAGTGGCGCAGCTAACGCATTAAGTCGACCG CCTGGGGAGTACGGCCGCAAGGTTAAAACTCAAATGAATTGACG GGGGCCCGCACAAGCGGTGGAGCATGTGGTTTAATTCGAAGCAA CGCGAAGAACCTTACCAGGCCTTGACATGCAGAGAACTTTCCAG AGATGGATTGGTGCCTTCGGGAACTCTGACACAGGTGCTGCATG GCTGTCGTCAGCTCGTGTCGTGAGATGTTGGGTTAAGTCCCGTA ACGAGCGCAACCCTTGTCCTTAGTTACCAGCACGTTAAGGTGGG CACTCTAAGGAGACTGCCGGTGACAA |
| 4 | 16S rRNA gene | Pseudomonas sp. LM8 | TAAGCGCGCGTTGTGAAAGCCCCGGGCTCAACCTGGGAATTGCA TCCAAAACTGTCTGACTAGAGTATGGCAGAGGGTGGTGGAATTTC CTGTGTAGCGGTGAAATGCGTAGATATAGGAAGGAACACCAGTG GCGAAGGCGACCACCTGGGCTAATACTGACACTGAGGTGCGAAA GCGTGGGGAGCAAACAGGATTAGATACCCTGGTAGTCCACGCCG TAAACGATGTCGACTAGCCGTTGGGATCCTTGAGATCTTAGTGGC GCAGCTAACGCATTAAGTCGACCGCCTGGGGAGTACGGCCGCAA GGTTAAAACTCAAATGAATTGACGGGGGCCCGCACAAGCGGTGG AGCATGTGGTTTAATTCGAAGCAACGCGAAGAACCTTACCAGGCC TTGACATGCAGAGAACTTTCCAGAGATGGATTGGTGCCTTCGGGA ACTCTGACACAGGTGCTGCATGGCTGTCGTCAGCTCGTGTCGTG AGATGTTGGGTTAAGTCCCGTAACGAGCGCAACCCTTGTCCTTAG TTACCAGCACGTTAAGGTGGG |

**Table 2.** Continued.

| 5 | 16S rRNA gene | *Pseudomonas pseudoalcaligenes 50018T* | DSM | CGGTCGAAGTTCACACATGCAAGTCGAGCGGTGAAGGGAGCTTG CTCCTGGATTCAGCGGCGGACGGGTGAGTAATG |
|---|---|---|---|---|
| | | | | CCTAGGAATCTGCCTGGTAGTGGGGGGATAACGTCCGGAAACGG GCGCTAATACCGCATACGTCCTGAGGGAGAAAGT |
| | | | | GGGGGATCTTCGGACCTCACGCTATCAGATGAGCCTAGGTCGGA TTAGCTAGTTGGTGGGGTAAAGGCCTACCAAGGC |
| | | | | GACGATCCGTAACTGGTCTGAGAGGATGATCAGTCACACTGGAA CTGAGACACGGTCCAGACTCCTACGGGAGGCAGC |
| | | | | AGTGGGGAATATTGGACAATGGGCGAAAGCCTGATCCAGCCATG CCGCGTGTGTGAAGAAGGTCTTCGGATTGTAAAG |
| | | | | CACTTTAAGTTGGGAGGAAGAGCAGTAAGTTAATACCTTGCTGTT TTGACGTTACCAACAGAATAAGCACCGGCTAACT |
| | | | | TCGTGCCAGCAGCCGCGGTAATACGAAGGGTGCAAGCGTTAATC GGAATTACTGGGCGTAAAGCGCGCGTAGGTGGTTC |
| | | | | AGCAAGTTGGATGTGAAATCCCCGGGCTCAACCTGGGAACTGCA TCCAAAACTACTGAGCTAGAGTACGGTAGAGGGTG |
| | | | | GTGGAATTTCCTGTGTAGCGGTGAAATGCGTAGATATAGGAAGG AACACCAGTGGCGAAGGCGACCAC |

**Table 3.** Comparative chart of potentiality of microorganisms from effluent and sediments against different concentration of TBP.

| S/No | Concentration of TBP | Growth observed on TBP | No. of colonies found after spread plate (Effluent- Sediment) | Colony identification | No. of colonies found after spread plate (Effluent) | Colony Identification |
|---|---|---|---|---|---|---|
| 1 | 10 ppm | +++ | Three colonies | *Pseudomnas* sp. LM8, *Pseudomonas pseudoalcaligenes* DSM 50018T, *Bacillus cereus strain* EK-15 | Two colonies | *Bacillus cereus* strain EK-15, *Pseudomonas* sp.N9-5 |
| 2 | 25 ppm | +++ | Two colonies | *Pseudomonas* sp. LM 8, *Pseudomonas pseudoalcaligenes* DSM 50018T | One colony | *Pseudomonas* sp. N9-5 |
| 3 | 50 ppm | ++ | Two colonies | *Pseudomonas* sp. LM8 *Pseudomonas pseudoalcaligenes* DSM 50018T | One colony | *Pseudomonas* sp. N9-5 |
| 4 | 75 ppm | ++ | One colony | *Pseudomonas pseudoalcaligenes* DSM 50018T | Nil | - |
| 5 | 100 ppm | + | One colony | *Pseudomonas pseudoalcaligenes* DSM 50018T | Nil | - |

## Conclusion

The present research study has identified the microbial consortium adaptability in industrial effluent disposal site. The potential micro-organism Pseudomonas *pseudoalcaligenes DSM 50018T* identified from microbial consor- tium by 16srRNA methods and confirmed on the basis of phylogenetic tree and sequencing. This organism has high potential to biodegrade the organic pollutant such as TBP present in low level nuclear waste.

**Table 4.** Biochemical characteristics shown by the most potential micro-organism isolated at 100ppm of TBP.

| S/ No | Characteristics | Result | S/No. | Characteristics | Result |
|---|---|---|---|---|---|
| 1. | ONPG | – | 15. | Esculin | – |
| 2. | Lysine decarboxylase | – | 16. | Arabinose | – |
| 3. | Ornithine decarboxylase | – | 17. | Xylose | + |
| 4. | Urease | – | 18. | Adonitol | – |
| 5. | Deamination | – | 19. | Rhamnose | – |
| 6. | Nitrate reduction | + | 20. | Cellobiose | – |
| 7. | H2S production | – | 21. | Melibiose | – |
| 8. | Citrate utilization | + | 22. | Saccharose | – |
| 9. | Voges Proskauer's | – | 23. | Raffinose | – |
| 10. | Methyl Red | – | 24. | Trehalose | – |
| 11. | Indole | – | 25. | Glucose | + |
| 12. | Malonate | + | 26. | Lactose | – |
| 13. | Gram Characteristic | – | 27. | Oxidase | + |
| 14. | Motility | Motile | | | |

## ACKNOWLEDGEMENTS

The present research work is a part of the Board of Research in Nuclear Sciences (BRNS) project entitled "Biotechnological approaches for the remediation of Low Level Nuclear Waste". The authors are grateful to the BRNS for giving the facility and financial grants to carry out the research work. The first author would like to extend thanks to the senior colleague Mrs. Anamika Singh for her technical assistance and support.

## REFERENCES

APHA (1989). Standard methods for examination of water and waste water. American Public Health Association, New York.

APHA (1979). Standard methods for examination of water and waste water. American Public Health Association, New York.

EPA (1992). Chemical information collection and data development tributyl phosphate test results. www.epa.gov/oppt/chemtest/tributph.htm

Fulekar MH (2005a). Bioremediation technologies for environment. Ind. J. Environ. Prot. 25(4): 358-364.

Fulekar MH (2005b). Environmental Biotechnology, Oxford and IBH publishing house, New Delhi, India. pp. 52-68.

Fulekar MH (2008). Bioinformatics: Applications in Life and Environmental Sciences. Capital & Springer publication, Germany. pp. 172-192.

Fulekar MH, Geetha M (2008). Bioremediation of chlorpyrifos by Pseudomonas aeruginosa using scale up technique. J. Appl. Biosci. 12: 657-660.

Fulekar MH, Jaya S (2008). Bioinformatics Applied in Bioremediation. Innovative Romanian Food Biotechnology 2:28-36.

He Z, Spain JC (1998). A Novel 2-Aminomuconate Deaminase in the Nitrobenzene Degradation Pathway of Pseudomonas pseudoalcaligenes JS45. J. Bacteriol. 180: 2502-2506.

Healy CE, Beyrouty PC, Broxup BR (1995). Acute and subchronic neurotoxicity studies with tri-N-butyl phosphate in adult Sprague_Dawley rats. Am. Ind. Hyg. Assoc. J. 56: 349-355.

Kumar H (2004). Recent trends in Biotechnology. Agrobios, India. pp. 55-58.

Kumar S, Mukerji KG, Lal R (1996). Molecular aspects of pesticide degradation by microorganisms. Crit. Rev. Microbiol. 22: 1-26.

McDonald AR, Odin M, Choudhury H (2002). Chronic risk assessment for tributyl phosphate (CAS NO. 126-73-8). In: Society of Toxicology Meeting, Nashville, TN, March 18-21.

Murphy CD, Quirke S, Balogun O (2008). Degradation of fluorobiphenyl by Pseudomonas pseudoalcaligenes KF707.FEMS microbiology letters (FEMS Microbiol Lett). 286(1): 45-90.

Raushel FM (2002). Bacterial detoxification of organophosphate nerve agents. Curr. Opin. Microbiol. 5: 288-295.

SIDS (2001). Initial Assesment Report for 12th SIAM Paris, France. pp. 2-13.

Singh D, Fulekar MH (2007). Bioremediation of phenol using microbial consortium in bioreactor. Innovative Romanian food Biotechnology. 1(30): 31-36.

Thomas R, Macaskie LE (1996). Biodegradation of tributyl phosphate by natural occuring microbial isolates and coupling to the removal of uranium from aqueous solution. Environ. Sci. Technol. 30: 2371-2375.

Xiao-Yu Chu, Ning-Feng Wu, Min-Jie Deng, Jian Tian, Bin Yao and Yun-Liu Fan (2008). Expression of organophosphorus hydrolase OPHC2 in Pichia pastoris: Purification and characterization. 49(1): 9-14.

# 2

# A geographic information assessment of exposure to a toxic waste site and development of systemic lupus erythematosus (SLE): Findings from the buffalo lupus project

Edith .M. Williams[1]*, Robert Watkins[2], Judith Anderson[3] and Laurene Tumiel-Berhalter[2]

[1]Institute for Partnerships to Eliminate Health Disparities, University of South Carolina, 220 Stoneridge Drive, Suite 208 Columbia, SC 29210, Columbia.
[2]Department of Family Medicine, State University of New York at Buffalo 173 CC, ECMC, 462 Grider Street, Buffalo, NY 14215, USA.
[3]Jericho Road Ministries, 318 Breckenridge Buffalo, NY 14213, USA.

**The Buffalo Lupus Project was a community-based participatory research partnership formed to address the relationship between an identified hazardous waste site and high rates of lupus and other autoimmune diseases in the surrounding community. Most cases identified began experiencing symptoms and were diagnosed in the periods when the site was inactive. Trends suggest that the impact of the site was more likely due to chronic exposure to waste rather than it being an acute trigger.**

**Key words:** Lupus, environmental exposure, buffalo lupus project, GIS.

## INTRODUCTION

Although we are all affected by environmental health threats, there are groups that bear a larger burden of impacts from higher exposure due to environmental degradation. Minorities and those who live in poverty often reside in areas characterized by differential exposure to health risks in the physical environment, poor quality housing, and deprivation of resources and facilities to adequately address the health related problems associated with such exposures (Cattell, 2001; Williams and Collins, 2001; Gee and Payne-Sturges, 2004). There is documented evidence of the over-abundance of contaminated sites located in minority and impoverished communities and the serious health risks such sites pose to residents (Harding and Greer, 1993; Dahlgren et al., 2007; Sapien, 2001). A number of studies have suggested that environmental exposure such as inhalation or ingestion of contaminants is related to the development of lupus (Balluz et al., 2001; Kardestuncer and Frumkin, 1997; Mongey and Hess, 2002). In lupus and other autoimmune diseases, the immune system loses its ability to differentiate between foreign substances and its own cells and tissues, causing the body to attack itself (Grossman and Kalunian, 2002). Literature suggests an overall mortality rate in lupus patients that is greater than two times the rate in the general population (Centers for Disease Control and Prevention, 2002; Bernatsky et al., 2006; Krishnan and Hubert, 2006).

There are approximately two million people in the United States who suffer from lupus, though it is likely that these estimates are low (Rus and Hochberg, 2002; Office of Minority Health, 2004; Calvo- Alen et al., 2005). Young women are most frequently affected by the disease, outnumbering male patient's ten to one (Cooper et al., 2002). The disease usually strikes between the ages of 15 and 40 years and African Americans are at particularly high risk for the disease. In the United States, blacks have three-fold higher incidence and prevalence rates of SLE, as well as cause-specific mortality rates, compared with whites (Rus and Hochberg, 2002; Office

*Corresponding author. E-mail: willi425@mailbox.sc.edu.

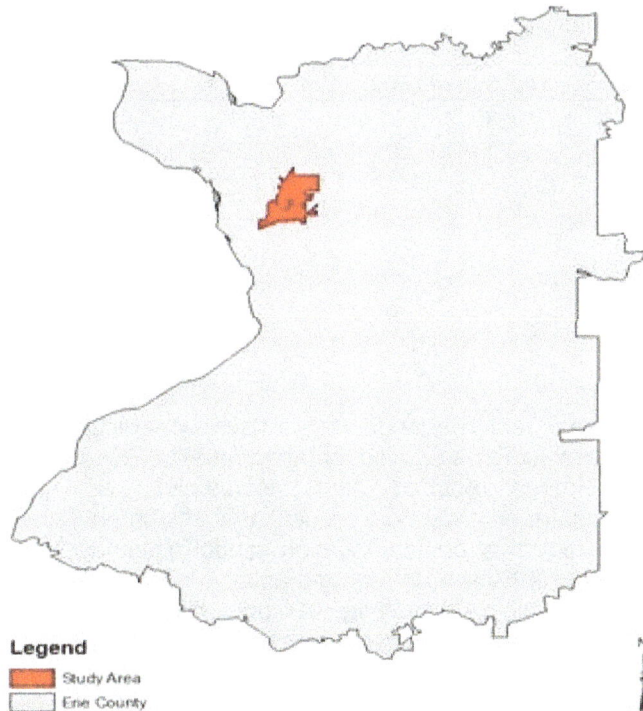

**Legend**
Study Area
Erie County

N

**Figure 1.** Study area in Erie county New York.

of Minority of Health, 2004; Alarcon et al., 2005; Oates et al., 2003). Environmental exposure is hypothesized to play an important role as either a trigger or modifying influence in the development of autoimmune diseases. Ultraviolet light and certain drugs have been identified as the only known environmental triggers. There are several toxic substances which have been implied in the development of lupus in ethnically and geographically vulnerable populations including mercury, presence of particulates, radiation, infectious agents, and other chemical factors and metals (Balluz et al., 2001; Kardestuncer and Frumkin, 1997; Mongey and Hess, 2002). In 1993, a cluster of women with SLE was identified in one eight-block area of the predominantly African American east side community of Buffalo, New York. There were also three toxic waste sites in the same neighborhood. Two had already been remediated. The focal point of contamination and largest of the three identified sites, 858 East Ferry Street, had been designated as a level 2 Superfund site that continued to sit uncontained (lack of fencing and signage indicating hazardous waste around the site) for several years. Community members were concerned that the high incidence of disease could be linked to the contaminants identified at the site caused by historical spill or dumping by the lead smelting plant that had once operated there.

The Buffalo Lupus Project was part of the five-year National Institute of Environmental Health Sciences (NIEHS) funded community-based participatory research

grant which was awarded in the fall of 2001. The purpose of the research partnership that became known as the Buffalo Lupus Project was to identify people in the community with lupus and other autoimmune disease and to see if exposure to the area toxic waste Superfund site was a trigger for high disease incidence in the area (Terrell et al., 2008). Buffalo Lupus Project research activities consisted of a registry to assess the city-wide prevalence of lupus and other autoimmune diseases, and a survey to investigate common environmental factors that could elucidate the complex causes of lupus and other autoimmune diseases. All eligible adults 18 years of age or older were invited to participate in research activities, and all participants provided their written informed consent on forms previously approved by the Health Sciences Institutional Review Board at the State University of New York at Buffalo. The Buffalo Lupus Project identified 185 cases of SLE in a population of 74,074 persons. This corresponds to an event or prevalence rate of 160 per 100,000 in the target population. These findings suggest that the observed number of events in the target population were greater than expected; it appears that there is approximately seven times greater risk of having SLE in the geographic study area relative to the general population estimates. The present study was designed to assess the geographical areas of highest risk in relation to the Superfund site, evaluate differences in disease clustering according to when the site was operational and after demolition, and gain insight into the potential impact of the site on affected community members.

**MATERIALS AND METHODS**

**Study area**

The study was conducted in the largely African American east side community of Buffalo, New York. The toxic waste sites of concern and identified lupus cluster were located in zip code tracking areas (ZCTAs) 14211 and 14215 of the 34th and 35th census tracts within this area (Figure 1). The City of Buffalo first identified the vacant property at 858 East Ferry Street as a hazardous site in 1997. Past operations of a zinc storage complex and lead smelter and refining facility, which operated from the 1920s through the early 1970s, was believed to be the source of contamination. Although state officials and interested groups usually refer to the Superfund site as 858 East Ferry Street, the original location of the lead smelting facility was the 2.3 acre lot at 856 East Ferry Street. The adjacent 3.32 acre empty lot (858 East Ferry Street) was used by the smelter to dump waste ash. The New York State Department of Environmental Contamination (NYSDEC) investigation in 1997-98 identified high concentrations of lead, mercury, arsenic, polychlorinated biphenyls (PCBs), volatile organic chemicals (VOCs), incinerator ash, and other metals in the soil on the site. Additionally, the water there was found to be contaminated with benzene and similar soil contaminants, at levels warranting remediation of the site (New York State Department of Environmental Conservation, 2005). Significant lead contamination was identified at surface levels between 149 - 11500 ppm and subsurface levels between 110 - 46700 ppm (New York State Department of Environmental Conservation, 2005).

**Table 1.** Comparison of buffalo lupus project survey participants to study area statistics (zip codes 14215 and 14211).

|                                     | Survey participants (n=66) | Zip code 14215 (n=43,569) | Zip code 14211 (n=29,039) |
| ----------------------------------- | -------------------------- | ------------------------- | ------------------------- |
| Race (%) - African American         | 82.8                       | 72.3                      | 71.5                      |
| Sex (%) - Female                    | 96.2                       | 54.8                      | 53.8                      |
| Median age                          | 49 years                   | 31.6 years                | 32.1 years                |
| Education (%)                       |                            |                           |                           |
| High school                         | 30.8                       | 79.8                      | 68.5                      |
| College                             | 47.7                       | 16.2                      | 7.4                       |

Source: US Census Bureau, 2000.

## Materials

Of the 309 Buffalo Lupus Project registrants who identified themselves as having SLE, 87 were eligible to participate in the survey due to having ever lived or currently living in the targeted zip codes and having a previous diagnosis of SLE. Trained interviewers completed surveys with 66 SLE cases, corresponding to a 76% response rate. Survey topics included demographics, diagnosis, health care utilization, residential history and exposures, occupational history and exposures, disease specific medical history, smoking history, family health history, and social support. As part of detailed residential histories, participants were asked to recall every address where they had ever lived and the specific years they resided there. All address information was de-identified and linked with other survey components by unique identifier. For each participant, residential data were categorized into ten-year increments, beginning when the smelting plant was active (1920 - 1929) and ending with the period of data collection (2000-2009). Each participant was assigned one address per ten-year period according to where they resided for the longest duration during that period. Addresses outside of Buffalo, New York were excluded.

## Methods

Information from completed surveys was entered into an electronic database, using the Microsoft Access program. A cross tabulation of participant ID and addresses was performed using the SPSS statistical package (SPSS Inc, Chicago, IL) to determine who lived where in which period. A geographic information system was constructed for the study using ArcMap software (Environmental Systems Research Institute, Redlands, CA) on a Windows-based work station. Locations of survey participants' residences within the study area were digitized into a GIS point file using the file containing participants' addresses that was matched, using ArcMap, against the U.S. Bureau of the Census TIGER (Topographically Integrated Geographic Encoding and Referencing system) line files. Incomplete records missing house numbers were assigned a random street number inside the range of highest and lowest valid street addresses in the zip code. The final GIS products for this study were databases from which locations of survey participants' residences and their respective disease status could be visually displayed. A total of nine maps were generated according to the nine ten-year periods, and each included participant residences in relation to the Superfund site of concern and whether the participant was pre-symptomatic, symptomatic without a diagnosis, or diagnosed during that period.

## RESULTS

Table 1 shows that 96.92% (n=63) of survey participants

were women. The median age of survey participants was 49 years. Table 1 also shows that 81.82% (n=54) of participants reported their race/ethnicity as African American. With regard to educational attainment, 30.77% of participants obtained a high school diploma or GED, while 47.69% (n=31) reported college to be their highest level of school completed. A cumulative percentage of 21.54% of participants reported participation in vocational/technical training and/or graduate or professional school. Buffalo Lupus Project survey participants were generally older and more highly educated than the general study area population residing in zip codes 14215 and 14211. Additionally, survey participants were comprised of more females and African Americans. Table 1 shows that African Americans make up approximately 72% of the population in zip codes 14215 and 14211, and approximately 54% of the population in both zip codes are female. The median age of residents of zip codes 14215 and 14211 is 32 years, and less than 20% in both zip codes reported graduating from college (CensusUSBot, 2006).

During the time period that the Superfund site of concern began operations as an active lead smelting plant (1920-1929), no survey participants reported residing in the area (Figure 2). In subsequent time periods (1930-1949), a small number (n=7) of participants reported residences that were clustered to the southwest of the site, but none reported experiencing disease symptoms during this time (Figures 3 and 4). In the following time periods that the site of concern was still an active industrial site (1950-1969), residential patterns of participants were more scattered, but still concentrated most densely to the west of the site (Figures 5 and 6). Few participants reported experiencing symptoms (n=5) or being formally diagnosed (n=3) during this time. During the time period that the smelting plant became inactive (1970-1979), residential patterns of survey participants were more dispersed, but most densely concentrated to the northeast of the site. An increased number of participants reported being formally diagnosed with SLE (n=9) during this period (Figure 7). During the time periods following (1980-1999), when the site of concern was inactive, but still being used as a waste dump, the majority of survey participants reporting residences in the area were formally diagnosed, and most densely

## Legend

⭐ 858 East Ferry

## Locations of Patients

◆ Pre-symptomatic

◇ Symptomatic

◆ Diagnosed

— Buffalo Streets

**Figure 2.** Residential patterns of Buffalo lupus project survey participants, 1920-1929 (n=0). Source: Author.

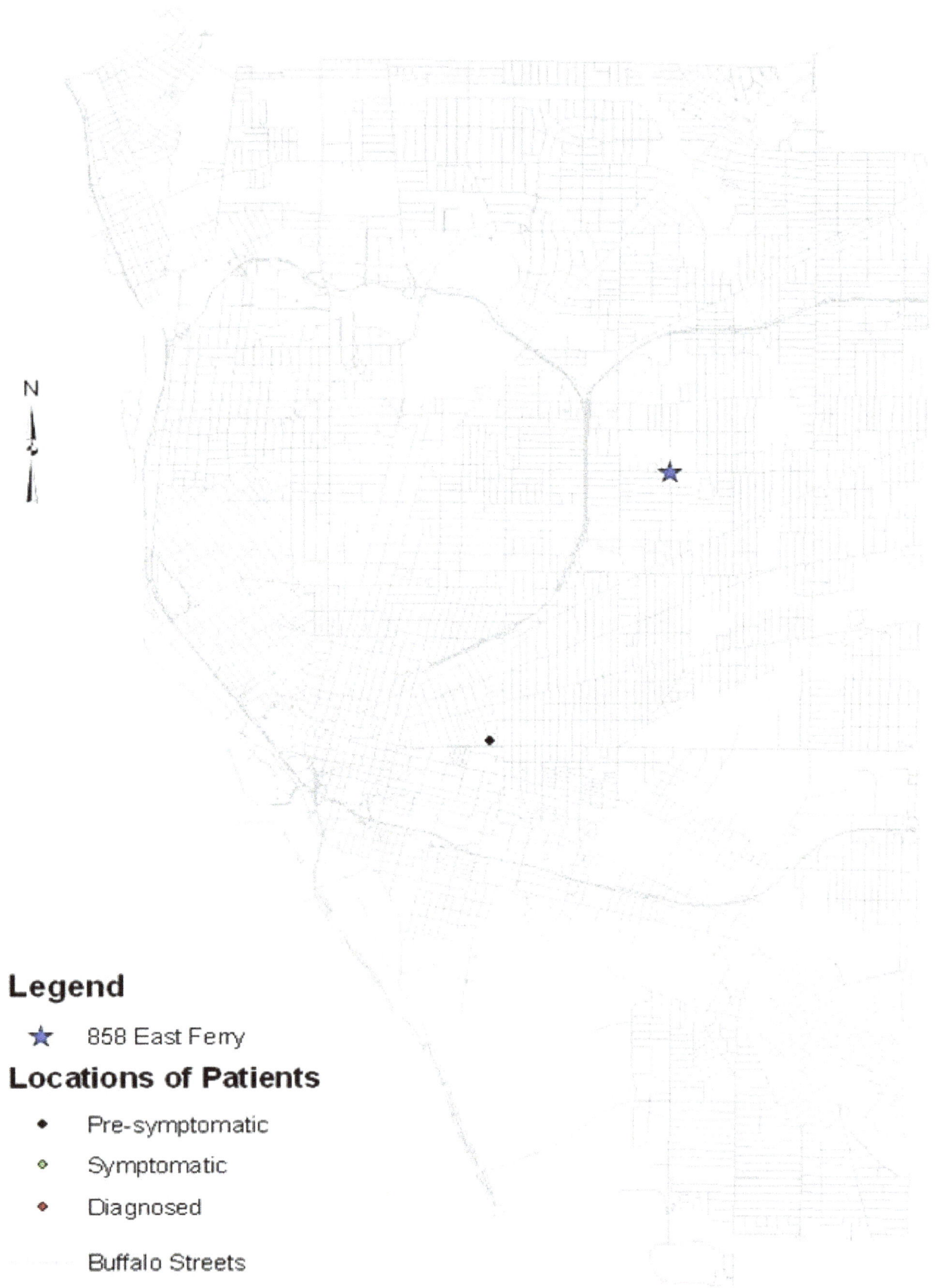

N

## Legend

⭐  858 East Ferry

## Locations of Patients

◆  Pre-symptomatic

◇  Symptomatic

◆  Diagnosed

      Buffalo Streets

**Figure 3.** Residential patterns of Buffalo lupus project survey participants, 1930-1939, n=2. Source: Author.

N

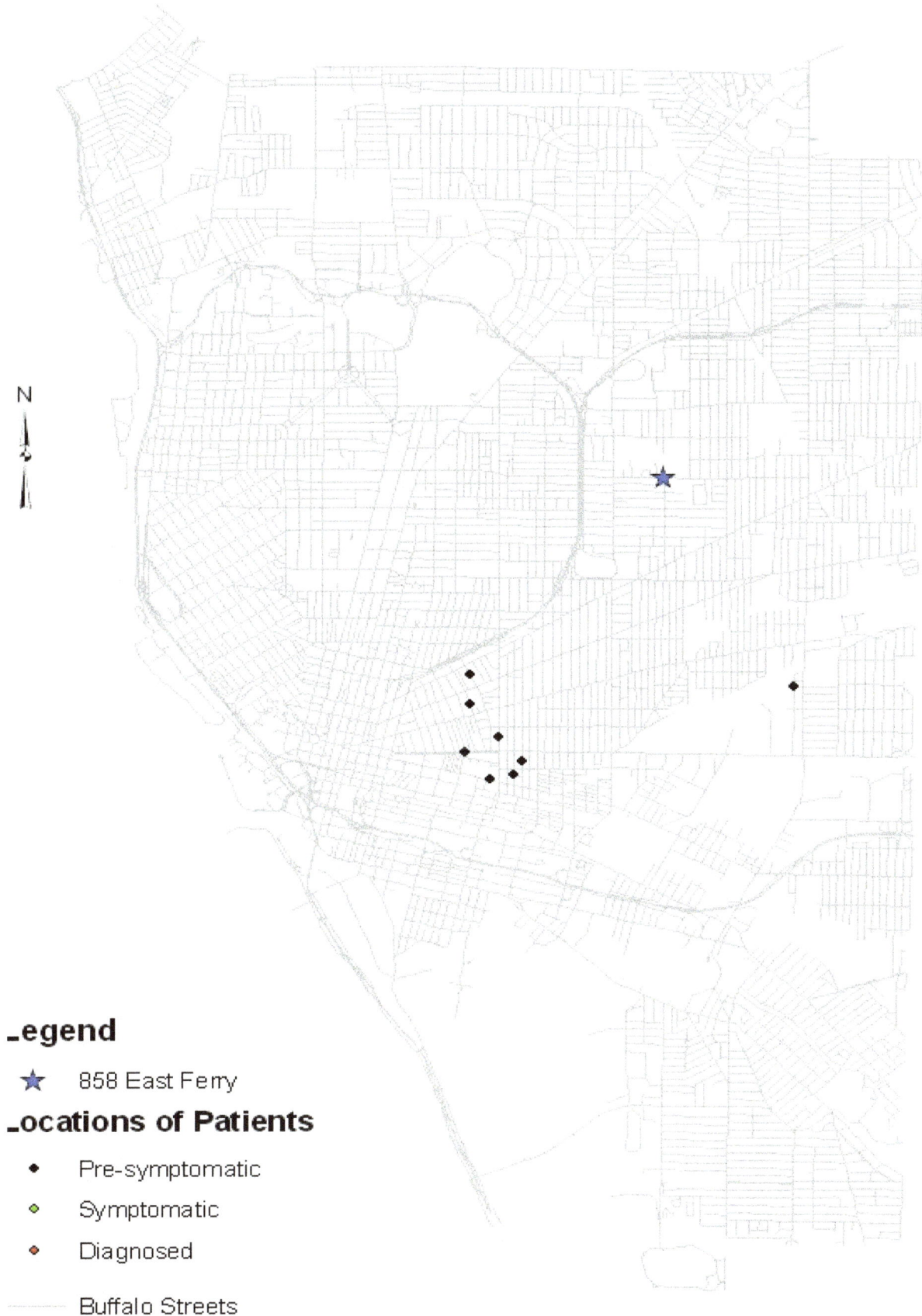

## Legend

⭐ 858 East Ferry

## Locations of Patients

◆ Pre-symptomatic

◇ Symptomatic

◆ Diagnosed

—— Buffalo Streets

**Figure 4.** Residential patterns of Buffalo lupus project survey participants, 1940-1949, n=14. Source: Author.

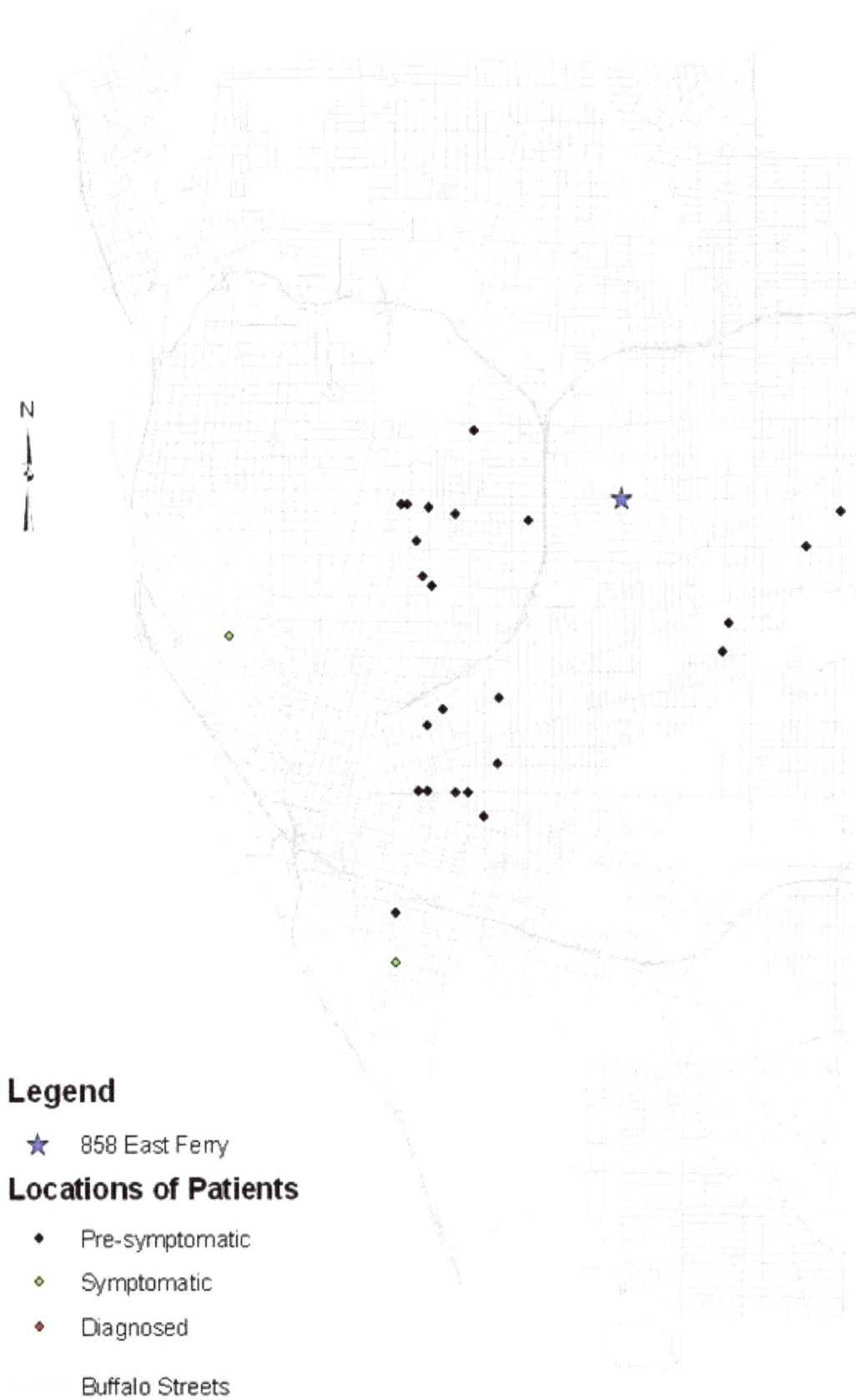

N

## Legend

⭐  858 East Ferry

## Locations of Patients

◆  Pre-symptomatic

◇  Symptomatic

◆  Diagnosed

      Buffalo Streets

**Figure 5.** Residential patterns of Buffalo lupus project survey participants, 1950-1959, n=33. Source: Author.

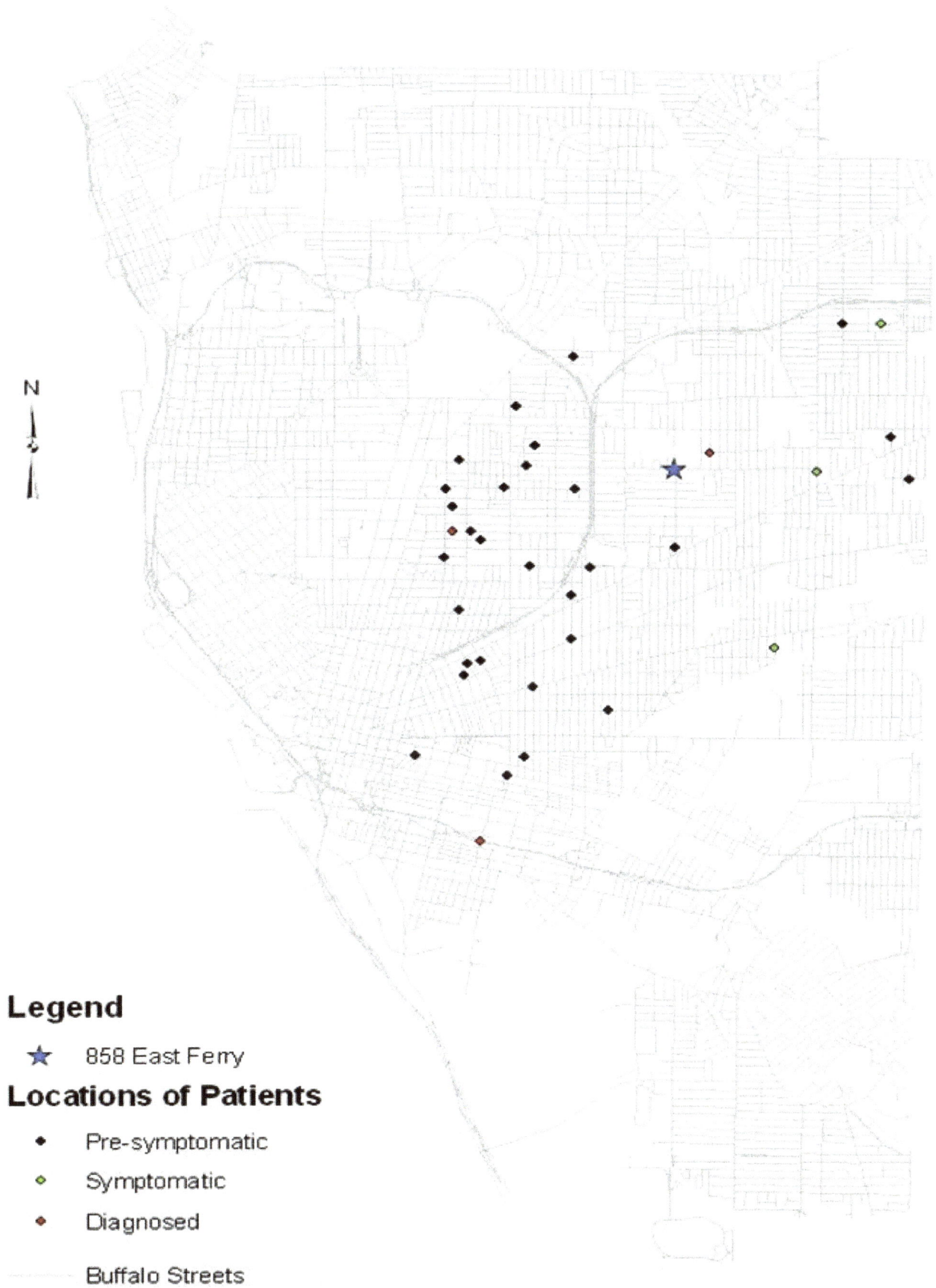

**Legend**

⭐  858 East Ferry

## Locations of Patients

◆  Pre-symptomatic

◇  Symptomatic

◆  Diagnosed

▭  Buffalo Streets

**Figure 6.** Residential patterns of Buffalo lupus project survey participants, 1960-1969, n=50. Source: Author.

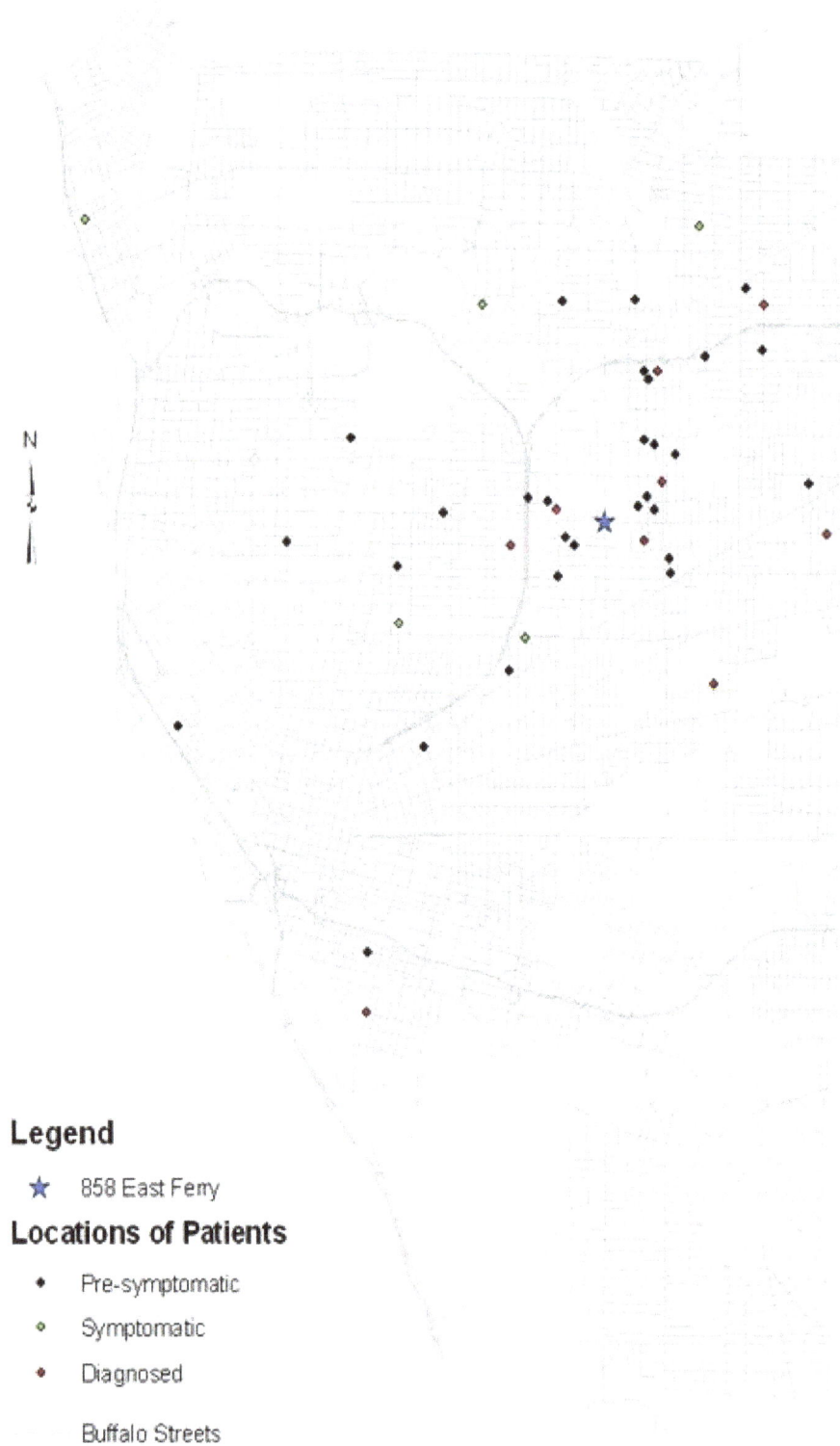

**Legend**

⭐   858 East Ferry

**Locations of Patients**

●   Pre-symptomatic

◇   Symptomatic

●   Diagnosed

    Buffalo Streets

**Figure 7.** Residential patterns of Buffalo lupus project survey participants, 1970-1979, n=57. Source: Author.

concentrated to the northeast of the site (Figures 8 and 9).

During the last reported time period (2000-2009), the site of concern was remediated. A total of 87,200 cubic yards of contaminated soil from the site and from off-site properties was stabilized on site and moved to a

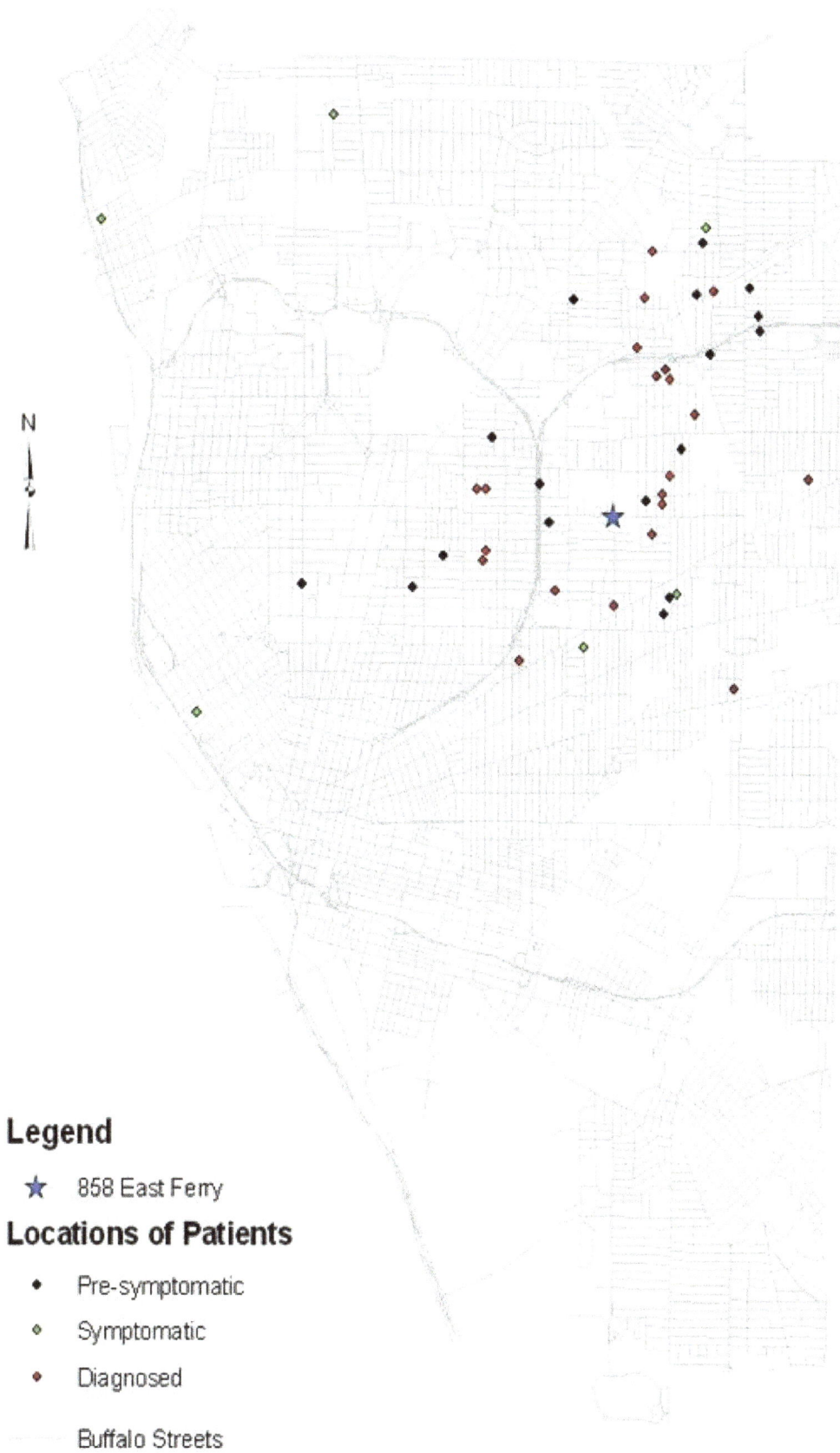

**Legend**

⭐  858 East Ferry

**Locations of Patients**

◆  Pre-symptomatic

◆  Symptomatic

◆  Diagnosed

Buffalo Streets

**Figure 8.** Residential patterns of Buffalo lupus project survey participants, 1980-1989, n= 59. Source: Author.

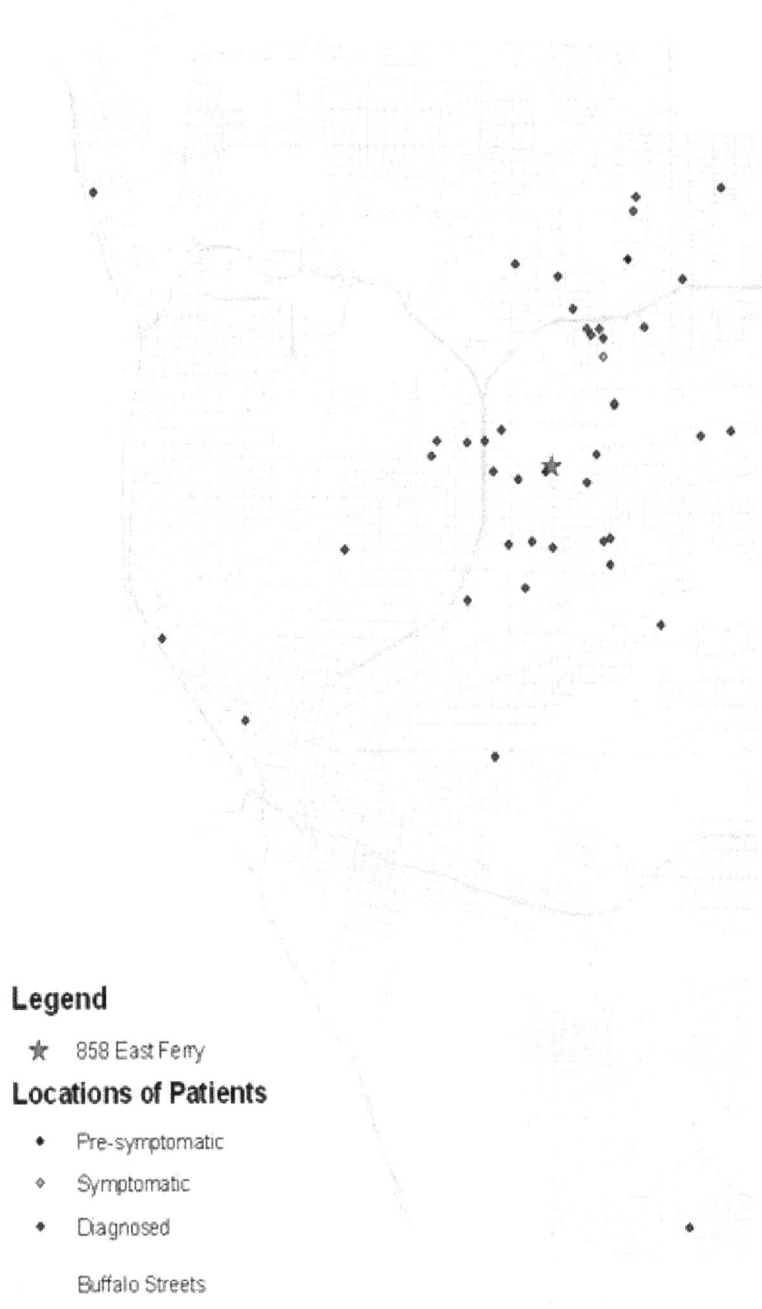

Legend

★   858 East Ferry

**Locations of Patients**

◆   Pre-symptomatic

◇   Symptomatic

◆   Diagnosed

    Buffalo Streets

**Figure 9.** Residential patterns of Buffalo lupus project survey participants, 1990-1999, n=61. Source: author.

hazardous waste landfill to achieve a clean up goal used for unrestricted future use on-site and industrial/commercial use off-site. At this time, all survey participants reporting residences in the area were formally diagnosed, and residential patterns were more dispersed, but still most densely concentrated to the northeast of the site (Figure 10).

**DISCUSSION**

The purpose of this study was to evaluate differences in SLE disease clustering according to activity of a nearby Superfund site. This study is similar to other anecdotal findings of higher rates of lupus in areas where dangerous environmental sites are located (Dahlgren et

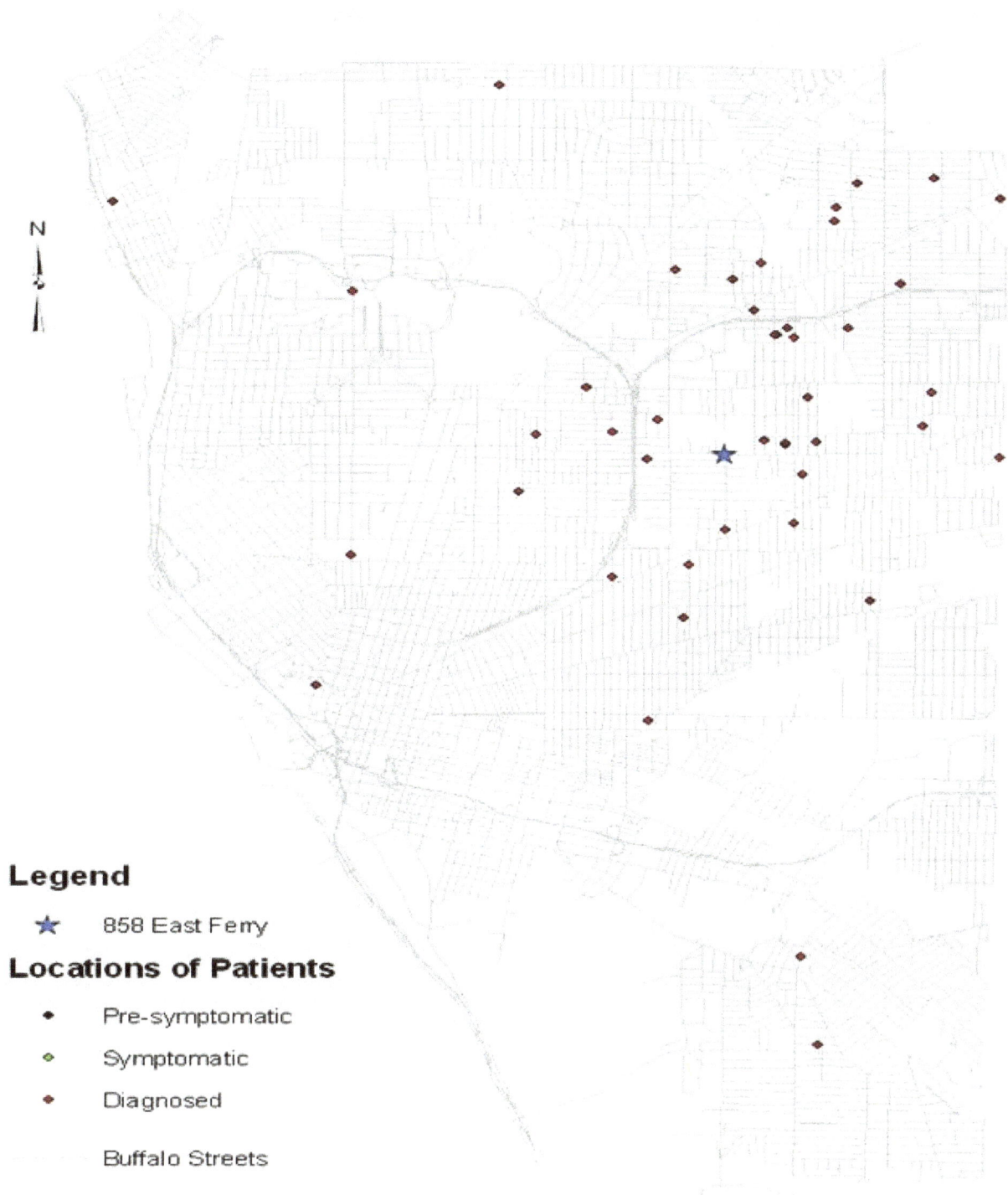

**Figure 10.** Residential patterns of Buffalo lupus project survey participants, 2000-2005, n=62. Source: author

al., 2007; Smith, 2007), but this study is unique in that it employed GIS technology to assess the geographical areas of highest risk in relation to the Superfund site. Our findings showed that in the early periods, when the site was active, all cases were clustered to the west of the site and few were symptomatic. Most cases began experiencing symptoms and were diagnosed in the periods when the site was inactive and resided more often to the northeast of the site. Such trends suggest that the impact of the site was more likely due to chronic exposure to waste rather than it acting as an acute trigger, although clustering does not prove a causal relationship between exposure and disease.

The primary advantage of the present study is that disease mapping was used to show changes in disease patterns over time in an identified cluster of SLE surrounding a nearby toxic waste site. However, small-area maps of disease are difficult to interpret in a meaningful way. With regard to the use of address geocoding for exposure assessment, residential location or any one location may not be the relevant site of exposure. It is possible that occupational location, for instance, may be as important as or more important than residential location. Therefore, the amount of time spent in different locations may need to be considered. Additionally, proximity of a residence to a source of contamination may not be synonymous with exposure. It is possible that wind direction or groundwater flow influence exposure levels.

## Conclusions

Even with these limitations, this project serves as an initial exploration of relationships between exposure and disease. More in-depth studies using GIS technology are recommended to more accurately estimate exposures within this geographic region.

**REFERENCES**

Alarcon G, Beasley T, Roseman JM, McGwin G Jr, Fessler BJ, Bastian HM, Vila LM, Tan F, Reveille JD (2005). Ethnic disparities in health and disease: the need to account for ancestral admixture when estimating the genetic contribution to both (LUMINA XXVI). Lupus, 14(10): 867-868.

Balluz L, Philen R, Ortega L, Rosales C, Brock J, Barr D, Kieszak S (2001). Investigation of Systemic Lupus Erythematosus in Nogales, Arizona. Am. J. Epidemiol., 154(11): 1029-1036.

Bernatsky S, Boivin J-F, Joseph L, Manzi S, Ginzler E, Gladman DD, Urowitz M, Fortin PR, Petri M, Barr S, Gordon C, Bae SC, Isenberg D, Zoma A, Aranow C, Dooley MA, Nived O, Sturfelt G, Steinsson K, Alarcon G, Senecal JL, Zummer M, Hanly J, Ensworth S, Pope J, Edworthy S, Rahman A, Sibley J, El-Gabalawy H, McCathy T, St Pierre Y, Clarke A, Ramsey-Goldman R (2006). Mortality in systemic lupus erythematosus. Arthritis Rheum., 54(8): 2550-2557.

Calvo-Alen J, Alarcon G, Campbell RJ, Fernandez M, Reveille J, Cooper G (2005). Lack of recording of systemic lupus erythematosus in the death certificates of lupus patients. Rheumatology, 44(9): 1186-1189.

Cattell V (2001). Poor people, poor places, and poor health: the mediating role of social networks and social capital. Soc. Sci. Med., 52: 1501-1516.

Census USBot. Buffalo city (2006). New York: Census 2000 Demographic Profile Highlights: Selected Population Group: Black or African American alone
http://factfinder.census.gov/servlet/SAFFIteratedFacts?_event=&geo _id=16000US3611000&_geoCo... Accessed 2006.

Centers for Disease Control and Prevention (2002). Trends in deaths from systemic lupus erythematosus-United States, 1979-1998. MMWR, 51(17): 371-374.

Cooper G, Dooley M, Treadwell E, St Clair E, Gilkeson G (2002). Hormonal and reproductive risk factors for development of systemic lupus erythematosus: Results of a population-based, case-control study. Arthritis Rheum., 46(7): 1830-1839.

Dahlgren J, Takhar H, Anderson-Mahoney P, Kotlerman J, Tarr J, Warshaw R (2007). Cluster of systemic lupus erythematosus (SLE) associated with an oil field waste site: a cross-sectional study. Environ. Health, 6: 8.

Gee G, Payne-Sturges D (2004). Environmental Health Disparities: A Framework Integrating Psychosocial and Environmental Concepts. Environ. Health Perspect., 112: 1645-1653.

Grossman J, Kalunian K. (2002). Chapter 2-Definition, classification, activity and damage indices. In D Wallace and B Hahn (Eds). Dubois' Lupus Erythematosus, 6 ed., pp. 19-31.

Harding A, Greer M (1993). The health impact of hazardous waste sites on minority communities: Implications for public health and environmental health professionals. J. Environ. Health, p. 55.

Kardestuncer T, Frumkin H (1997). Systemic Lupus Erythematosus in relation to environmental pollution: An investigation in an African-American community in North Georgia. Arch. Environ. Health, 52(2): 85-90.

Krishnan E, Hubert H (2006). Ethnicity and mortality from systemic lupus erythematosus in the USA. Ann. Rheum. Dis., 65(11): 1500-1504.

Mongey A-B, Hess E (2002). Chapter 3-The Role of Environment in Systemic Lupus Erythematosus and Associated Disorders. In D Wallace and B Hahn (Eds). Dubois' Lupus Erythematosus, 6 ed., pp. 33-64.

New York State Department of Environmental Conservation (2005). Focused Feasibility Report: East Ferry Street Site Work Assignment, pp. 1-31.

Oates J, Levesque M, Hobbs M, Smith EG, Molano I D, Page GP, Hill BS, Weinberg, B, Cooper GS, Gilkenson GS (2003). Nitric oxide synthase 2 promoter polymorphisms and systemic lupus erythematosus in african-americans. J. Rheumatol., 30(1): 60-67.

Office of Minority Health. Burden of Lupus. <http://www.cdc.gov/omh/AMH/factsheets/lupus.htm> Accessed 2004, pp. 1-4.

Rus V, Hochberg M (2002) . Chapter 4-The Epidemiology of Systemic Lupus Erythematosus. In D Wallace and B Hahn (Eds). Dubois' Lupus Erythematosus. 6 ed., pp. 65-83.

Sapien J (2001). Human exposure 'uncontrolled' at 114 Superfund sites EPA secrecy about sites' toxic dangers extends even to senators' inquiries: Center Public Integrity, pp. 1-5.

Smith S (2007). Study finds Boston lupus rates highest in Roxbury, Mattapan. Globe.2007, pp. 1-2.

Terrell J, Williams E, Murekeyisoni C, Watkins R, Tumiel-Berhalter L (2008). The Community-Driven Approach to Environmental Exposures: How a community-based participatory research program analyzing impacts of environmental exposure on lupus led to a toxic site cleanup. Environ. Justice, 1(2): 87-92.

Williams D, Collins C (2001). Racial residential segregation: A fundamental cause of racial disparities in health. Public Health Rep., 116: 404-416.

# 3

# Microbiological assessment and some physico-chemical properties of water sources in Akungba-Akoko, Nigeria

## A. O. Ajayi and T. O. Adejumo*

Department of Microbiology, Adekunle Ajasin University, P.M.B. 01, Akungba-Akoko, Ondo state, Nigeria.

A comparative study was carried out to determine the quality of three water sources: borehole, well and stream in Akungba-Akoko, Ondo State. The water sources were assessed for microbiological quality and physico-chemical properties (temperature, odor, color and pH). The stream had the highest plate count of $40 \times 10^5$ cfu/ml, while those of borehole and well water had the least of $12 \times 10^5$ cfu/ml and $14 \times 10^5$ cfu/ml respectively. The isolated organisms were identified as *Escherichia coli, Staphylococcus* sp., *Bacillus* sp., *Streptococcus* sp., *Klebsiella* sp., *Pseudomonas* sp., *Flavobacterium* sp., *Enterobacter* sp., *Proteus* sp. and *Pseudomonas* sp. The most probable number (MPN) of the water sample sources ranged from 8 to 120 coliforms per 100 ml, which signified undesirable level of water pollution in the area covered. Water samples from the boreholes had a coliform range of 32 to $38 \times 10^0$ cfu/ml, and were adjudged to be less prone to contamination and potable than the well and stream sources, which recorded relatively higher coliform load of 44 to $70 \times 10^0$ cfu/ml.

Key words: Bacteriology, borehole, stream, well, water assessment.

## INTRODUCTION

Water is one of the most abundant resources on which life on earth depends; in some places, availability of water is critical, limited and renewable. Shortage of water could lead to disease outbreak and economic loss, hence water is a necessity, it is a unique liquid and without it life is impossible. Water plays a vital role in the proper functioning of the earth's ecosystem. Man uses water for various purposes which include drinking, transportation, industrial and domestic use, irrigation in agriculture recreation, fisheries, and waste disposal among others (Shittu et al., 2008; Ajayi and Akonai, 2005). Water that is of a good drinking quality is important to human physiology, and man's continued existence depends so much on its availability (Lamikanra, 1999; FAO, 1997). The quality of water for drinking deteriorates due to inadequacy of treatment plants, direct discharge of untreated sewage into rivers and stream, and inefficient management of piped water distribution system (UNEP,

2001). The contaminated water therefore has critical impact on all biotic components of the ecosystem and this could affect its use for other purposes.

Water receives its bacteria spores from air, sewage, organic waste, dead plants and animal, at times almost all microorganisms may be found in water, but bacteria appeared to be the major water pollutants. Majority of the bacteria found in nature live on dead decaying organic matter as saprophytes (Peter and George., 1989). Bacteria also helps in the digestion of poisons from food and water. Presence of other species could cause various diseases to man and other animals. Water obtained from wells, boreholes, streams and river are never chemically pure, even rain water contains dissolved materials from the air as well as suspended dust intermixed with microorganisms (Prescott et al., 2008). Impurities in water may be floating as suspended matter consisting of insoluble materials of greater density than water which could be removed by sedimentation and in the form of bacteria. The bacteriological examination of water is performed routinely by microbiologists, and this will ensure a safe supply of water for drinking, bathing,

---

*Corresponding author. E-mail: toadejumo@yahoo.com.

**Table 1.** Sample sources and their physical parameters.

| Sample source | pH | Temperature (°C) | Odor | Color |
|---|---|---|---|---|
| Ilale Well | 7.50 | 28.3 | Odorless | Colorless |
| Ilale Borehole | 6.52 | 28.7 | Odorless | Colorless |
| Akua Well | 7.10 | 28.5 | Odorless | Colorless |
| Akua Borehole | 6.40 | 28.5 | Odorless | Colorless |
| Okusa Well | 7.80 | 28.3 | Odorless | Colorless |
| Okusa Borehole | 6.90 | 28.3 | Odorless | Colorless |
| Ibaka Well | 7.44 | 28.6 | Odorless | Colorless |
| Ibaka Borehole | 7.00 | 28.9 | Odorless | Colorless |
| Okeoko Stream 1 | 8.59 | 28.0 | Odorless | Colorless |
| Okeoko Stream 2 | 8.10 | 28.2 | Odorless | Colorless |

swimming and other domestic and industrial uses. Microbiological examination is usually intended to identify water sources which have been contaminated with potential disease-causing microorganisms. Such contamination generally occurs either through improperly treated sewage or improperly functioning sewage treatment system. Chemical analysis can however determine whether water is polluted and provides other useful information (APHA, 1998).

In order to determine whether water is contaminated or contain any microorganism known to be pathogenic or indicative of faecal pollution, it is necessary to carry out a bacteriological examination (analysis) on it. This study therefore aims to investigate the quality of water from boreholes, wells and streams in Akungba-Akoko with a view to determining the presence and levels of pathogenic microorganisms that are indicative of faecal pollution. Similarly, the physico-chemical parameters of the water sources were also determined because it also serves as water quality determinant factor.

## MATERIALS AND METHODS

The facilities used for this study like petri-dishes, pipettes, inoculation loop and culture media were adequately sterilized using appropriate laboratory methods (DIFCO, 1984).

## Study site and sample collection

Water samples were collected from well, borehole, and stream at different sites from Akungba-Akoko (Table 1). The samples were obtained as follows according to the practice of the inhabitants: For the boreholes, the water was allowed to flow out for 2 min before the samples were collected. The stream water was collected by dipping the sample bottle into 2 m depth of the water body, while well water samples were obtained by using a clean fetching bowl to draw out water. The first three bowls fetched were discarded, while the fourth sample was aseptically poured into a sterile sample bottle.

## Assessment of physico-chemical parameters

### (i) *pH readings*

The pH of the samples was determined using the Fisher Accument pH meter (Model 600 fisher scientific co, U.S.A). 10ml of each of the samples was poured into a sterile beaker and the anode of the pH meter was deep into it and readings were obtained when it was stable.

### (ii) *Temperature*

A simple thermometer in centigrade scale (500, 0.5 divisions) was used to measure the water temperature of each sample. The thermometer was inserted into the water sources to determine the approximate temperature.

### (iii) *Odor and color*

The odor and the color of the water samples were observed after collecting the samples by physical observation.

## Microbiological tests

### Media used and their preparation

The media used for these analyses were Nutrient agar (NA), Eosine Methylene Blue (EMB), and MacConkey broth which is a differential medium for isolation of gram negative bacteria and screens them (isolates) for lactose fermentation. Sugars used for fermentation test were Glucose, Mannitol, Lactose, Fructose and Galactose. Similarly, Motility indole ornithine fluid media was used for motility and indole test. All the media were prepared according to the manufacturer's instruction and adequately sterilized in an autoclave at 121°C for 15 min.

## Enumeration and detection of bacteria

Pour plate techniques: The aliquot of the specimens to be cultured was placed in the bottom of an empty, sterile Petri dish and melted. A cooled agar was poured over it, the plate was swapped to allow proper mixing. The agar was allowed to gel (solidified) after which the plate was incubated in an incubator at 37°C for 24 h.

Sub-culturing of isolates and stock cultures Nutrient Agar (NA) was poured aseptically into plates and allowed to solidify.

**Table 2.** Total bacterial and coliform counts of water samples.

| Sample source | Total bacterial count | | Coliform count | |
|---|---|---|---|---|
| | cfu ×$10^{-3}$ ml$^{-1}$ | cfu × $10^{-5}$ ml$^{-1}$ | cfu × $10^{0}$ ml$^{-1}$ | cfu ×$10^{2}$ ml$^{-1}$ |
| Ilale Borehole | 50 | 16 | 38 | 10 |
| Akua Well | 54 | 14 | 44 | 12 |
| Akua Borehole | 52 | 18 | 38 | 10 |
| Okusa Well | 70 | 20 | 30 | 10 |
| Okusa Borehole | 45 | 12 | 40 | 11 |
| Ibaka Well | 62 | 17 | 40 | 15 |
| Ibaka Borehole | 53 | 14 | 32 | 8 |
| Okeoko Stream 1 | 20 | 40 | 70 | 30 |
| Okeoko Stream 2 | 02 | 32 | 68 | 24 |
| Control | 0 | 0 | 0 | 0 |

Key: NA = Nutrient Agar and EMB= Eosine Methylene Blue Agar.

Specific colonies on the samples obtained were sub-cultured by streaking on the NA plates incubated at 37°C for 24 h. When primary isolation of the plates has been properly streaked, individual colonies was picked and incubated on fresh NA. Subsequent sub-culturing was carried out until pure cultures of the different isolates were obtained. These pure isolates were transferred onto agar slants in McCartney bottles and kept in the refrigerator as stock culture for subsequent tests during identification.

Total plate counts: The heterotrophic plate count (HPC)/ total count was carried out to provide an estimate of the total number of bacteria in each of the samples that would develop into colonies during the period of incubation on Nutrient agar and Eosine methylene blue Agar plates. This test detects a broad group of bacteria including the pathogens, non pathogenic and opportunistic pathogens. The laboratory procedure involves making serial dilution of the sample in sterile distilled water and cultivating $10^{-3}$ and $10^{-5}$ then $10^{0}$ and $10^{-2}$ dilution factor into the center of Petri dish. The prepared media were allowed to cool to about 45°C before they were added to the dilution factors. The plates were incubated at 37°C for 24 h in inverted position to prevent condensation from the lid to the agar, after which the number of the colonies formed was counted. The acceptable value of the total number of Colony Forming Units (CFU) during the plate count for potable water was a total of less than 102 per ml.

Coliform count: Most probable number (MPN) method: This was done as recommended by standard method (APHA, 1998). The materials and media used for the analysis consisted of the followings. Fermentation tubes with aluminum caps, Durham tubes, MacConkey Broth (Single and double strength) inoculating loop, Bunsen burner, syringes (10, 5 and 2 ml). The most probable number tube fermentation technique is performed in three stages: Presumptive test, confirmative test and completed test.

Presumptive tests: With sterile pipettes, 1 ml of water samples was dispensed into two sets of 5 tubes containing 5 ml of sterile single strength MacConkey broth: 10 ml of each sample was also dispensed into a set of test tubes containing 5 ml of sterile double strength broth. Each fermentation tube contains an inverted Durham tube. A tube containing single strength broth was inoculated with 1 ml of the sterile distilled water to serve as control. The procedures were carried out aseptically and the tubes were recorded by for 48 h. After incubation, the results were recorded by looking for the presence of trapped gas bubble inside the Durham

indicates positive results. The MPN of coliforms in 100 mL of the water sample was estimated by the numbers of positive tubes and the results were checked on the MPN tables.

Confirmative tests: A loop full of the sample from positive tubes were transferred into a plate containing Eosine Methylene blue agar (EMB) by streaked method and incubated at 37°C for 24 to 28 h. The agar inhibits Gram positive organisms and allows the Gram negative coliforms to grow. Coliforms produce colonies with dark centers, green metallic sheen and large pinkish colonies.

The completed tests: The organisms that grew on the confirmed test media were inoculated into nutrient agar slants and tubes of MacConkey broth. After incubation at 37°C for 24 h, the broth was checked for production of gas and a Gram's stain was made from organisms on the nutrient agar slant. A positive test indicates that coliforms were present in the water sample when the tests showed Gram negative, non spore- forming rod with gas production on MacConkey Broth.

**Biochemical tests and identification of microbial isolates**:

Morphological and biochemical characteristics of the microbial isolates were used for the identification of the isolates according to Baron et al., 1990, Benson (1990) and Bitton (1994). The Bergey's Manual of determinative bacteriology by Buchanan and Gibbons (1974) was used to compare the characteristics with the results obtained.

## RESULTS AND DISCUSSION

The temperature of the water samples ranged between 28.0 and 28.9°C, while the pH ranged from 6.40 to 8.59. All the samples were colorless and odorless as shown in Table 1. The water samples from the boreholes had a coliform range of 32 to 38 × $10^{0}$ cfu/ml were adjudged to be less prone to contamination and potable than the well and stream sources which recorded relatively higher coliform load of 44 to 70 × $10^{0}$ cfu/ml (Table 2).

This study showed a comparative nature of the bacteria isolates obtained from different water sources in Akungba-Akoko. The assessment of microbiological quality of water from different sources was essential for

**Table 3.** Morphological and biochemical characteristics of isolates.

| Morphological characteristics of isolates on solid media | Gram's staining | Cell morphology | Catalase | Motility | Indole | Starchy hydrolysis | Glucose | Lactose | Fructose | Galactose | Mannitol | Identification |
|---|---|---|---|---|---|---|---|---|---|---|---|---|
| Circular Creamy Flat Entire Moderate Smooth | − | Short Rod | + | + | + | + | AG | AG | − | + | + | Escherichia coli |
| Irregular White Flat Entire Moderate Smooth | − | Cocci in chains | − | − | − | + | AG | + | AG | − | AG | Streptococcus sp. |
| Circular Creamy Low convex Smooth Moderate Smooth | − | Rod | + | − | − | + | AG | + | − | + | + | Enterobacter sp. |
| Circular Creamy hite Raised Rhizoid Moderate Dull | − | Motile rod | + | + | − | + | AG | AG | − | − | − | Pseudomonas sp. |
| Irregular Pale White Flat Smooth Moderate Smooth | + | Cocci in clusters | + | − | − | + | + | − | − | + | + | Staphylococcus sp. |
| Circular Yellow Flat Undulated Moderate Dull | − | Rod | − | − | − | − | AG | AG | AG | + | + | Flavobacterium sp. |
| Circular Creamy hite Raised Rhizoid Moderate Dull | − | Rod | + | + | − | − | AG | − | − | − | + | Pseudomonas sp. |
| Irregular Creamy Flat Undulated Moderate Rough | + | Motile -rod | − | + | AG | − | − | AG | − | + | + | Proteus sp. |
| Circular Opaque Raised Entire Moderate Smooth | − | Rod | − | − | − | − | AG | AG | − | + | + | Klebsiella sp. |
| W10- Circular White Flat Entire Moderate Smooth | + | Rod | − | + | − | + | AG | − | + | + | + | Bacillus sp. |

**Key:** AG= Production of gas and acid, + = Positive tests, − = Negative tests.

detecting the presence or absence of organisms that might constitute health hazards in water, which could be used as a guide to monitor and protect the water sources. The total bacteria counts for all the samples were generally high, exceeding the limit of $1.0 \times 10^2$ cfu/ml which was the standard limit of heterotrophic count for drinking water (EPA, 2002).

The high total plate counts observed in stream water indicated the presence of high organic matters and related nutrient sources. The primary sources of bacterial contamination might include the surface runoff, sewage treatment facilities, natural soil/plants bacteria and improper management activities of the inhabitants like washing, refuse dumpage, faecal droppings, dipping of different materials inside the water sources. The stream water also had the highest number of coliform. Various groups of microorganisms were isolated and identified during the study. They include *Escherichia coli*, *Streptococcus* sp., *Enterobacter* sp., *Pseudomonas* sp., *Staphylococcus* sp., *Flavobacterium* sp., *Pseudomonas* sp., *Proteus* sp., *Klebsiella* sp., and *Bacillus* sp. (Table 3).

The presence of some of these organisms is indicative of water contamination.

Bacteriological tests are extremely sensitive and specifically designed to reveal the evidence of water pollution. A great majority of quality problem with community drinking water are related to fecal contamination. Although, a significant number of serious probes may occur as a result of chemical contamination from variety of natural and man- made sources.

Water from the wells, boreholes, streams and other sources may look clean and have no undesirable odor or taste. However, it is unfortunate that the pathogens found in these water sources can be harmful by causing serious illnesses. In this study, the boreholes had the lowest bacteria count which may be due to its depth. In some cases, improper construction of boreholes, its pipes, proximity to toilet facilities and various human activities around the borehole could contaminate the water. Well water could also be contaminated due to shallowness, animal waste, closeness of refuse dump sites, proximity to toilet facilities (closeness to latrines), improper placement of well covers, using different drawing bowls and plunging of bowls/bucket directly from the soil into the well as well as human activities around it (Bitton, 1994).

The total coliform for samples examined during this study were exceedingly high as against the EPA maximum contamination level (MCL) for coliform bacteria in drinking water of zero total coliform per 100ml of water (EPA, 2002). The high coliform count obtained in the samples may be an indication that the water sources were faecally contaminated (EPA, 2002; Osunide and Enuezie, 1999). None of the water sources in this study complied with the EPA standard for coliforms in water. The microorganisms generally isolated in this study include *Escherichia coli, Streptococcus* sp., *Enterobacter* sp., *Pseudomonas* sp., *Staphylococcus* sp., *Flavobacterium* sp., *Pseudomonas* sp., *Proteus* sp., *Klebsiella* sp., and *Bacillus* sp. (Table 3). The presence of some this organisms signifies contamination of water from some domestic sources. The *Staphylococcus* species is known to produce enterotoxin (Okonko et al., 2008). *Proteus* species is an intestinal flora, but also widely distributed in soils and water (Schlegel, 2002). P*seudomonas aeruginosa* is an example of non-faecal coliforms, while *E. coli* are a fecal coliform. Other organisms encountered include *Pseudomonas* species, *Klebsiella* species, *Streptoccocus* species and *Bacillus* species.

## Conclusion

This study showed that most domestic water sources are potential health risks for consumers. Among the strategies to adopt to combat water pollution problem is the promotion of household water storage, and the need to improve on personal behavior and hygiene practices to reduce microbial load in water supply. The key to avoiding bacterial contamination of water include proper well and borehole location and construction, control of

human activities to prevent sewage from entering the water bodies. It is evident that water-borne diseases are due to improper disposal of refuse, contamination of water by sewage and surface runoff. Appropriate programmes must be put in place to educate the general populace on the need to purify water to make it fit for drinking and other domestic purposes.

## REFERENCES

Ajayi AO, Akonai KA (2005). Distribution pattern of enteric organisms in the Lagos Lagoon. Afr. J. Biomed. Res., 8(3): 163-168.

American Public Health Associated (APHA) (1998). Standard method for examination of water and waste water, 20th Ed. American Public Health Associated Inc, New York. American Water Work Associated (AWWA). Research Foundation, 1993. Bacterial examination of water. West Quincy Avenue Denver, pp. 81-85.

Baron EJ, Sydney MF (1990). Bailey and Scott's Diagnostic Microbiology. The C.V. Mosby Co., St. Louis, p. 861.

Benson HJ (1990). Microbiological Applications: A Laboratory Manual in General Microbiology. Wm. C. Brown Publishers, Dubuque, p. 459.

Bitton G (1994). Waste Water Microbiology. Gainesville, New York Wiley – Liss, p. 118.

Buchanan RE, Gibbons NE (eds) (1974). Bergey's Manual of Determinative Bacteriology, 8th Edition. The Williams and Wilkins Co., Baltimore, p. 124.

DIFCO (1984). Dehydrated Culture Media and reagents for Microbiology. Tenth Edition. DIFCO Laboratories Detroit Michigan 48232 U.S.A.

Environmental Protection Agency US (EPA) (2002). Safe Drinking Water Act Amendment, http://www.epa. gov/safe/mcl/Html.

Food and Agriculture Organization (FAO) (1997). Chemical Analysis Manual for Food and Water. 5th Ed. FAO, ROME, 1: 20-26.

Lamikanra A (1999). Essential Microbiology for Students and Practitioners of Pharmacy, Medicine and Microbiology. 2nd Edition. Amkra books, p. 406.

Okonko IO, Adejoye OD, Ogunusi TA, Fajobi EA, Shittu OB (2008). Microbiological and Physicochemical analysis of different water samples used for domestic purposes in Abeokuta and Ojota, Lagos State, Nigeria. Afr. J. Biotechnol., 7(5): 617-621.

Osunide MI, Eneuzie NP (1999). "Bacteriological analysis of ground water" Nig. J. Microbiol., 13: 47-54.

Peter HR, George BJ (1989). General Microbiology 2nd Edition. Water Treatment. George Brooks, Time mirror/Mosby College publishing. St Louis, USA, p. 231.

Prescott LM, Harley JP, Kleins DA (2008). Microbiology 11th Edition. Treatment of Water McGram Hill Company New York. P. 1088

Schlegel HG (2002). General Microbiology. 7th Edition Cambridge University Press. P 480.

Shittu OB, Olaotan JO, Amusa TS (2008). Physicochemical and Bacteriological Analysis of water used for Drinking and Swimming Purposes in Abeokuta, Nigeria. Afr. J. Biomed. Res., 2: 285-290.

UNEP (2001). State of Water Sanitation and Health Programme Nepal. United Nation Environment Programme (UNEP).

# 4

# Influence of pollutants on bottom sediment of sewage collecting Kalpi (Morar) River, Gwalior, Madhya Pradesh (M. P.)

### Avnish K. Verma* and D. N. Saksena

Limnology Research Unit, Aquatic Biology Laboratory, SOS in Zoology, Jiwaji University, Gwalior-474011 (M. P.), Madhya Pradesh, India.

**To investigate the nutrient value of bottom soil pH, conductivity, potassium, exchangeable potassium, total phosphorus, available phosphorus, organic carbon, total nitrogen, available nitrogen, calcium and magnesium were analyzed from monthly samples collected from six sampling stations during two year of study, that is, April 2002-March 2004 of study. Soil from station A was found unpolluted due to low level of pollution load and less human activities. While at other stations sewage and municipal wastes increased the amount of calcium, magnesium and other nutrient.**

**Key words:** River, sediment pollution, nutrient level.

## INTRODUCTION

The quality of soil of aquatic environment plays a major role in determining the fertility of water body. Soil acts as a reservoir of nutrients and several biogeo-cycling taken places at bottom and exchange of nutrients takes at bottom water interface. Nutrients are also derived from the drainage water and mineralization of organic matter. The influence of soil type and nutrients, on the benthos is well known. The role of sediment nutrients is very much useful in determining nutrients of the river ecosystem.

Nutrients are also derived from the drainage water and mineralization of organic matter. The bottom sediment also provides the shelter for various life forms including both micro and macrozoobenthic organisms. Sreenivasan (1976) reported that, the bottom sediment is the main source for different forms of nitrogen. Jhingran (1991) insisted that the organic carbon of sediment is a common constituent of all organic matter and this can also be a measure of bacterial activity. Kumar and Ramachandra (2003) have observed low values of conductivity in Sharavathi river Karnataka may result in least nutrient transfer, complexation and exchange of elements.

A survey repot on Mithi river (M.P.C.B., 2004) showed that the river had pH value between 6.5-8.5 and sulphate between 1500-1400 mgl⁻¹ due to high load of municipal and domestic sewage. Due to sediment contamination, several changes take place in the sediment including that of degradation of bottom-feeding invertebrate communities, increased incidence of fish tumors and other abnormalities (Crane, 2005). Padmaja et al. (2008) studied the sediment quality of water tank in near Medak District A.P. and suggested its suitability in agriculture due to rich in nutrients.

## MATERIALS AND METHODS

The sediment samples have been collected from different sampling stations by Ekman's dredge, scoop and direct by hand wherever possible and various physical, chemical parameters have analyzed after Trivedy and Goel (1986) and Chopra and Kanwar (1999).

The samples were kept in polythene bags and transported to laboratory as early as possible. Apparent density, pH and water holding capacity were analyzed immediately. For other parameters, soil sample were dried in air under the shadow in natural conditions. Soon after drying, stones and similar objects are picked up and the soil is grounded in a mortar to break up aggregates or lumps, taking care not to break actual soil particle. The soil was then passed with 2 mm sieve. The mesh size allows all the nutritionally important factors to pass through. Approximately, 4-5 gm of soil was grounded to get more fine particles which can pass through mesh size of 0.5 mm.

---
*Corresponding author. E-mail: ak_water79@yahoo.com.

**Table 1.** The mean values of Soil Characteristics in Kalpi (Morar) river during April 2002 - March 2004.

| S/No. | Parameter | A | B | C | D | E | F |
|-------|-----------|-----|-----|-----|-----|-----|-----|
| 1 | Apparent density ( g.cm$^{-3}$) | 1.93 | 1.34 | 1.85 | 1.76 | 1.59 | 1.65 |
| 2 | Water holding capacity (%) | 25.39 | 26.53 | 26.69 | 26.06 | 28.57 | 26.85 |
| 3 | Soil pH | 7.36 | 7.51 | 7.43 | 7.48 | 7.60 | 7.52 |
| 4 | Conductivity (mS.cm$^{-1}$) | 0.85 | 1.63 | 1.72 | 1.76 | 1.00 | 0.94 |
| 5 | Total phosphorus (%) | 0.067 | 0.069 | 0.071 | 0.079 | 0.069 | 0.068 |
| 6 | Available phosphorus (%) | 0.017 | 0.018 | 0.020 | 0.024 | 0.022 | 0.016 |
| 7 | Total nitrogen (%) | 0.034 | 0.052 | 0.053 | 0.054 | 0.036 | 0.035 |
| 8 | Available nitrogen (mg..100 gm$^{-1}$ ) | 23.99 | 31.23 | 32.41 | 33.56 | 31.31 | 26.80 |
| 9 | Organic carbon (%) | 1.06 | 2.06 | 2.23 | 2.37 | 1.46 | 1.19 |
| 10 | C and N ratio | 30.94 | 41.53 | 40.68 | 45.37 | 45.21 | 35.20 |
| 11 | K (mg.100 g$^{1}$ ) | 62.10 | 74.06 | 74.44 | 77.76 | 67.80 | 66.69 |
| 12 | Exchangeable (K mg.100 g$^{1}$) | 33.12 | 40.10 | 42.77 | 43.40 | 36.46 | 33.29 |
| 13 | Ca (m.e.100 g$^{-1}$) | 1.25 | 1.43 | 1.50 | 1.53 | 1.30 | 1.24 |
| 14 | Mg (m.e. 100 g$^{-1}$) | 0.40 | 0.45 | 0.51 | 0.52 | 0.44 | 0.43 |

**Physical characteristics**

Apparent density, water holding capacity and specific conductivity.

**Chemical characteristics**

pH, total phosphorus, available phosphorus, total nitrogen, available nitrogen, organic carbon, carbon and nitrogen ratio, potassium, exchangeable potassium, calcium and magnesium.

## RESULTS AND DISCUSSION

The physical and chemical properties of soil from all six stations are presented in the Table 1 and Figures 1 - 14. Soil of river were noticed with higher value of nutrients, this was occurred due to sedimentation through sewage and waste pouring.

## Apparent density

Apparent density is a weight measurement in which the entire soil volume is taken in to consideration (Buckman and Brady, 1950) and which allows knowing the nature of bottom. In a study by Kumar and Ramchandra (2003) apparent density of Sharavathi river in Kerala was found between 0.783 and 1.475 g. cm$^3$ indicating rocky bottom condition. Physico-chemical analysis of bottom soil reveals minimum density of 0.923 and maximum of 1.092 g.m$^{-3}$ in Khandaleru reservoir, Tumayya (Asadi et al., 2008). In the present study Kalpi (Morar) river was having apparent density (1.34 to 1.93 g.cm$^{-3}$) similar to those of other workers (Figure 1).

**Water holding capacity**

The water holding capacity is determined to know the water content of soil. The moisture content of sediment samples was found up to 37.702% from Sharavathi river (Kumar and Ramchandra, 2003). In the present study, the water holding capacity was found 38.61% at station D during October, 2003 (post manson) the minimum of 25.39% value was evident at station A (Figure 2).

**pH**

The pH of soil is known to control both biogenic and abiogenic reactions and has a direct influence on the sediment nutrient status as well as the condition of the overlying water. The vertical concentration gradient of ions in the water, which controls the benthic communities, is closely related to pH (Banerjea, 1967). Hosetti et al. (1995) observed soil pH of 7.2 in Tunga river at Shimoga. Pathak et al. (2001) observed pH ranging from 6.7 to 7.2 of Mahanadi river from Durgapalli to Narsinghpur. Kumar and Ramchandra (2003) showed pH value varying between 6.37 and 7.39 in the sediment of Sharavathi river.

The pH values observed by above workers are least favorable for the bottom decomposition. pH of the sediments was to be found high in Mithi river of Mumbai due to domestic and municipal sewage pouring in to the river (MSPCB, 2004). The pH of sediment in Ase river, Nigeria was quite high during summer due to high concentration of heavy metals (Iwegbue et al., 2006) the similar trend of pH were noticed at Niger Delta by (Iwegbue et al., 2007). The pH of Kalpi river was noticed always moderately alkaline and varying from 7.1 – 8.14 during the study (Figure 3).

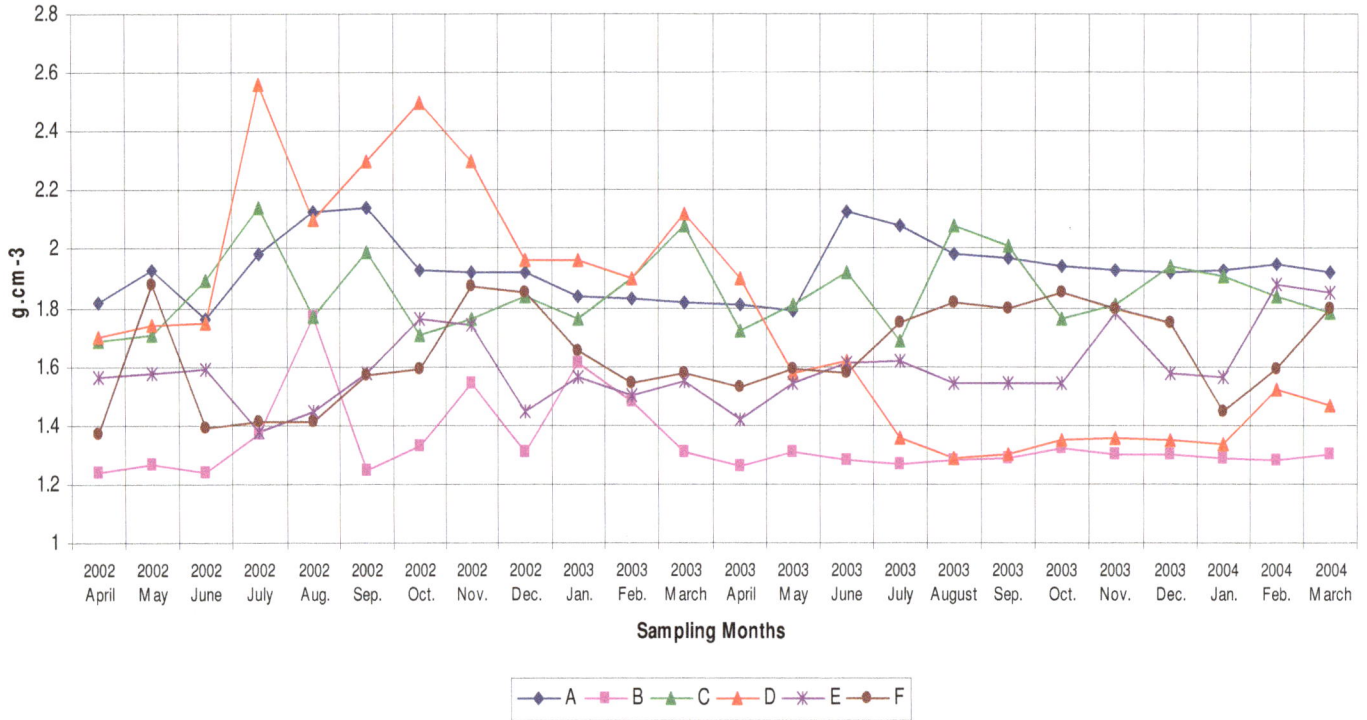

**Figure 1.** Showing sediment apparent density at various stations on Kalpi (Morar) river.

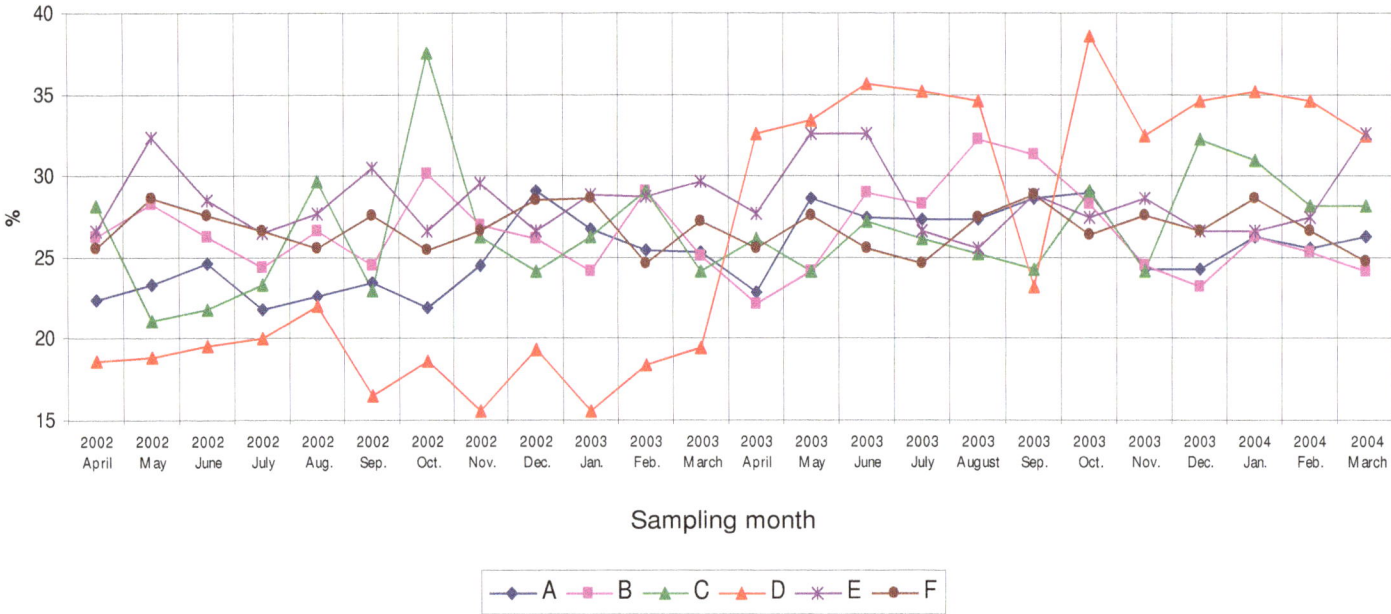

**Figure 2.** Showing water holding capacity at various stations on Kalpi (Morar) river.

## Specific conductivity

Specific conductivity of sediment has an influence on the water medium. The fluctuations in conductivity of soil and that of water are quite proportional and ahs become a common trend. Ayyappan and Gupta (1985) have observed an increase in dissolved organic matter that affected the values of conductance. The specific conductivity values were lower in Sharavathi river, Kerala which resulted from least nutrient transfer, complexation and exchange

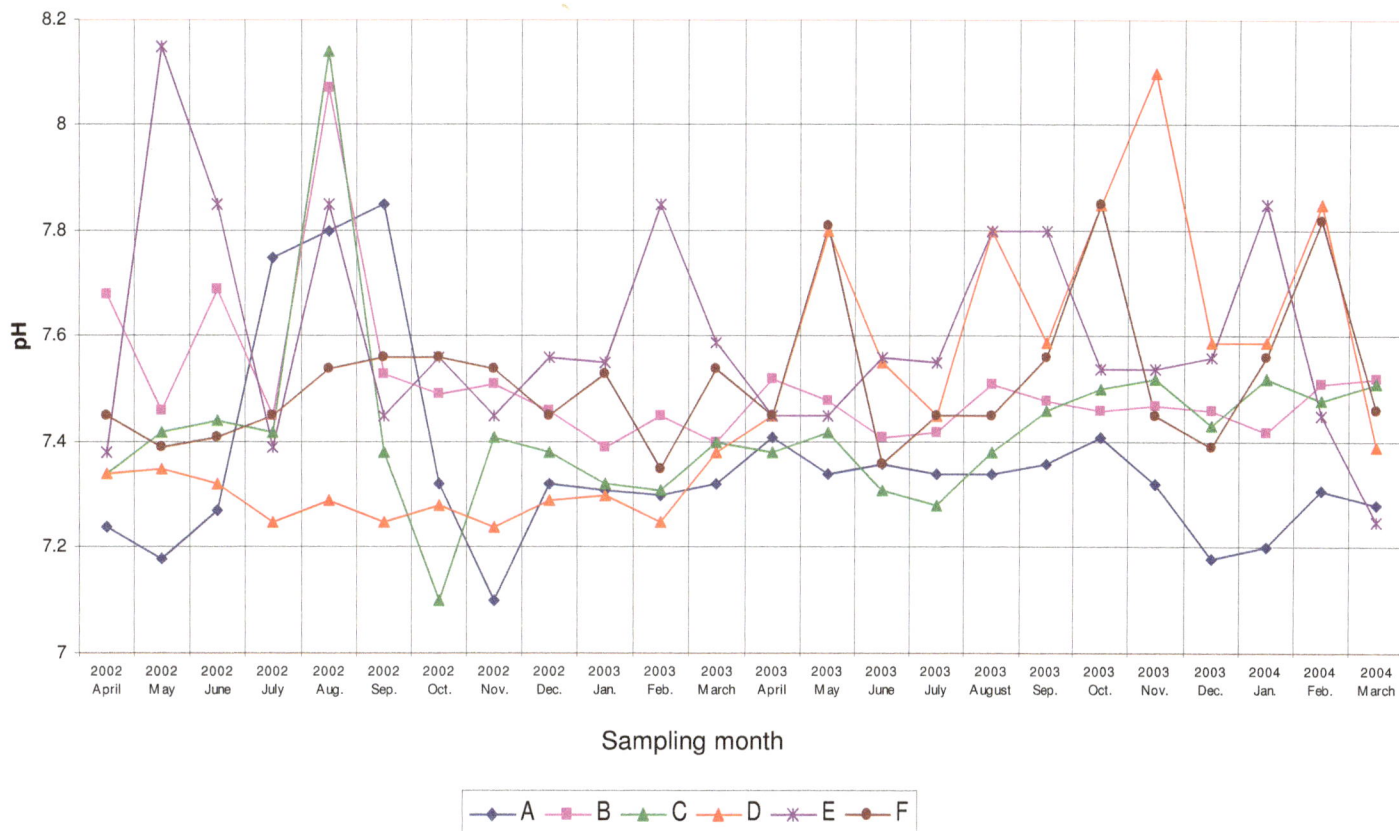

**Figure 3.** Showing soil pH at various stations on Kalpi (Morar) river.

of elements (Kumar and Ramchandra, 2003).

According Assadi et al. (2008), the specific conductivity values of the soil were within permissible limit and do not indicative of contamination. In the Kalpi (Morar) river, conductivity was minimum (0.81 m.Scm$^{-1}$) at station A in the month of February, 2003, where the sediment was almost without contamination, while higher of value of specific conductivity were noted at station D (2.51 m.Scm$^{-1}$) followed by station C (2.14 m.Scm$^{-1}$), and station B (2.01 m.Scm$^{-1}$). It was because of the fact that the municipal sewage started pouring at these points (Figure 4).

## Phosphorus

The high phosphorus content in all the soil profiles may be attributed to continuous addition of fertilizers, manures and sewage sedimentation. Plants in the sediments of river continuously absorb available phosphorus. Analysis of phosphorus revealed a maximum value at Ramapuram area due to surrounding agricultural lands in the downstream of Khandaleru reservoir at reservoir (Assadi et al., 2008). In the present study, the phosphorus was minimum in the sediment at station A while maximum amount was observed at station D (Figure 5).

## Available phosphorus

Ghosh (1989) observed higher amount of available phosphorus in Hoogly estuary near haldia port due to seepage of municipal sewage and industrial effluents. In Kalpi (Morar) river also, the sediments were having good amounts of available phosphorus in the areas of the river with appreciale organic matter load (Figure 6).

## Nitrogen

Nitrogen being a major nutrient plays an important role in determining the fertility. The carbon and nitrogen (C: N) ratio is also an important factor in soil fertility because it is an indicator of decomposition of organic matter. Sreenivasan (1976) reported that the bottom sediment is the main source for different forms of nitrogen. Ayyappan and Gupta (1985) observed a negative correlation between the nitrate–nitrogen content and available nitrogen content of water and the available nitrogen content sediment. Purusothama (1985) stated that increased water level cause enhancement in nitrogen content of the sediment during rainy season. In Kalpi (Morar) river, the total Nitrogen was high at places where organic pollution was higher. The high value of Nitrogen content was at

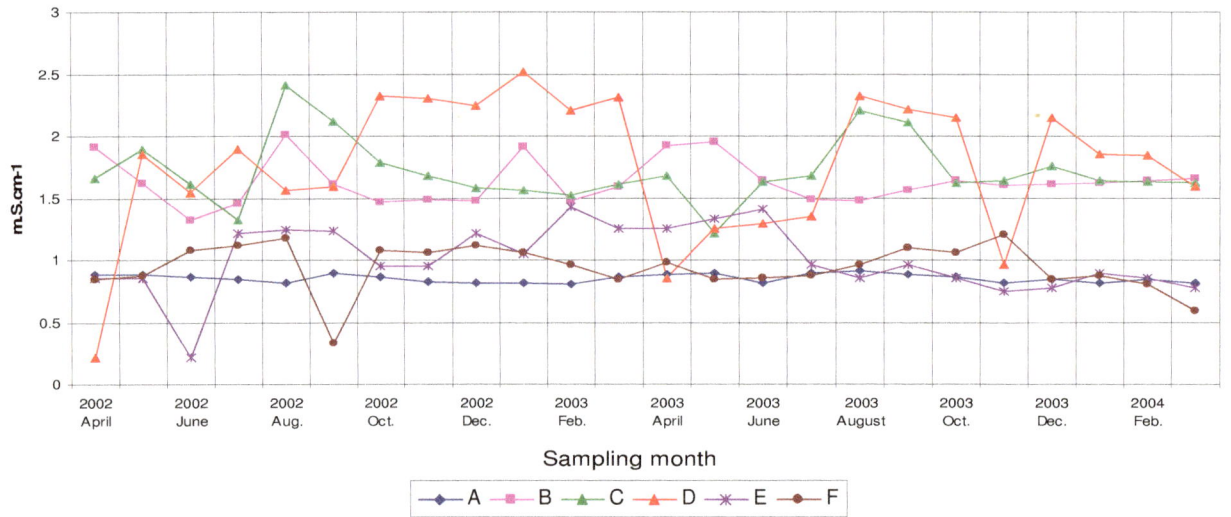

**Figure 4.** Showing specific conductivity m.S/cm at various station on Kalpi (Morar) river.

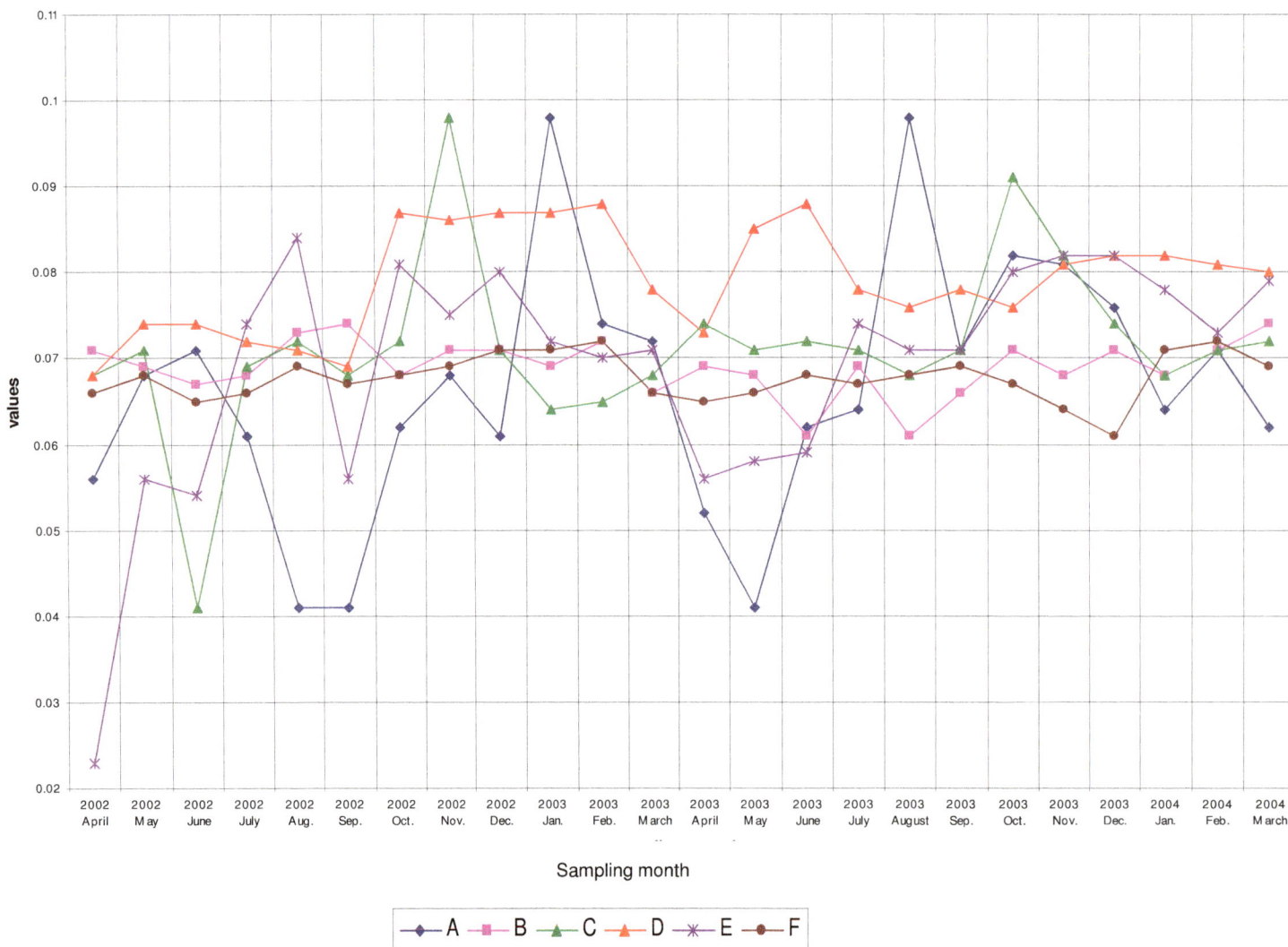

**Figure 5.** Showing total phosphorus at various stations on Kalpi (Morar) river.

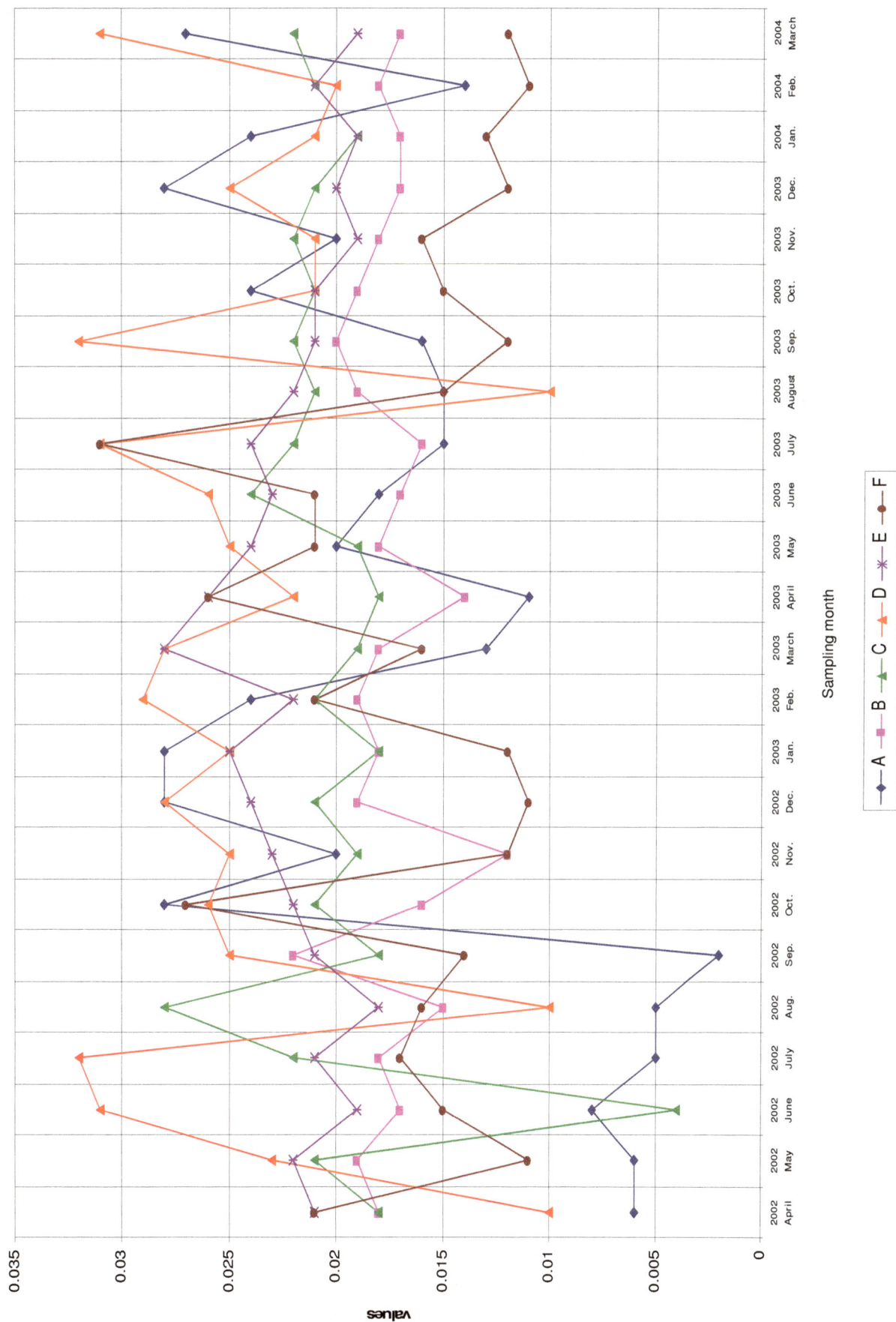

**Figure 6.** Showing available phoshphorus at various stations on Kalpi (Morar) river.

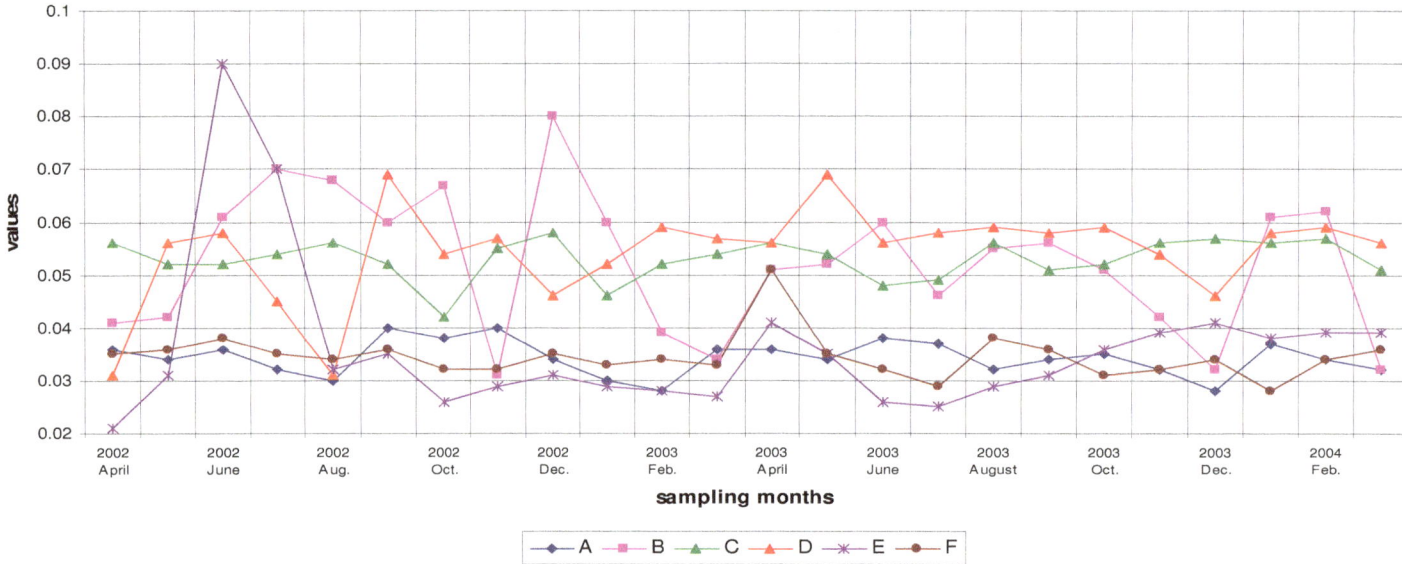

**Figure 7.** Showing total nitrogen at various stations on Kalpi (Morar) river.

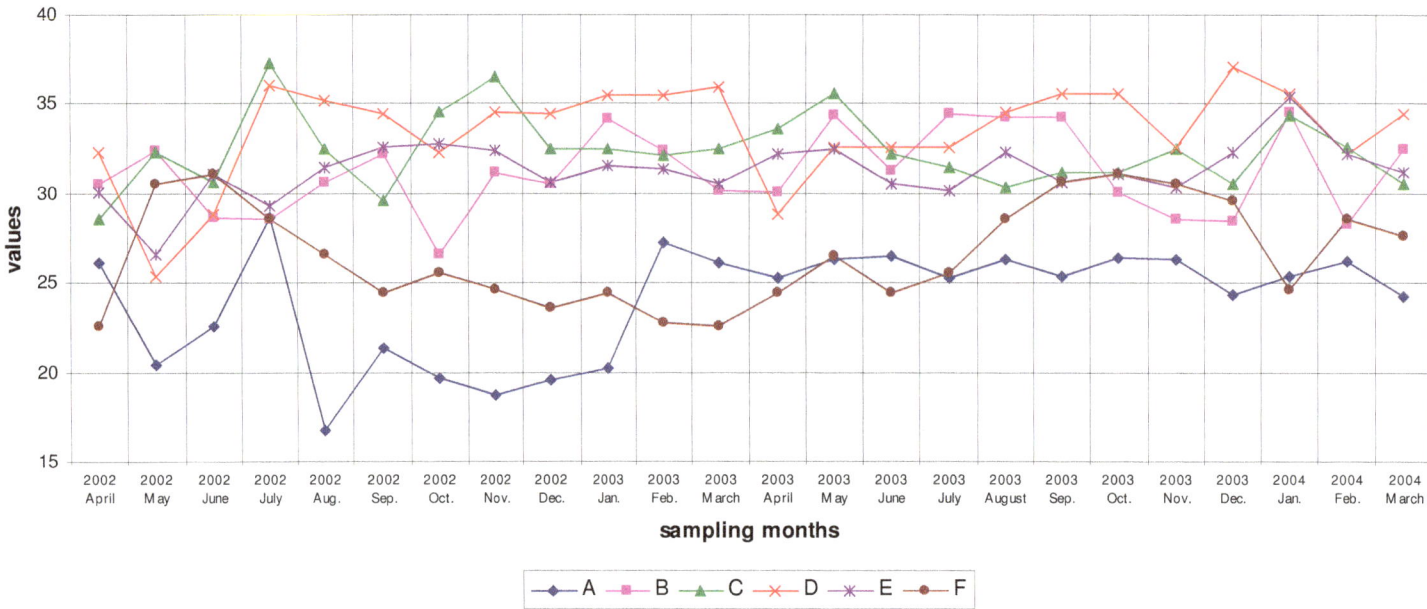

**Figure 8.** Showing available nitrogen at various stations on Kalpi (Morar) river.

station B, C and D probably due to addition organic matter in the form of sewage and a low value was obtained at station A where no organic pollution was observed (Figure 7).

## Available nitrogen

Banerjee and Pakrasi (1986) have shown a low concentration of available nitrogen in newly constructed impoundments at Sunderban. Ghosh and Chaudhary (1989) found high values of available nitrogen in Haldia

port at Hoogly estuary were due to dumping of organic wastes. Low level of available nitrogen in ponds along chilka lake may be due to the least human activity in that region (Gupta et al., 1999). Nasnolkar et al. (1996) reported higher sediment nitrogen content in Mandovi estuary.

In the present study also, the available nitrogen was obtained with its lower value at station A 16.8 mg.100 g$^{-1}$ with less organic material input and higher value of 37.10 mg.100g$^{-1}$ was obtained at station D with greater influx or sewage (Figure 8).

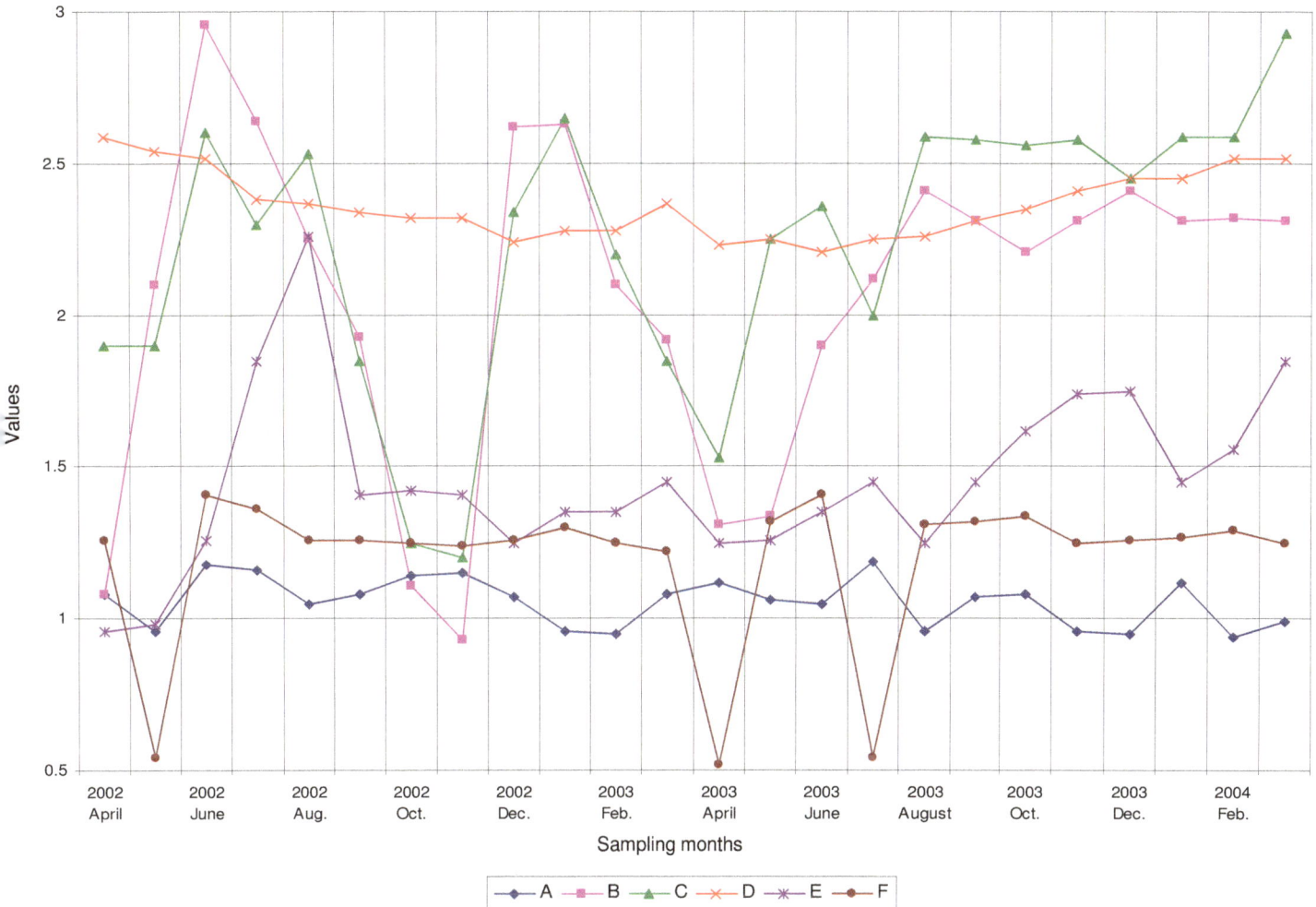

**Figure 9.** Showing organic carbon at various stations on Kalpi (Morar) river.

## Organic carbon

The organic carbon represents the organic matter in the sediment. The dead organic matter gets deposited in the bottom and undergoes chemical and bacterial decomposition. Estimation of organic carbon can serve as an important tool in determining the status of food available to the benthic fauna and also indicates the extent to which the bottom soil is fertile for the subsistence of benthic fauna. The organic carbon also exerts an influence on the available phosphorus level in the soil. Jhingran (1991) described that, the carbon is the common constituent of all organic matter and is a measure of bacterial activity. Kumar and Ramchandra (2003) observed the organic carbon is rich condition in Sharavathi river due to large amount of organic matter. The organic matter content of Ase river sediments are generally low (Iwegbue et al., 2007).

Similar low organic matter content has been reported for some river system in the Niger Delta (Horsfall et al.,

1999) due to less anthropogenic activities on both the side. Higher organic matter was observed in the soil of Khandaleru reservoir due to unplanned irrigation practices which contributed input of fertilizer into the reservoir (Assadi et al., 2008). The organic carbon content is generally high in the sediment of river Kalpi (Morar) ranging 0.93 - 2.59%, showing higher contamination. The highest values were obtained from station D (April, 2002), while the minimum values were evident 1.06% at station A, where least human activities were observed. Station D is the highly polluted point of the river (Figure 9).

## Carbon and nitrogen ratio

The carbon to nitrogen (C:N) ratio is also an important factor in soil fertility because it is indicative of the rate of decomposition of organic matter. The application of nitrogenous fertilizers cause reduction in the population of both nitrogen fixing and denitrifying bacteria and

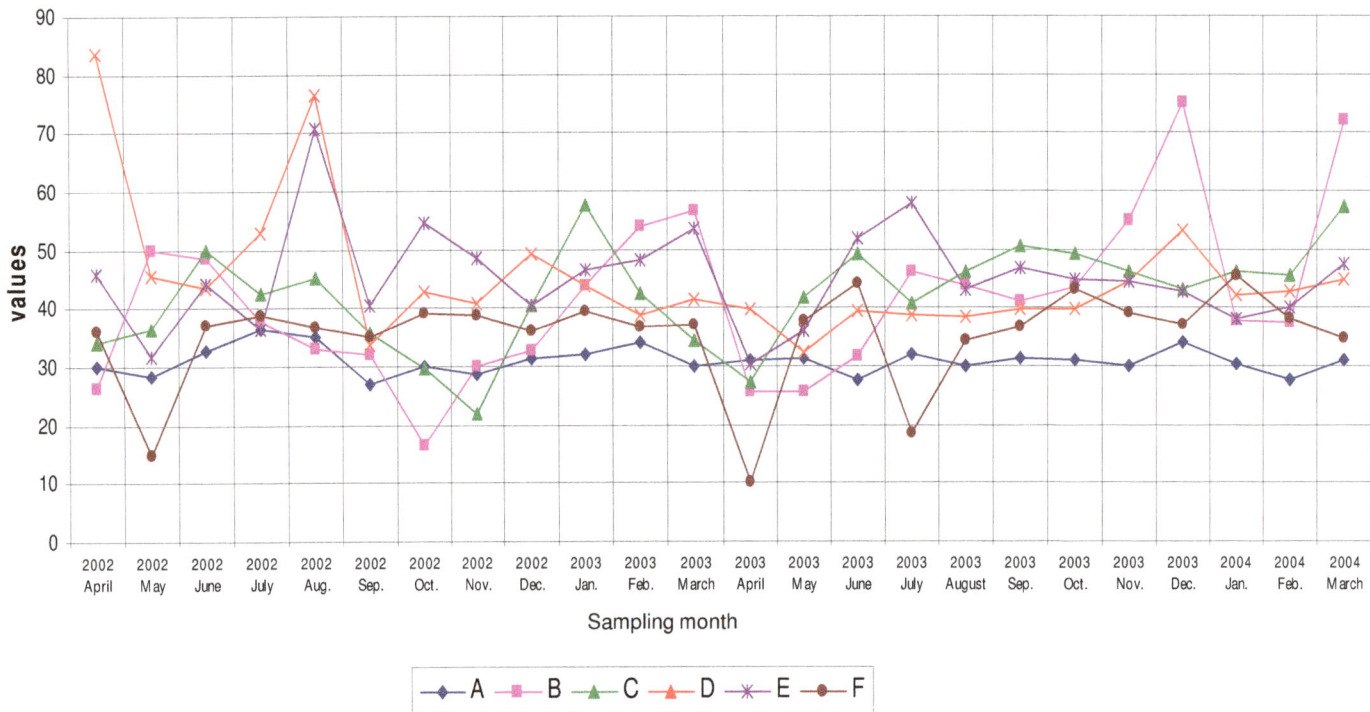

**Figure 10.** Showing carbon and nitrogen ratio at various stations on Kalpi (Morar) river.

further influenced by various factors (Hepher, 1952). In the present study the carbon and nitrogen ratio were fluctuated with the range of 30.94 to 45.37 (Figure 10).

## Potassium

The values of potassium in sediment were found ranging from 0.013 to 0.89 mg.gm$^{-1}$ in Sharavathi river which considered as low because of least pollution and human activities (Kumar and Ramchandra, 2003). In the present study, the potassium was found in good amount with 42 showing higher concentration at station D which is highly polluted (Figure 11).

## Exchangeable potassium

The mean exchangeable potassium was observed to be at station A and it was found in increasing order from B to D. Again, the exchangeable potassium was reduced at station E and F as these stations are pollution free areas (Figure 12).

## Calcium

Calcium promotes the activity of soil bacteria concerned with the fixation of the free nitrogen or the formation of nitrates from organic forms of nitrogen. Calcium deficiency is commonly associated with the acidity, which will lead to the accumulation of toxic salts of iron, aluminium and manganese in the sediments. In the present study, the amount of calcium in soil was higher in contaminated soils in comparison to uncontaminated soil, probably due to the sedimentation of water pollutants. The higher amount of calcium in the water itself reflects the high calcium content in the bottom soil. The highest mean value of calcium was in soil 1.53 m.e.100 g$^{-1}$ at station D while lowest mean was 1.25 m.e.100 g$^{-1}$ at station A in Kalpi (Morar) river (Figure 13) mg.100 g$^{-1}$ at station A to 85.3 mg.100 g$^{-1}$ at station D.

## Magnesium

In Sharavati river in Kerala was highly dependent on the parent materials or rock (Kumar and Ramchandra, 2003). Usually, amount of magnesium is higher in organic matter rich soils than in other soils. At station D the highest value of magnesium was obtained due to high sewage load at this point the lowest value of this element was evident as this point is pollution free (Figure 14).

## Conclusion

Apparent density and water holding capacity of river

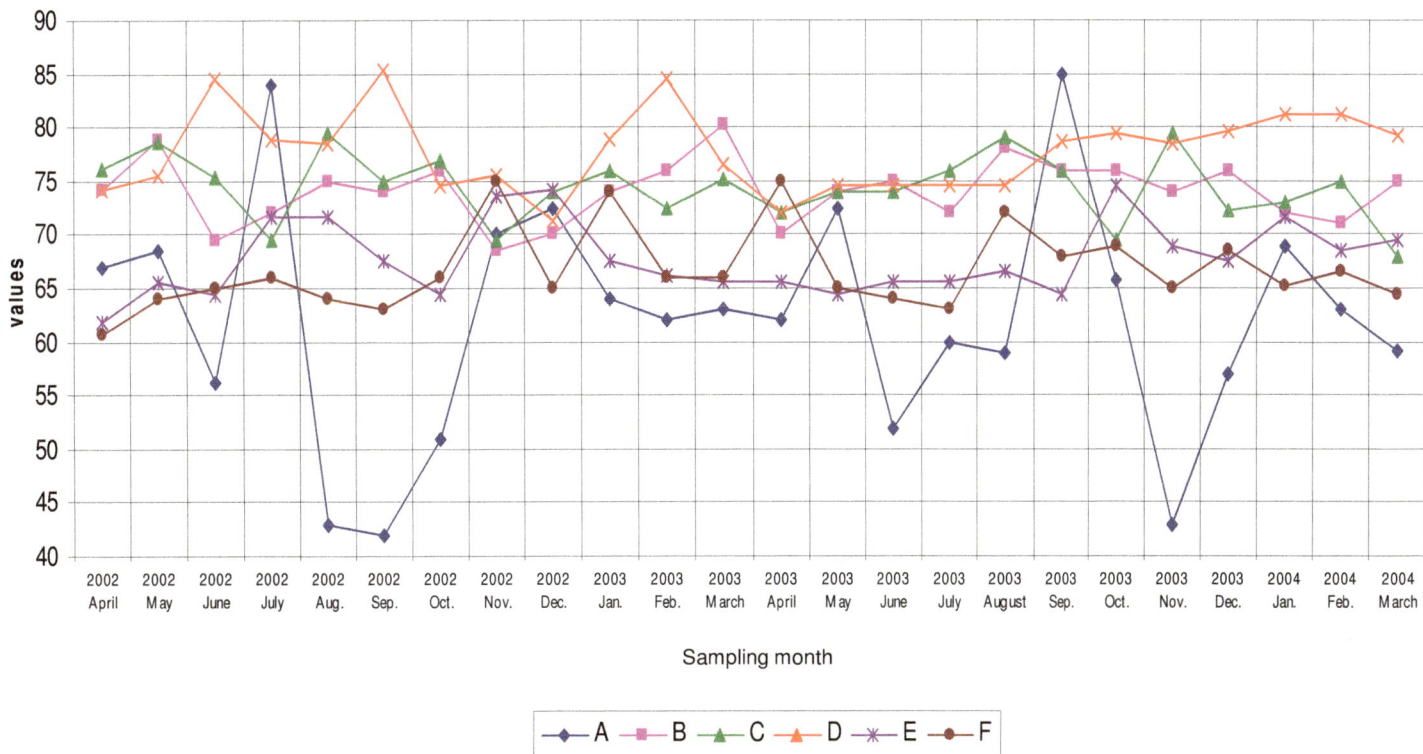

**Figure 11.** Showing potassium (mg/100 g) in sediment at various stations on Kalpi (Morar) river.

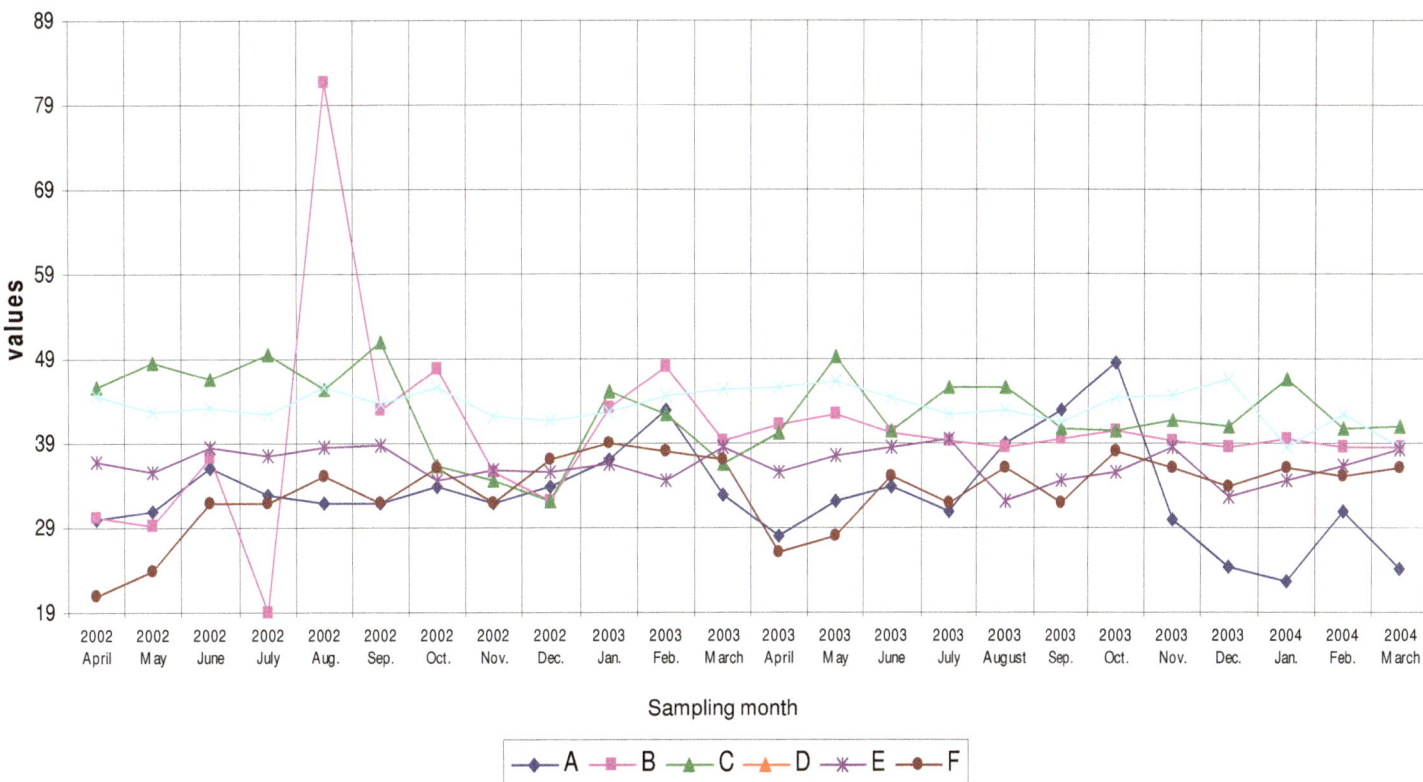

**Figure 12.** Showing exchangeable potassium at various stations on Kalpi (Morar) river.

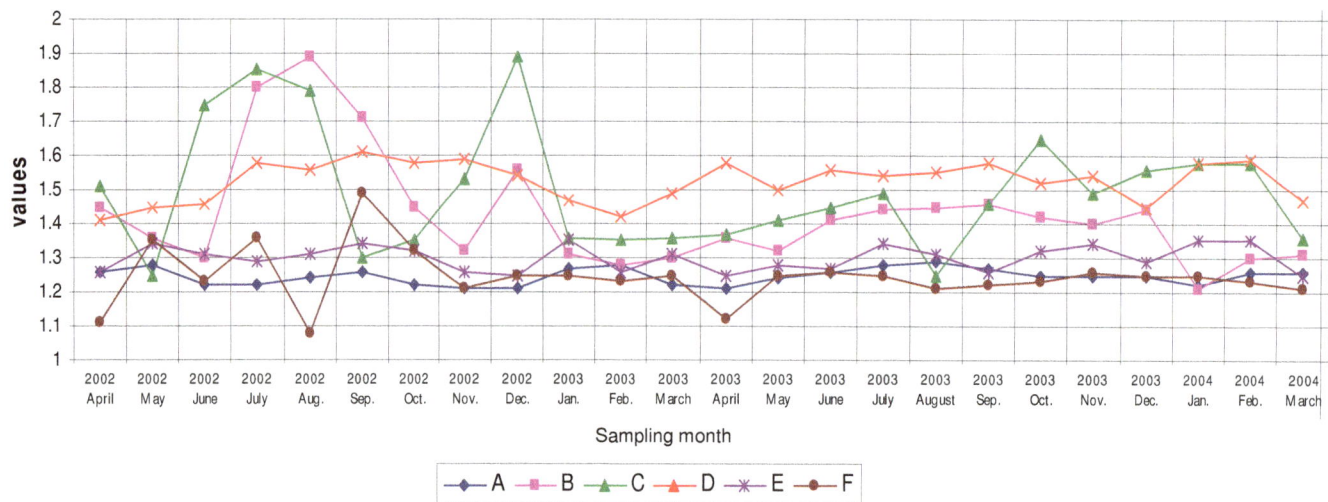

**Figure 13.** Showing calcium at various stations on Kalpi (Morar) river.

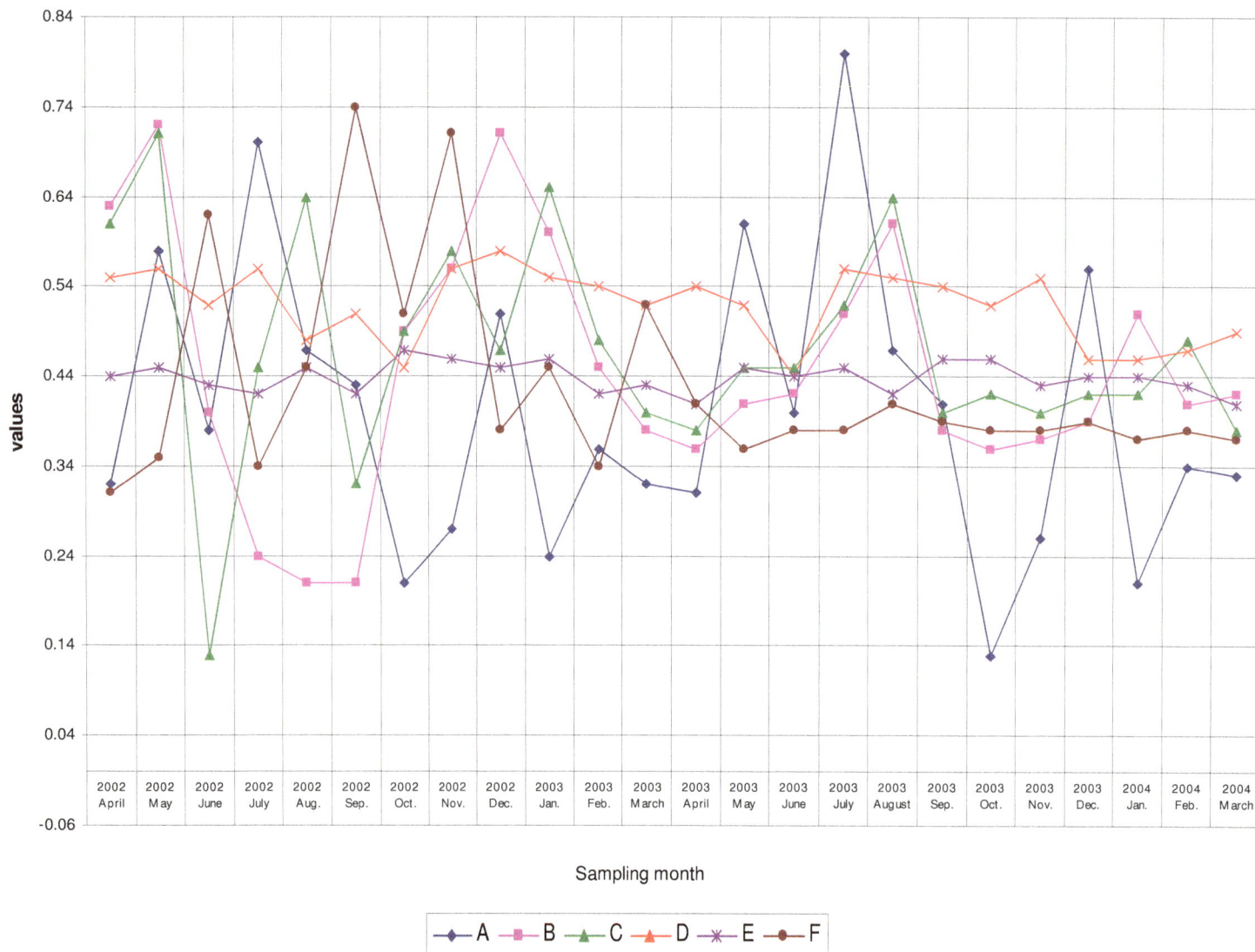

**Figure 14.** Showing magnesium at various stations on Kalpi (Morar) river.

sediment ranged between 1.24 to 2.56 and 15.16 to 38.61%, respectively. The pH of Kalpi (Morar) river was noticed always moderately alkaline varying from 7.1 – 8.14 during the study that has direct bearing on biogenic and abiogenic reactions and vice versa. The specific conductivity of sediment ranged form 2.51 mScm$^{-1}$ at station D during, 2.14 mScm$^{-1}$ at station C during and 2.01 mScm$^{-1}$ at station B which are indicative of high organic load at station B, C and D. It also showed that station A, E and F lower values of specific conductivity. The organic carbon content is generally high in the sediment of river ranging high in the sediment of river ranging from 0.93% at station A to 2.52%n at station D, showing high contamination. Higher organic carbon at polluted sites is due to chemical and bacterial decomposition and sewage contamination. Nitrogen and available nitrogen were also high at contaminated sites while less at less contaminated sites, other nutrients like phosphorus and available phosphorus were also evident high at polluted sites were in increase order from station A to station D.

Carbon and nitrogen ratio were range from 31.28 to 49.36, higher at station D indicating higher contamination at this point. Potassium, exchangeable potassium, magnesium and calcium were found high at station B, C and D and low at station A, while moderate values at station E and F. Higher contents of these nutrients shows higher organic level at the river sediment. The high input of sewage has increase the amount of Ca, Mg and other nutrients at B, C and D stations. Station A, E and F are having less concentration of these cations. The soil parameters are also indicating the increasing pollution in stations B, C and D in Kalpi (Morar) river.

## ACKNOWLEDGEMENT

The authors would like to give a special vote of thanks to Madhya Pradesh Council of Science and technology for financial assistance.

## REFERENCES

Asadi SS, Azeem S, Prasad AVS, Anji RM (2008). Analysis and mapping of soil quality in Khandaleru catchment area using remote sensing and GIS. Curr. Sci., 95(3): 391-396.

Ayyappan S, Gupta TRC (1985). Limnology of Ramasamudra tank-bottom sediments Mysore. J. Agric. Sci., 14: 408-420.

Banerjea SM (1967). Water quality and Soil condition of fish ponds in some state of India in relation to fish production. Indian J. Fish., 14: 115-144.

Banerjee RK, Pakrasi BB (1986). Physico-chemical nature and bimass production of newly constructed brackish water impoundments (Nonagheri) in lower Sunderban. Proc. Sympo. Coastal Aquaculture, Part-4: 1103-1106.

Buckman HO, Brady NC (1950). The Nature and Properties of soil. Fifth edition, The Macmillian Co., Newyork, p. 57.

Crane JL, Richards CD, Breneman SL, Schuldt JA (2005). Evaluating methods for assessing sediment quality in a Great Lakes embayment. Aqu. Ecosy. Health Manage., 8: 323-349.

Chopra SL, Kanwar JS (1999). Analytical Agricultural Chemistry. Kalyani Publishers, New Delhi.

Ghosh PB, Choudhary A (1989). The nutrient status of the sediments of Hooghly estuary. Mahasagar., 22(1): 34-41.

Gupta BP, Muralidharan M, Joseph KO, Krishnani KK (1999). Suitability of coastal saline soils of Gopalapuram Nellore for shrimp farming. Indian J. Fish., 46(4): 391-396.

Horsfall M, Horsfall MN, Spiff AI (1999). Speciation of heavy metals in inter-tidal sediment of the Okirina river system, River state, Nigeria. Bulla. Chem. Soc. Ethiop., 13: 1-9.

Hosetti BB, Natraj S, Chandrashekhar AS, Kamalkar SB, Patil SR (1995). Ecological studies on River Tunga at Shimoga with special reference to water pollution and Land use. J. Notcon., 7: 111-117.

Iwegbue CMA, Nwajei GE, Isirimah NO (2006). Characteristic levels of heavy metals in sediments and dredge sediment of a municipal creek in the Nigeria. Environmentalist, 26: 139-141.

Iwegbue CMA, Nwajei GE, Isirimah NO (2007). Assessment of contamination by heavy metals in Sediment of Ase river, Niger Delta, Nigeria. Res. J. Environ. Sci., 1(15): 220-228.

Kumar R, Ramachandra TV (2003). Water and soil and sediment investigation of aquatic ecosystem. Nati. Sem. on River Conse. and Man., St. Thomas's college Thrissur, Kerala. Maharashtra state Pollution Control Board (MPCB) (2004). A survey report on Mithi river water pollution and recommendation for its control. A Report submitted to Govt. of Maharashtra.

Nasnolkar CM, Shirodkar PV, Singbal SYS (1996). Studies on organic carbon, nitrogen and phosphorus in the sediment of Mandovari estuary. Indian J. Mar. Sci., 25: 120-124.

Padmaja VK, Wani SP, Sahrawat KL, Jangawad LS (2008). Economic evaluation of sediments as a source of plant nutrients. Curr. Sci., 95(8): 1042-1050.

Pathak V, Mahavar LR, Sarkar A (2001). Ecological status and production dynamics of a stretch of river Mahanadi. J. Inland Fish. Soc. India, 33(1): 25-31.

Trivedy RK, Goel PK (1986). Chemical and Biological methods for water pollution studies. Env. Publishing, Karad, India.

# 5

# Oil spills and community health: Implications for resource limited settings

Albert SALAKO[1], Oluwafolahan SHOLEYE[2]* and Sunkanmi AYANKOYA[1]

[1]Obafemi Awolowo College of Health Sciences, Olabisi Onabanjo University, Sagamu, Nigeria.
[2]Department of Community Medicine and Primary Care, Olabisi Onabanjo University Teaching Hospital, Sagamu, Nigeria.

Oil spillage has become of increased relevance in recent times because of the magnitude with which it occurs, the effects it has on the environment, and the quality of life of people residing in the affected areas. The World Health Organization (WHO) has drawn the attention of many member states to the burden of non-communicable diseases on the limited resources available to health. Much emphasis has been placed on environmental factors influencing the health of populations. Oil spillage affects both living and non-living components of the environment which in turn affect man directly and indirectly. This paper discusses the health implications of oil spillage, including the gastro-intestinal dermatologic and neurological effects associated with it. It advocates for better environmentally-friendly and sustainable techniques, particularly in resource-limited communities, where the socioeconomic, physical and invariably the health consequences will be least mitigated and most severe. Effective response and mitigation are necessary to cushion the adverse effects of inadvertent spills on the populations around effected regions.

Key words: Oil spillage, health, environment, non-communicable disease, pollution.

## INTRODUCTION

The World Health Organization (WHO) defines health as a state of complete physical, mental and social well being and not the mere absence of disease or infirmity (WHO, 1948). This definition gives a holistic dimension to the issues concerning the health of individuals as well as whole communities (Jekel et al., 2007). It provides a template for defining various dimensions to health and its determinants, in spite of the criticisms put forward against this definition.

The environment has been recognized as an important contributor to the attainment of good health or ill health. The environment includes both living and non-living components. The quality of the environment affects man, influencing his actions while man's actions and inactions influence his environment (Roche, 2003). The increased burden of disease exerted by chronic non-communicable diseases has made it necessary for all factors influencing

health to be studied. A lot of work has been done in the area of genetics, nutrition, lifestyle modification and environmental factors. Insightful epidemiological studies have linked specific health conditions with exposures to hazardous substances in the environment. Thus, environmental health deals with the impact the environment has on a population (WHO, 1993). A growing concern in the field of environmental health is the impact of oil spillage on the well being of individuals, residing close to the affected areas, as well as the degradation of the environment resulting from such spills.

Oil spills have been reported over several decades in many parts of the world, from Africa to North America. The Niger Delta region of Nigeria has been greatly affected by oil spillage in recent years, with its devastating effects on the environment and the populace. The federal government set up agencies to supervise the efforts of various stakeholders in combating the problems arising from oil spillage. While success has been recorded in a few quarters, a lot still needs to be done, in order to achieve a healthy environment.

*Corresponding author. E-mail: folasholeye@yahoo.com.

This paper discusses the implications of oil spillage on the well being of the public and their environment.

## THE OCCURRENCE OF OIL SPILLS

A lot of oil is used and needs to be transported all over the world. Many countries depend on crude oil to power their economics. It has been estimated that the world uses about 2.73 billion gallons per day. States go to any length to acquire oil production capacity or to be assured access to free flow of oil. This has resulted in some form of conflict or the other in certain areas which may only possibly contain oil reserves. Oil production and trade fuel the economics of the world, making countries rely on the oil market supply chain. Some countries, like the United States, use up to 70 million gallons of oil daily. Crude oil is used as fuel for transportation and in factories; it is useful for electric power generation, asphalt production, wax, plastics, pesticide, fertilizers, heating, paints production and varnishes.

When oil spills, it is as a result of various contributing factors. It is estimated that 31.5 billion gallons of oil are at sea being transported, making it easy for spillage to occur. But not all spills come from tankers; some are from pipelines, oil wells, storage tanks and vessels cleaning out tanks. When oil of any type is released into the natural environment, the result is termed an "Oil Spill". Most people think of marine oil spills when spillage is mentioned, forgetting that the escape of oil into the natural environment is a problem on land as well.

The causes of oil spills can be broadly classified as follows:

1) Accident: This could be due to negligence or otherwise. This is the most widely acknowledge causes of spillage, involving very large cargo ships on sea;
2) Natural disasters includes earthquakes, adverse weather conditions, hurricanes, etc., it is also responsible for dumping of oil and gas in very large amount, into seas and oceans.
3) Intentional spills: This could be as a result of vandalism, terrorism, wars and washing of tanks and vessels into the sea.

A number of processes take place when oil spills. Oil generally floats because it is lighter than water. It needs to adhere to heavier particles like sand, algae and silt in order for it to sink. The following processes take place when spillage occurs:

1) Spreading, evaporation, emulsification
2) Petro/photo-chemical oxidation, and bacterial degradation.

Spreading takes place immediately oil is released on water. Oil usually floats and spreads out into a very thin film on the water surface known as "sheen". Sheens are often seen as silvery or rainbow-colored in puddles in parking lots. This process stands to be the most significant. Spreading is affected by gravitational forces which reduce film thickness and surface tension as well as inertia forces.

Bacterial degradation also occurs following spillage. Petroleum as well as its refined products, is a mixture of a wide range of hydrocarbon fractions with sulphur, oxygen, and nitrogen compounds. These fractions which include straight or short-branched alkanes, cyclo-alkanes, aromatic hydrocarbons, and heterocyclic compounds, form a potential carbon and energy source for microbial activities and hence are biodegradable. When spillage occurs in land, micro organisms make use of it in the soil, as energy source, with such additional nutrients as nitrogen, phosphorous, potassium and an adequate supply of oxygen. This microbial activity causes the disappearance of crude oil from the soil, after a considerable time.

Accumulation of carbon dioxide, water, ammonia and some other intermediates are unavoidable and the formation of waxy solids, by unknown biosynthetic processes are all hazardous and significantly long testing (Cole, 1997).

## OIL SPILLAGE AND THE ENVIRONMENT

The environmental cost of oil spills has been extensive. They include the destruction of wildlife and biodiversity loss of fertile soil, pollution of air and drinking water, degradation of farmland and damage to aquatic ecosystems, all of which have caused serious health problems for the inhabitants of areas surrounding oil production and spill.

Oil can have a significant impact on marine larvae, birds and mammals in particular, and to a lesser extent on fish. Some components of oil are toxic if exposure occurs within the first two days of a spill. Oil on feathers hinders the water-repellency of birds. Oil on fur reduces its insulating capacity.

Several studies have been conducted on the effects of crude oil and its products on the environment. The acute and chronic consequences of oil spills deplete environmental resources which may result in death of living organisms in the environment, either immediately or with time (Hofer, 1998; Cole, 1997; MacFarland, 1998). The consequences oil spillage in the oceans is dreadful. The world's coastlines become polluted, fisheries and marine life get contaminated, colonies of sea birds are annihilated and other affected animals, leading to dead. The effects of oil on marine life are caused by either the physical nature of the oil (physical contamination and smothering) or its chemical components (toxic effects and accumulating) (Cole, 1997).

The environmental damage that results from oil retraction and production can also directly affect human life in the region. Damage can include pollution of water resources and contamination of the soil. Environmental devastation is damaging to vegetation, livestock, and invariably to the health of the humans. Oil spills can interfere with desalination plants which require a continuous supply of clean seawater (Awobajo, 1979). The impact of oil spills on living components of the environment may be lethal, sometimes affecting reproductive ability, physiologic processes and behavior of the organisms. These could be attributed to a number of factors, which are:

i) Type of the oil spilled, amount of the oil spilled, topography of the affected area, weather condition at the time of the spills, previous exposure of the area to oil, exposure to other pollutants, geomorphology of the coast, ii) State of the oil (whether fresh or not).

Coastline erosion is another problem associated with the environmental degradation due to oil spillage (Ibe, 1998). This in turn poses a lot of problems for humans residing close to the affected areas.

## HEALTH IMPLICATIONS

The emphasis of environment health has been the promotion, protection and sustenance of the health of people, as it relates to environmental conditions. It is not new that many epidemiological studies have been conducted to ascertain the effects of certain exposures on the health and well being of the concerned populations (Moeller, 1997). Hydrocarbon compounds contain substances that are hazardous to human health, causing a wide range of symptoms in various systems of the human body (Hofer, 1998). Oil spills are therefore harmful to human health in a multiplicity of ways, as extensively documented in scientific literature (Aquilera et al., 2010; Ha et al., 2008; Levy et al., 2011).

### Oil smoke

Oil smoke from evaporation or burning is inhaled as volatile organic compounds. Oil vapors have been noted to cause headaches, dizziness, nausea, vomiting, eye and throat irritations, as well as breathing difficulties. People who inhale large amounts of fumes are in danger of chemical poisoning called hydrocarbon pneumonia.

Hofer (1998) discussed the environmental and health effects of crude oil contamination of the sea, during transportation. Oil contains a lot of particulate matter, polycyclic aromatic hydrocarbons, hydrogen sulphide, acidic aerosols, and volatile organic compounds which are detrimental to human health. These particles have been noted to be harmful to the heart and lungs. Volatile organic compounds can lead to respiratory problems,

allergic reactions and weakened immune systems. They are also associated with harmful effects on the gastro-intestinal tract and liver.

The Colorado Air Quality Control Commission (2006) report stated that oil and gas developments were the primary sources of the Denver region's air pollution. In the Rocky mountain region, the emission of sulphur (iv)oxide has increased by 147%, as a by-product of petroleum production. The chemical aggravates heart and lung diseases and is poisonous at high levels.

Meo et al. (2009) conducted a study involving subjects that took part in the oil cleanup operators following spillage from a Creek tanker 'Tasman Spirit'. They studied 50 exposed subjects matched with an equal number of controls, all non-smoking. They found that subjects involved in the cleanup exercise had higher health complaints as compared to their matched control. This led to the assertion that even occupational exposure to oil-contaminated water could lead to respiratory and general health complaints in workers.

A retrospective cohort study examined the acute health effects of the sea express oil spills; Lyons et al. (1999) concluded that residence in oil-contaminated area (as a result of spillage) are significantly associated with greater anxiety and depression scores, worse mental health, self-reported headache, sore eyes and sore throat after adjusting for age, sex, smoking status, anxiety and the belief that oil had affected their health. People from the exposed areas reported higher rates of physical and psychological symptoms than the control areas. The authors said the symptoms were those expected from the known toxicological effect of oil, suggesting a direct health effect on the exposed population (Lyons et al., 1999).

A similar study, which considered the psychological dimensions of the health sequelae of a community exposure in Texas, showed that adverse effects on the mental health of the people were recorded, as a result of exposure to hazardous chemical (Dayal et al., 1994).

Neurological effects have also been reported in several studies. Akpofure et al. (2000) in a study on integrated grass root post-impact assessment of acute damaging effects of continuous oil spills in the Niger Delta (January 1998 to January 2000) found increased phobic and anxiety disorders among the exposed. Some subjects were observed to have been perplexingly neurotic and schizophrenic. Agoraphobic disorder and tendencies were also noted. However, not all studies support an increased incidence of mental health problems following an oil spill, among people exposed to the pollutant.

A cross-section study examining the health-related quality of life and mental health in the medium-term aftermath of the "prestige oil spill" in Spain, found out significant increase in mental health problems due to oil spills. It concluded that almost 18 months after the 2002 sinking of the oil tanker, off    the Calician coast, worse health-related quality of life (HRQOL) and mental health levels were not in evidence among subjects exposed to the spill. Nevertheless, some scales suggest the possibility of the slight impact on the mental health of residence in unaffected areas (Carrasco et al., 2007).

Gastro-intestinal symptoms, allergic reaction and generalized discomfort have also been reported as consequences of oil spillage. A retrospective cohorts study in England, after water contaminated incident near Worcester in April 1994, reported diarrhea, nausea, headache, abdominal pains, skin irritations and itchy eyes at higher rates among the exposed subjects when compared with the unexposed (Fowle et al., 1996).

A similar study following oil pollution on a beach near Karachi, Pakistan, measured enzymes and certain parameters in blood samples taken from study participants living or working in the vicinity of the polluted area. Lymphocytes and eosinophils were slightly raised, while some people had raised serum glutamic pyruvic transaminase levels (Khursid et al., 2008).

An important cause for concern is the cumulative effect of the crude oil contents on human health over a long period of time (Khursid et al., 2008). A study on the natural radioactivity and trace elements in crude oils implications for health, noted that Niger Delta oil has low metallic contents but the cumulative effects of the radio-activity and metallic properties on human health is of concern (Ajayi et al., 2009). An incident that also readily comes to mind, showing the effect of environmental pollution on human health is the Minamata Bay incident in Japan, where more than 100 cases of central nervous system damage fish, as a result of toxic waste disposal on the bay by an industrial outside. The compound in questions was methyl mercury (Kurtand et al., 1960; Jekel et al., 2007).

In essence, the health risks associated with oil spillage are real and require urgent attention. Preventive efforts are better coordinated with excellent results than pallia-tive measures. Indeed, conditions like post-traumatic stress disorder, obsessive-compulsive disorder, somato-form stress disorder, hypochondrasis and affective stress disorders are better avoided than managed, by preventing spillage and if need be, immediately curtailing the spread of any accidental spills.

## SOCIAL DIMENSIONS OF OIL SPILLS

Many communities in the Niger Delta region of Nigeria have been worst hit by the problem of oil spillage, with its attendant environmental, social and economic depriva-tion. The mangrove forests, intertidal shores and other natural habitats have been gravely affected by oil exploration and exploitation (Osuji et al., 2010; Omorodion, 2004; Amnesty International, 2009; Nwilo and Badejo, 2005). The people, who are mainly rural farmers, fishermen and traders, have lost their means of livelihood due to the menace, resulting in food shortages and unemployment, increasing their vulnerability to HIV/AIDS (Udonwa et al., 2004; Udoh et al., 2008).

Migrant workers in the oil industries, operating in the oil-rich regions have been found to engage in high-risk behaviours, over the years, though, many workplace programs are now addressing such and related issues (Nwauche and Akani, 2006; Macilwain, 2006). Weak public sector health delivery and educational systems, poverty, migration, concurrent sexual partnerships and food insecurity have been identified as the enabling factors for HIV transmission in the region (Udoh et al., 2009).

## MANAGEMENT OF OIL SPILLS

The effective and full clean-up of spills, remains an intricate problem, all over the world. Technical methods have been characterized by various levels of success and challenges. The spill size and prevailing weather conditions affect the choice of methods. Common clean-up techniques include:

1) Containment and recovery, which involves the use of booms and skimmers for oil recovery, use of sorbents, dispersants, burning.
2) Bioremediation.
3) Shoreline clean-up.
4) Masterly inactivity, particularly, in open ocean spills where clean up is difficult and not efficient (Atlas, 1995; Hoff, 1993).

## INTERNATIONAL HEALTH REGULATIONS

The 58th World Health Assembly IHR (WHA 55.16) is on global health response to natural occurrence, accidental release/deliberate use of biological and chemical agents/radio nuclear material that affect health.

Article 12 of IHR states "public health emergency of international member states can be determined by the process of deciding a public health emergency affecting members states by the Director-General (DG) of WHO". The roles of the DG and the affected state parties are clearly outlined. Oil spillage could also constitute a public health emergency when it occurs in a large great dimen-sion, the guidelines stipulates by WHO must be followed.

## THE NIGERIAN EXPERIENCE

Several laws and policies have been formulated to aid in management of oil spills in the country. Some other laws guide operations in the petroleum industry. As is charac-teristic of many other aspects of the country's public activities, the problem is not with the laws and policies, but with implementation. Some of these laws include:

1) Oil Pollution Act (OPA) of 1990.
2) Federal Environmental Protection Agency Act Cap 131

LFN, 1990.
3) Endangered Species Decree Cap 108 LFN, 1990.
4) Harmful Waste Act Cap 165 LFN, 1990.
5) Petroleum (drilling and production) Regulations, 1969.
6) Mineral Oil (Safety) Regulations, 1963.
7) International Convention on the Establishment of an International Fund for compensation for Oil Pollution Damaged, 1971.
8) Convention on the Prevention of Marine Pollution damage, 1972.
9) African Convention on the Conservation of Nature and Natural Resources, 1968.

The Federal Government in 1962 promulgated the Environmental Impact Assessment (EIA) decree No. 86 of 1992, to protect and sustain the ecosystem. The law makes EIA compulsory for any major project that may have adverse effects on the environment (Olagoke, 1996; Ntukekpo, 1996). A number of federal and state agencies deal with the problem of oil spillage in Nigeria, including the Department of Petroleum Rouses (DPR), the Federal Ministry of Environment, the State Ministries of Environment and the National Maritime Authority.

A major issue that has remained a challenge is effective implementation of all documented guidelines, as well as coordination of all efforts of various stakeholders including civil society groups.

## GEOGRAPHIC INFORMATION (GIS) SYSTEM FOR OIL SPILLS MANAGEMENT

A successful combating operation for a marine oil spill is dependent on a rapid response from the time of spill is reported until it is fully controlled (Wardley-Smith, 1977). In order to improve on the timely response and decision-making process, GIS is applied as an essential operational tool. Information on the exact position and size of the spill can be plotted on maps in a GIS environ-ment. It offers opportunities for integration of oil drift forecast models in the computer program framework (Milaka, 1995). Required information for oil spill sensitivity mapping can be depicted on a set of thematic maps using GIS even though they can in theory be depicted on a single sheet. GIS could also be used to assess the adequacy of any given oil spill contingency plan (Parthiphan, 1994; Smith and Loza, 1994).

## PREVENTIVE STRATEGIES

It is a known fact that clean-up exercises, following oil spills, are not totally effective and neither do they fully restore the initial properties of the affected areas. Preventive plans are therefore more appropriate for health promotion and protection. Strategies to be employed include:

1) Improved piloting: training of ship and tanker crews.

2) Training of storage and pipeline facility crews.
3) Enforcing pollution rules at sea.
4) Building spill-resistant vessels.
5) Maintaining pipeline and vessels.
6) Preparing for spills response via effective training and (practice) drills.

Awareness creation is also an important strategy for the prevention of oil spills and associated health effects. The more enlightened people are, the easier it is for them to prevent, recognize and promptly report any case of oil spillage in their vicinity. In Nigeria, much of the incidents of oil spillage have also been attributed to sabotage. Political upspring and youth restiveness have greatly contributed to the destruction of oil pipelines. These could have been avoided if only the youth were educated about the adverse effect of spills on the environment and human health.

The various agencies, ministries and department, at local, national and international levels, need to be more pro-active in carrying out their stipulated duties as well as think of innovative ways to ensure the protection of the ecosystem and the health of the public.

Environmental health as a discipline needs to be taught in more details, with appropriate emphasis on contemporary practical issues facing the world today, including climate change and pollution (IoM, 1995). The problem of oil spills must be of high priority, so that more focused research may be conducted in this area.

## CONCLUSION

Oil spillage is a major challenge in today's world. Everyone is affected in one way or the other. The resources of the environment are depleted and the ecosystem disrupted. So much can still be done in preventing further spills, as well as in the management of existing cases.

The Government health institutions, agencies and the general public all have a role to play. Policy implementation with effective monitoring and evaluation is needed. The time to act is now.

### REFERENCES

Ajayi TR, Torto N, Tchokossa P, Akinlua A (2009). Natural Radioactivity and Trace Elements in Crude Oils: Implication for health. Environ. Geochem. Health 31(1):61-69.
Akpofure EA, Efere ML, Ayawa P (2000). The Adverse Effects of Crude Oil Spills in the Niger Delta, Nigeria. Urhobo Historical Society. Available at: (http://www.waado.org/environment/petrolpolution/oilspills/OilSpills_AdverseEffects.html)
Amnesty International (2009). Nigeria: Petroleum, pollution and poverty in the Niger Delta. London, United Kingdom Amnesty International Publications p 10.
Atlas RM (1995). Petroleum Biodegradation and Oil Spill Bioremediation. Marine Pollut. Bull. 31:178-182.
Awobajo (1979). "Environmental Aspect of the Petroleum Industry in the

Niger Delta: Problems and Solutions" in Proceedings of Petroleum Industry and the Environment of the Niger Delta p 145

Carrasco JM, Perez-Gomez B, Lope V (2007). Health-related quality of life and mental health in the medium-term aftermath of the prestige oil spills in Galiza (Spain): A cross-sectional study. BMC Public Health 17:7:245.

Cole HA (1997). Pollution of the Sea and its Effects. Proc. R. Soc. Lond. B. Biol. Sci. 205(1158):17-30.

Dayal HH, Baranowski T, Li YH, Morris R (1994). Hazardous Chemicals; Psychological Dimension of the health Sequelae of a Community Exposure in Texas. J. Epidemoil. Community Health 8(6):560-568.

Fowle SE, Constantine CE, Fone D, McCloskey B (1996). An epidemiological study after a water contamination incident near Worcester, England in April 1994. J. Epidemiol. Community Health 50(1):18-23.

Hofer J (1998). Environmental and Health effects resulting from marine bulk liquid transport. Environ. Sci. Pollut. Res. Int. 5(4):231-7.

Hoff R (1993). Bioremediation: An Overview of its development and use for Oil Spill cleanup. Marine Pollut. Bull. 26:476-481.

Ibe AC (1998) .Coastline Erosion in Nigeria. Ibadan University Press, Ibadan p 217.

Institute of Medicine (1995). Environmental Health: Integrating a Missing Element into Medical Education. National Academy Press Washington, DC. pp. 44-51.

Jekel JF, Katz DL, Elmore JG, Wild DM (2007). Introduction to Preventive Medicine in: Epidemiology, Biostatistics and Preventive Medicine. 3rd Edition. WB Saunders, Philadelphia. p 225.

Khursid R, Sheikk MA, Igbal S (2008). Health of People Working/Living in the Viccinity of an Oil-polluted beach, near Karachi, Pakistan. East Mediterr. Health J. 14(1):179-82.

Kurtand LT (1960). Minamata Disease. World Neurol. 1:370-395.

Lyons RA, Temple JM, Evans D (1999). Acute Health Effects of the Sea Empress Oil Spills. J. Epidemiol. Community Health 53(5):306-10.

MacFarland HN (1998).Toxicology of Petroleum Hydrocarbons. Occup. Med. 3(3):445-454.

Macilwain C (2006). Oil firms back AIDS project in the Niger Delta. Nature 7:444(7120):663.

Meo SA, AL-Dress AM, Rasheed S, Meo IM (2009). Health complaints among subjects involved in Oil Clean up Operations during Oil Spillage from a Greek tanker "Tasmin Spirit". Int. J. Occup. Med. Environ. Health 22(2):143-148.

Milaka K (1995). Use of GIS as a tool for Operational Decision Making, Implementation of a National maritime Oil Spills Contingency Plan for Estonia, Carl Bro International a/s. Glostrup, Denmark.

Moeller DW (1997). Environmental Health. Revised Edition. Cambridge Mass Harvard University Press.

Ntukekpo R (1996). Spillage: Bane of Petroleum. Ultimate Water Technology and Environment. Environ. Res. J. 6(2):100-105

Nwauche CA, Akani CI (2006). An assessment of high-risk sexual behavior and HIV transmission among migrant oil workers in the Niger Delta area of Nigeria. Niger. J. Clin. Pract. 9(1):48–51.

Nwilo PC, Badejo OT (2005). Oil spill problems and management in the Niger Delta. International Oil Spill Conference. Miami, Florida, USA. Euro. J. Sci. Res. 52(4):592-603

Olagoke W (1996). Niger Delta Environmental Survey: which way forward? Ultimate Water Technology and Environment. Environ. Res. J. 6(2):100-105

Omorodion FI (2004). The impact of petroleum refinery on the economic livelihoods of women in the Niger Delta region of Nigeria. JENDA. J. Cult. Afr. Women Studies 6:1-15.

Osuji LC, Erondu ES, Ogali RE (2010). Upstream petroleum degradation of mangroves and intertidal shores: the Niger Delta experience. Chem. Biodivers. 7(1):116-128.

Parthiphan K (1994). Oil Spill Sensitivity mapping using a Geographical Information System. University of Aberdeen. EGIS Foundation.

Roche N (2003). Environmental Health in: Luaccs AO, Gilles HM (Eds). Short Textbook of Public Health Medicine for the Tropics. 4th Edition. London Book Power pp. 337-351.

Smith LA, Loza L (1994). Texas Turns to GIS for Oil Spill Management. Geo. Info. Syst. p 48.

Udoh IA, Mantell JE, Sandfort T, Eighmy MA (2009). Potential pathways to HIV/AIDS transmission in the Niger Delta of Nigeria: Poverty, migration and commercial sex. AIDS Care 21(5):567-574.

Udoh IA, Stammen RM, Mantell JE (2008). Corruption and oil exploration, expert agreement about the prevention of HIV/AIDS in the Niger Delta of Nigeria. Health Educ. Res. 23(4):670-681.

Udonwa NE, Ekpo M, Ekanem IA, Inem VA, Etokidem A (2004) . Oil doom and AIDS boom in the Niger Delta Region of Nigeria. Rural Remote Health 4(2):273.

Wardley-Smith J (1977). The Control of Oil Pollution on the sea and Inland waters. Graham and Trotman Limited, London p 181.

World Health Organization (1948). Constitution of the World Health Organisation. Geneva: World Health Organization.

World Health Organization (1993). Consultative Meeting on Environmental Health Sofia, Bulgaria. October.

# 6

# Genotoxic damage in oral epithelial cells induced by fluoride in drinking-water on students of Tula de Allende, Hidalgo, Mexico

Patricia VÁZQUEZ-ALVARADO[1], Arcelia MELÉNDEZ-OCAMPO[2], Rosa María ORTIZ-ESPINOSA[3], Sergio MUÑOZ-JUÁREZ[3] and Alejandra HERNANDEZ-CERUELOS[3]*

[1]Dentistry Academic Area, Science Health Institute, Autonomous University of the State of Hidalgo, Mexico.
[2]Faculty of Dentistry, National Autonomus University of México, Mexico.
[3]Medicine Academic Area, Science Health Institute, Autonomous University of the State of Hidalgo, Mexico.

Fluoride (F⁻) compounds are present on the earth's surface, water, volcanoes and are also a product of petrochemical and cement industries. Little amounts of F⁻ are required for the formation of bones and enamel, however, according to World Health Organization (WHO), ingestion of over 1.5 mg/L of F⁻ may be a health hazard due to the toxic effects on the kidney, liver and it may also cause dental or skeletal fluorosis. The aim of this study is to compare the genotoxic damage in cell of oral epithelium detected by (the) comet assay, through a population of scholars from two communities located in the city of Tula; one consuming high concentration of F⁻ from drinking-water San Miguel Vindhó (SMV) and the other with levels under the limit La Malinche (LaM). 113 (students) teenagers between the ages of 12 to 15 were selected to obtain epithelial cells by internal cheek brushing. 30 of them were selected (from) by the dental clinic service of the Health Science Institute (UAEH, Mexico) as negative and positive control groups before and after professional appliance of sodium fluoride (NaF) (2%). 31 from LaM and 52 from SMV. 200 cells per person were analyzed to measure tail moment, tail length. Visual scoring was used to classify results according to degrees of damage. These results showed a significant difference between the populations, indicating very low basic damage in both the negative control group and LaM and severe damage in both the exposed population and the positive control group. These results indicate that high levels of F⁻ may be the cause of genotoxicity in oral epithelial cells.

**Key words:** Epithelial oral cells, DNA damage, fluoride, comet assay, Tula.

## INTRODUCTION

Fluoride (F⁻) is a trace essential element for bones and teeth development in human beings and animals (Ling and Jian, 2006; Griffin et al., 2007). It is also potentially toxic with a low margin of safety (ASTDR, 2003). The human's consumption from drinking water to prevent dental decay is advised to be between 0.7 to 1.2 mg/L (Kaseva, 2006). Chronic exposure to F⁻ over the limit established by the World Health Organization (WHO) (1.5 mg/L) is the cause of dental and skeletal fluorosis. It also increases the probability of renal diseases (WHO, 2004), liver and parotid glands damage (Shanthakumari et al., 2004), as well as brain damage, and is also linked to decreasing IQ in school children (Wang et al., 2007). Toxic effect over cells of soft tissues such as endothelium, gonads and neurological system has also been reported, including modifications in cell proliferation, membrane permeability and induction of apoptosis (National Research Council, 2006; Yan et al., 2007).

Studies in China and India reported an increased prevalence of skeletal fluorosis linked to the absorption of

---

*Corresponding author. E-mail: alejandra.ceruelos@gmail.com.

drinking water containing F⁻ above the level of 1.4 mg/L (Jolly et al., 1968; Choubisa et al., 1997; Xu et al., 1997). However, such studies suffer from limitations: the diagnostic criteria are not always specified or consist of self-reported symptoms and only drinking water is considered as a source of exposure. If the intake of F⁻ totals is between 6 to 14 mg/day, there is a clear risk of skeletal adverse effects (Li et al., 2001).

F⁻ can be found naturally in mineral deposits of non-metallic compounds, in sedimentary, hydrothermal, metamorific and also volcanic soils, which determine the nature of the water source (Hernández, 1997). In Mexico, high levels of F⁻ in water is chronic and endemic in the zone called "fluorite belt" that includes the states of Coahuila, Durango, Zacatecas, San Luis Potosí, Guanajuato and Queretaro (Ortega, 2009). The State of Hidalgo is not in this area; nevertheless, the geological properties of Tula de Allende which are characterized by natural deposits of limestone, kaolin, gypsum, dolomite, quartz, clay and silica are associated with very high levels of F⁻ in underground water (Geological Monograph, 1992).

Tula de Allende is located over the geological volcanic fault of El Doctor (Segerstrom, 1961), constituted principally by limestone deposits. Those deposits are used in the production of cement, which is one of the main economic activities of this municipality, as well as, thermo electrical, the petroleum industry and agricultural production. Tula is located north 20°10′, south 19° 57′ latitude, longitude east 99° 15′, and west 99° 30′ and at an altitude of 2477 m above sea level. In a previous report, Vazquez-Alvarado et al. (2010) showed high levels of F⁻ found in Ex-Hacienda well which provides drinking water to the community of San Miguel Vindhó (SMV), with an average yearly concentration of 1.41 mg/L (April, 2008 to April, 2009), and a maximum of 1.99 mg/L observed in April, 2009. The control well Manzanitas-I, had an average of 0.62 mg/L with a maximum level of 0.78 mg/L on April of the same year, this well supplies tab water to the community of La Malinche (LaM). A correlation between the high concentration of F⁻ found in the water of SMV and the presence of dental fluorosis with 85% of prevalence in its community was established. Meanwhile, a prevalence of only 4% was observed in the community of LaM.

Currently, lymphocytes of peripheral blood are the most used biomarkers in human; nevertheless, it is also important to use other tissues such as oral epithelium which is located in one of the most common site of malignant transformation (Szeto et al., 2005). Comet assay is useful to evaluate genotoxcity. It is a fast and simple technique, which analyses and quantifies with accuracy DNA damage in nucleated cells (Singh et al., 1988; Horváthová et al., 1998). Cell damage can be mea-sured individually, and DNA breaks can be determined on cells obtained from diverse tissues of a single organism (Obwald et al., 2003; Tovalin et al., 2006). Comet assay

has been used in humans (Lee et al., 2004) to establish the genotoxic effect in individuals exposed to different damaging contributors including radiation, chemical and oxidative stress (Tice et al., 2000; Dusinka and Collins, 2008). The first place in contact with xenobiotics, including the food, are the epithelial cells of the mouth (Winning and Townsed, 2000), making these cells an attractive target with the potential to determine in vivo DNA damage (Glei et al., 2005; Landi et al., 2003; Kleinsasser et al., 2006). The aim of this study is to determine if the F⁻'s concentration in drinking water is a significant factor in the induction of genotoxicity on the exposed community of SMV, by the comet assay on oral epithelium cells.

## MATERIALS AND METHODS

### Chemicals

Low melting point agarose (LMPA), dimethyl sulfoxide (DMSO), ethylene diamine tetraacetic acid (EDTA) disodium salt, Triton 100X, Trizma, ethydium bromide and blue trypan solution 0.4%, were obtained from Sigma-Aldrich (St. Louis MO., USA.). Agarose, trypsine 1:250 of porcine pancreas from Gibco In vitrogen were obtained from Carisbad Ca. USA. Sodium chloride, potassium phosphate, sodium phosphate, sodium hydroxide and methanol, were obtained from JT Baker Mexico city; deionized water was purchased from Hycel (Mexico city). Saline isotonic solution and distillated water were purchased from PiSA Laboratory (Mexico City). Interdental brushes with no additives were obtained from Oral B (Braun, Mexico city). Commercial sodium fluoride gel (2%) for professional application was supplied by Medicom (Montreal, Quebec, Canada).

### Geographical characteristics

SMV is an urban location in the south of the municipality of Tula de Allende lying at an altitude of 2,100 m. It has 10,488 inhabitants. LaM is in the north of the city, 2030 m above sea level, it has 2,000 inhabitants. Both communities studied are located in the same municipality and have very similar geographic and socio-economic conditions (Delimitation of metropolitan areas in Mexico, 2005).

### Water sample

The wells of Ex- Hacienda and Manzanitas-I, were sampled for this study on March, 2009 according to Mexican Official Norm (NMX by its abbreviations in Spanish) (NMX-AA-051-SCFI-2001) by personal of Development and Research Laboratory of Studies of Quality of the Water (IDECA, S.A. de C.V., México, city). One litre (1 L) of water was taken in polyethylene bottles previously washed with HCl 10% and rinsed with distilled water, the recipients were labeled with identification codes including: date, time, location, air temperature, water temperature, name and signature of the person that performed the sample.

Temperature was measured with a check temperature pocket thermometer (HANNA instruments) in situ, calibration number H198501. pH was taken with a digital meter (OAKTON 1234251). Samples were transported in an icebox to maintain the temperature at 4°C (39.2 °F). Once in the laboratory, water samples were distillated according to Standard Methods (APHA-WPCF-AWWA, 1998). This took place before the start of the procedure to determine procedure to determine F⁻'s concentration in order to eliminate any interference due to color, turbidity and presence of other substances. Measurements were performed with a spectrophotometer by SPANDS method in acid media, based on Mexican regulations (NMX-AA-077-SCFI-2001), and concentrations of F⁻ were determined by comparing with a standard calibration curve. In addition, quantification of carbonates (NMX-AA-036-SCFI-2001) and aluminum (NMX-AA-051-SCFI-2001) were realized to avoid possible interferences in F⁻ determination.

## Populations of study

The population of study was 113 scholar children and the criteria of inclusion for the pupils were:

(1) No gender discrimination
(2) Age group between 12 and 15 years old.
(3) Must be born and must live in the community of which water supply was being sampled.
(4) Have informed and signed parental consent.
(5) Voluntary participation in the study.
(6) Have not undergone orthodontic treatments.

Social-demographic data was obtained from a survey completed by the population of the study. Volunteers were questioned about smoking and alcohol consumption habits, viral diseases, history of exposure to X-rays and recent vaccinations. In addition, each volunteer was examined to evaluate the presence of lesions or infections in the oral cavity to rule out any possibility that could interfere with the results.

Ethical aspects were strictly followed to protect the dignity and the wellbeing of the subjects involved, ensuring respect, confidentiality and the protection of human rights. The ethical considerations were based on the General Law of Health of Mexico (2007) which deals with studies related to the inspection of human beings.

Thirty (30) volunteers were selected as the control group from the dental clinic service of the Health Science Institute (Autonomus University of Hidalgo State), for professional application of sodium fluoride (NaF) (2%) to prevent tooth decay. Samples of epithelial cells were obtained before and after application of the F⁻ to provide negative and positive control groups, respectively. F⁻ was applied in special trays for the duration of 4 min. Once the trays were removed, patients were asked to spit and at the same moment samples were taken with the interdental brush on the left and right side of the mucosa of the cheek, avoiding contact with teeth, tongue and gums. This process was carried out in less than 60 s. Concerning the communities, 31 students were sampled from the non-exposed community (LaM) and 52 from the exposed area (SMV).

## Comet assay

Samples were obtained in the morning before recess at a time when students had not done any previous physical activity. Epithelial cells of buccal mucosa were collected from each student according to Besarati-Nia et al. (2000) by repeatedly washing out the mouth with tepid distillated water to remove exfoliated death cells and then gently brushing the internal part of both the right and the left cheeks with an interdental brush. Two samples per volunteers were taken. The brushes were immersed in 200 µl phosphate-buffered saline (PBS) 37°C and stirred in a vortex for 5 s to obtain the cell suspension. Viability of the cells was determined by Trypan blue dye exclusion, with 95% of cell survival. 10 µl of cell suspension were mixed with 75 µl of LMPA (0.5%). The mixture was expanded with a cover slide over a microgel of normal agarose

(1%) and placed in-between microscope slides previously coded and left over ice for gelation. Five minutes (5 min) later, a second layer of LMPA was added. The micro-slides were immersed for 30 min in a solution of trypsine 0.25% at 37°C (Szeto et al., 2005) before being rinsed in saline isotonic solution and immersed in solution of lysis (2.5 M NaCl, 100 mM Na₂ EDTA,10 mM Tris, Triton 100X 1%, DMSO 10% pH 10) at 4°C for 4 h. Slides were then rinsed and placed immersed in buffer (NaOH 300 mM, 1 mM Na₂ EDTA, pH > 13) in an electrophoresis chamber for 30 min, after what electrophoresis was performed for 20 min (300 mA, 25 V). Slides were washed three times with 0.4M Tris pH = 7.5, and dehydrated with absolute methanol for 5 min for their preservation (Tice et al., 2000).

## Qualitative evaluation

Slides were dye with 50 µl of ethidium bromide (0.02 mg/ml). 100 cells per sample were observed and classified in a visual scale according to Browne (2009). The criteria to determine the percentage of damaged nucleus and to calculate the damage index (DI) in order to obtain a numeric value for statistical analysis following this formula:

DI = % of nucleus grade 0 (0) + % of nucleus grade 1 (1) + % of nucleus grade 2 (2) + % of nucleus grade 3 (3)+ % of nucleus grade 4 (4)

(#): number of times the size of the head appeared to migrate on the tail.

## Quantitative evaluation

From every sample, 100 cells were measured to determine tail moment (TM) and tail length (TL) using Metasystem Image Analyzer and the COMET 2.0 software, with fluorescence Microscope Carl Zeiss Axioimager using 20 X/0.065 dry, exciter filter 515 to 560 nm and barrier filter 590 nm.

## Statistical analysis

Data were analyzed with Instat 3.0 software using non-parametric Wilcoxon test for damage index. Student T-test was used for comparing TM and TL among populations and controls, for both, were considerate $P < 0.05$, CI of 95% and standard error (SE).

## RESULTS

## Water quality

Concentration of F⁻ was equal to 1.67 mg/L in the Ex-Hacienda well which is the source of drinking water to the SMV community. This can be considered over the maximum advisable limit of Mexican regulation (NMX-AA-077-SCFI-2001) for allowed level of F⁻ (1.5 mg/L). Water temperature was 20°C, air temperature 29.2°C, pH *in situ* 7.69, aluminum < 0.2 mg/L, alkalinity 287 mg/L. For the Manzanitas-I well which provides drinking water to LaM community, the level of F⁻ was equal to 0.69 mg/L, water temperature 24.1°C, air temperature 29.2°C, pH 7.4, aluminum < 0.2 mg/L and alkalinity 217 mg/L.

**Table 1.** Comet categories for visual scoring grade 0 to 4 in scholar aged 12 to 15.

| School population | Grade 0 | Grade 1 | Grade 2 | Grade 3 | Grade 4 | Average mean and SE | Confidence intervals (CI) |
|---|---|---|---|---|---|---|---|
| Negative control | 95.2 | 0.3 | 1.0 | 0.8 | 2.7 | 15.36 ± 2.52 | (10.20, 20.53) |
| Positive control | 41.7 | 3.3 | 3.0 | 7.0 | 45.0 | 210.97 ± 11.09[a] | (188.27, 233.66) |
| LaM (control population) | 97.0 | 0 | 0 | 0.1 | 2.7 | 12.83 ± 1.76 | (9.22, 16.43) |
| SMV (exposed population) | 40.0 | 7.0 | 2.0 | 5.0 | 46.0 | 190.50 ± 9.71[ab] | (170.97, 210.03) |

The damage index was determined using non-parametric Wilcoxon test. [a]Statistical difference when compared to negative control versus all groups, P < 0.0001; [b]Statistical difference when La Malinche is compared with San Miguel Vindhó, P < 0.0001.

**Table 2.** Measurement of TM and TL in controls and both communities of Tula.

| School population | Average mean and SE | Confidence Intervals 95% | P value |
|---|---|---|---|
| **Negative control** | | | |
| Tail moment (TM) | 6.54 ± 1.40 | (3.66, 9.42) | |
| Tail length (TL) | 10.29 ± 2.25 | (5.69, 14.89) | |
| | | | |
| **Positive control** | | | |
| Tail moment | 118.37 ± 6.58[a] | (104.91, 131.83) | |
| Tail length | 128.47 ± 7.25 | (113.64, 143.30) | |
| | | | <0.0001* |
| **LaM (control population)** | | | |
| Tail moment | 4.81 ± 0.64 | (3.53, 6.10) | |
| Tail length | 3.89 ± 0.79 | (2.29, 5.48) | |
| | | | |
| **SMV (exposed population)** | | | |
| Tail moment | 104.01 ± 4.90[ab] | (94.27, 113.75) | |
| Tail length | 111.45 ± 5.14 | (101.23, 121.67) | |

The measurement of the Tail Moment and Tail Length were considered with the test of Student-Newman-Keuls. [a], Presents statistical difference when compared to negative control versus all groups, P < 0.0001; [b], presents statistical difference when compared La Malinche with SMV, P < 0.0001.

## Comet assay

Table 1 shows the results of the percentage of damaged nucleus observed for the negative control, positive control, LaM and SMV, graded according to Browne's scale. Grade zero represents intact nucleus; Grade one minimum damage, Grade two moderate damage, Grade three severe damage and Grade four highly damaged. From those results, we can observe that in the negative control and non-exposed population most of the nucleuses are intact with 95.2 and 97%, respectively graded zero. On the other hand, the positive control 41.7% and SMV 40% could be considered intact and 45 and 46% highly damage with a classification on grade four, respectively. Statistical analysis of DI showed significant differences between positive control and SMV when they were compared to the negative control, indicating an increase in DNA damage. For LaM, there are not significant differences in this parameter.

Results in Table 2 show average values and standard error of TM and TL for controls and communities. TM 118.37 and TL 128.47 nm of the positive control had a significant increase when compared with the negative control TM 6.54 and TL 10.29 nm. For LaM, the basal genotoxic damage with TM average equal to 4.81 and TL 3.89 nm, was similar to the one observed in the negative control, nevertheless, for the exposed community, values of TM reached 104.01 and TL 111.45 nm, similar to those observed in the positive control group; this data indicates that the induction of genotoxic damage in the epithelial cells may be directly related to the high concentration of fluoride in water from the Ex-Hacienda well consumed by SMV.

## DISCUSSION

In all the statistical tests that were conducted in this study, the confidence intervals (CI) denoted a strong statistical power and all P values (<0.0001) were statistically significant. Dusinska and Collins (2008) mentioned that when an individual measurement of comets is used for a population, SE value must be low. The number of cells evaluated per person (200 cells) gave a total of 6,000 for control groups and 6.200 for non-exposed community (LaM); 10,400 for exposed population (SMV), giving enough statistical creditability to establish a possibility of genotoxic effect of $F^-$ under the experimental conditions.

Basal damage of epithelial cells in normal conditions is very low as observed in the negative control group where most of the nucleus were classified with no damage, and had low values on TM and TL. This indicates that the manipulation of the cells as biomarkers during the sample and experimental procedure was correct, and avoided false positive results. For human studies, cells such as peripherial lymphocytes (Yañez et al., 2003; Jasso et al., 2007) or spermatozoids (Codrington et al., 2004) are used as biomarkers of genotoxicity caused by environmental xenobiotics, nevertheless, the use of other type target cells is needed. Other studies using oral epithelial cells have indicated that they can be used as a good biomarker since they are the first in contact with ingested xenobiotics, the sample is easy to obtain, and the sensitivity to detect genotoxicity is high (Szeto et al., 2005; Ribeiro et al., 2004; Winning and Townsend, 2000).

The increase on DNA damage of epithelial cells when they were in contact with NaF made evident the high genotoxic potential of $F^-$, since most of the cells of this group were classified with the maximum degree of damage (45%). Considering that the samples were taken from the same students used for negative control, we can infer that the topical appliance of fluoride gel on teeth is able to induce a very significant DNA damage on the oral epithelial cells in an acute exposure.

For LaM community, genotoxic damage is as low as the one observed in the negative control group, and even though the city of Tula has an important pollution problem due to its industrial activities (ATSDR, 2003; Shashia, 2003; Wania, 2004). The life style and socio-demographic conditions of the students did not influenced the basal level of DNA damage of oral epithelium cells, validating this community as a good population of reference.

A marked genotoxic effect was found in the exposed community with similar distribution observed in the positive control group, for grade of damage, damage index, TM and TL. This indicates that it is very possible that the induction of DNA damage observed in the SMV is directly related with the presence of $F^-$ in the water consumed in the area; since social-demographic conditions were very similar to those observed at LaM. On March 2009, levels of $F^-$ at the Ex-Hacienda well (1.67 mg/L) were above the level permitted by the Mexican regulation. This was enough to have a very similar response to the DNA damage found in the positive control where the concentration of $F^-$ was more higher, indicating the low security margin of $F^-$. Genotoxicity of $F^-$ evaluated with comet assay using hippocampus neuronal cells *in vitro*

was reported by Zhang et al. (2008), there was an increase in TM for exposition to 2.1 and 4.2 $F^-$ mM for 24 h. Also in fetal teeth exposed to $F^-$ (10 and 20 mM) for 48 h there was a decrease in cellular proliferation and an increase on apoptosis (Yan et al., 2007).

Exposure to $F^-$ affects cells of hard tissues such as teeth and bones, as well as in soft tissues such as gonads, kidney, endothelium (National Research Council, 2006). Exposure to $F^-$ increases the production of superoxide as a consequence of hydrogen peroxide metabolism, inducing peroxinitrate, hydroxyl radicals increasing oxidative stress (Barbier et al., 2010; Garcia-Montalvo et al., 2009; Chinoy, 2003), and inducing lypo-peroxidation of membranes, apoptosis and DNA damage (Wang et al., 2007). The clinical application of NaF induced genotoxicity on the oral epithelium in acute exposition according to Zeiger et al. (1993). Tiwari and Rao (2010) considered $F^-$ as a mutagenic agent that even in low concentrations can increase the induction of chromosomal aberrations. In a previous study, the exposed community was observed to have dental fluorosis and a correlation with the high concentration of $F^-$ in the well water (Vazquez-Alvarado et al., 2010). However, the high concentration of $F^-$ also affected DNA of epithelial cells, making them susceptible to mutations, transformation and neoplasia development (Guachalla and Ascarrunz, 2003), if the DNA damage is unable to be repaired. Evidence that cytogenetic biomarkers are correlated with cancer risk has been strongly validated in recent results from both cohort and nested case-control studies, showing that Comet assay is a marker of cancer risk. Reflecting both, the genotoxic effects of carcinogens and individual cancer susceptibility (Bonassi et al., 2007; Smerhovsky et al., 2001; Rekhadevi et al., 2007)

In conclusion, $F^-$ in water for human consumption over the limit of 1.5 mg/L is able to induce a strong genotoxic effect; therefore, other epidemiological studies must be done to establish the toxic impact over the health of an open population chronically exposed to this agent.

### REFERENCES

APHA (American Public Health Association), AWWA, (American Water Works Association), WEF (WaterEnvironment Federation) (1998). Standard Methods for the examination of water and wastewater, American Public Health Association, 20th Edition.

ATSDR (Agency for Toxic Substances and Disease Registry) (2003). Toxicological Profile for Fluorides, Hydrogen Fluoride and Fluorine. Department of Health and Human Services, Public Health Services, Atlanta, GA, US.

Barbier O, Arreola-Mendoza L, Del Razo LM (2010). Molecular mechanisms of fluoride toxicity. Chemico-Biologic Interact 188:319-333.

Besarati-Nia A, Van Straaten HWM, Godschalk RWL, Van Zandwijk N, Balm AJM, Kleinjans JCS (2000). Immunonoperoxodidase detection of polycyclic aromatic hydrocarbon-DNA adducts in mouth floor and buccal mucosa cells of smokers and non-smokers. Environ. Mol. Mutagen 36:127-33.

Bonassi S, Znaor A, Ceppi M, Lando C, Chang WP, Holland N, Kirsch-Volders M, Zeiger E, Ban S, Barale R, Bigatti MP, Bolognesi C, Cebulska-Wasilewska A, Fabianova E, Fucic A, Hagmar L, Joksic G, Martelli A, Migliore L, Mirkova E, Scarfi MR, Zijno A, Norppa H,

Fenech M (2007). An increased micronucleus frequency in peripheral blood limphocytes predicts the risk of cancer in humans. Carcinogenesis 28:625-631.

Browne M (2009). Imaging and Image Analysis in the Comet Assay. In: Dhawan A and Anderson D (Eds.), The Comet Assay in Toxicology. Royal Society of Chemistry, Cambridge, U.K. p. 408.

Chinoy NJ (2003). Fluoride stress on antioxidant defence systems. Fluoride 36:138-141.

Choubisa SL, Choubisa DK, Joshi SC, Choubisa L (1997). Fluorosis in some tribal villages of Dungapur district of Rajasthan, India. Fluoride, 30: 223-228.

Codrington A, Hales B, Robaire B (2004). Spermiogenic Germ Cell Phase-Specific DNA Damage Following Cyclophosphamide Exposure. J. Androl. 25(3):354-362.

Delimitation of metropolitan areas in Mexico (2005). Kindly provide the English version www.inegi.org.mx/prod_serv/contenidos/espanol/bvinegi/productos/g eografia/publicaciones/delimex05/dzmm-2005_20.pdf.

Dusinska M, Collins A (2008). The comet assay in human Biomonitoring: gene-environment interactions. Mutagenesis 23(3):191-205.

García-Montalvo EA, Reyes-Pérez H, Del Razo LM (2009). Fluoride exposure impairs glucose tolerance via decreased insulin expression and oxidative stress. Toxicology 263:75-83.

General Law of Health of Mexico (2007). Title Fifth: Investigation for the Health. Unique chapter. Article 100. Mexico.

Geologic Monograph Mining of the State of Hidalgo (1992). Secretariat of Energy, Mining and Industry, Mininig and Basic Industry Department, Mineral Resources Council. pp. 69-73.

Glei M, Habermann N, Obwald K, Seidel C, Pool-Zobel BL (2005). Assessment of DNA damage and its modulation by dietary and genetic factors in smokers using the Comet assay: a biomarker model. Biomarkers 10 (2-3):203-217

Griffin SO, Regnier E, Griffin PM, Huntley V (2007). Effectiveness of fluoride in preventing caries in adults. J. Dent. Res. 86(5):410-415.

Guachalla L, Ascarrunz M (2003). Genetic Toxicology: a science in constant development. Biofarbo 11:75-82.

Hernández A (1997). El flúor y el Abastecimiento del Agua, Tecnología Internacional del agua. Madrid, España. pp. 26-239.

Horváthová E, Slamenová D, Hlinciková L, Mandal T.K, Gábelová A, Collins AR (1998). The nature and Origin of DAN single-strand breaks determined with the comet assay. Mutag. Res. 409:163-171.

Jasso Y, Espinoza G, González D, Razo I, Carrizales L, Torres A, Mejía J, Monroy M, Ize A, Yarto M, Díaz F (2007). An Integrated Health Risk Assessment Approach to the Study of Mining Sites Contaminated With Arsenic and Lead. Int. Environ. Assess. Manag. 3(3):344-350.

Jolly SS Singh BM, Mathur OC, Malhotra KC (1968). Epidemiological clinical and biochemical study of endemic dental and skeletal fluorosis in Punjab. Br. Med. J. 4:427-429.

Kaseva ME (2006). Contribution of trona (magadi) into excessive fluorosis-a case study in Maji ya Chaid ward, Northern Tanzania. Sci. Total Environ. 366: 92-100.

Kleinsasser N, Schmid K, Sassen A, Harréus U, Staudenmaier R, Folwaczny M, Glas J, Reichl F (2006). Cytotoxicity and genotoxicity effects of resin monomers in human salivary gland tissue and lymphocytes as assessed by the single cell microgel electrophoresis (comet) assay. Biomaterials 27:1762-1770.

Landi S, Naccarati A, Mathew R, Hanley N, Daily L, Devlin R, Vásquez M, Pegram R, DeMarini D (2003). Induction of DNA strand breaks by trihalomethanes in primary human lung epithelial cells. Mutag. Res. 538: 41-50.

Lee E, Oh E, Lee J, Sul D, Lee J (2004). Use of the Tail Moment of the Lymphocytes to Evaluate DNA Damage in Human Biomonitoring Studies. Toxicol. Sci. 81:121-132.

Li Y, Liang C, Slemenda CW, Ji R, Sun S, Cao J, Emsley CL, Ma F, Wu Y, Ying P, Zhang Y, Gao S, Zhang W, Katz BP, Niu S, Cao S, Johnston Jr CC (2001). Effect of long-term exposure to fluoride in drinking water on risks of bone fractures. J. Bone Mineral Rev. C15(2):123-138.

Ling FH, Jian GC (2006). DNA damage, apoptosis and cell cycle changes induced by fluoride in rat oral mucosal cells and hepatocytes.

World J. Gastroenterol. 12:1144-1148.

Mexican Official Norm (2001): NMX-AA-036-SCFI-2001. It establishes the maximum permissible limits of polluting agents in the unloadings of waste waters and national goods.

Mexican Official Norm (2001): NMX-AA-051-SCFI-2001. Waste waters. Sampling. Declaration of use. Published in the Official Newspaper of the Federation, 25 of March of 1980.

Mexican Official Norm (2001): NMX-AA-077-SCFI-2001. Water analysis; Determination of natural, residual and residual water fluorides treated. Official newspaper of the Nation (it cancels to NMX-AA-077-1982).

National Research Council (NRC) (2006). Fluoride in drinking-water, a scientific review of EPA´s standards, Washington DC.

Obwald K, Mittas A, Glei M, Pool BL (2003). New revival of an old biomarker: characterization of buccal cells and determination of genetic damage in the isolated fraction of viable leucocytes. Mutagen. Res. 544:321-329.

Ortega AM (2009). Presencia, distribución, hidrogeoquímica y origen de arsénico, fluoruro y otros elementos traza disueltos en agua subterránea, a escala de cuenca hidrológica tributaria de Lerma-Chapala, México. Revista Mexicana de Ciencias Geológicas 26(1):43-161.

Rekhadevi PV, Sailaja N, Chandrasekhar M, Mahboob M, Rahman F, Paramjit G (2007). Genotoxicity assessment in oncology nurses handling anti-neoplastic drugs. Mutagenesis 22:395-401.

Ribeiro D, Bazo A, DaSilva C, Alencar M, Favero D (2004). Chlorhexidine induces DNA damage in rat peripheral leukocytes and oral mucosal cells. J. Periodont. Res. 39:358-361.

Segerstrom K (1961). Southeastern Geology of the State of Hidalgo and northeastern Mexico. Bol. Assoc. Petr. Geol., 3 and 4: 147-168. Cited in: Geological-Mining Monograph of the State of Hidalgo, 1992.

Shanthakumari D, Srinivasalu S, Subramanian S (2004). Effect of fluoride intoxication on lipidperoxidation and antioxidant status in experimental rats. Toxicology 204:219-228.

Shashia A (2003). Fluoride an adrenal gland function in rabbits. Fluoride 36:241-251.

Singh NP, McCoy MT, Tice RR, Schneider EL (1998). A simple technique for quantification of low levels of DNA damage in individual cells. Exp. Cell Res. 175:184-191.

Smerhovsky Z, Landa K, Rossner P, Brabec M, Zudova Z, Hola N, Pokorna Z, Mareckova J, Hurychova D (2001). Risk of cancer in an occupationally exposed cohort with increased level of chromosomal aberrations. Environ. Health Perspect. 109:41-45.

Szeto YT, Benzie IFF, Collins AR, Choi SW, Cheng CY, Yow CMN, Tse MMY (2005). A buccal cell model comet assay: Development and evaluation for human biomonitoring and nutritional studies. Mutat. Res. 578:371-381.

Tice R, Aguerrí E, Anderson D, Burlinson B (2000). Single cell gel/comet assay: Guidelines for in vitro and in vivo genetic toxicology testing. Environ. Mol. Mutagen. 35:206-221.

Tiwari H, Rao MV (2010). Curcumin supplementation protects from genotóxic effects of arsenic and fluoride. Food Chem. Toxicol. 48:1234-1238.

Tovalin H, Valverde M, Morandi MT, Blanco S, Whitehead L, Rojas E (2006). DNA damage in outdoor workers occupationally exposed to environmental air pollutants. Occupational Environ. Med. 63:230-236.

Vázquez-Alvarado P, Prieto-García F, Coronel-Olivares C, Gordillo–Martínez A, Ortiz-Espinosa R, Hernández-Ceruelos A (2010). Fluorides and dental fluorosis in students from Tula de Allende, Hidalgo, México. J. Toxicol. Environ. Health Sci. 2:24-31.

Wang SX, Wang ZH, Cheng XT, Li J, Sang ZP, Zhang XD, Han LL, Qiao SY, Wu ZM, Wang ZQ (2007). Arsenic and fluoride exposure in drinking water: children´s IQ and growth in Shanyin County, Shanxi. Chin. Environ. Health Perspect. 115:643-647.

Wania F (2004). Transport and fate of chemicals in the environment. In: Encyclopedia of Physical Science and Technology, 3ed Edition Elsevier pp. 89-105.

Winning T, Townsend G (2000). Oral Mucosal Embriology and Histology. In: Elsevier Science. Clinic Dermatol. 18:499-511.

World Health Organization (WHO) (2004). Fluoride in drinking-water. Background document for development of WHO Guidelines for drinking-water quality p. 9.

Xu RQ, Wu DQ, Xu RY (1997). Relations between environmental and endemic fluorosis in Hohot region, Inner Mongolia. Fluoride 30:26-28.

Yan Q, Zhang Y, Li W, DenBesten PK (2007). Micromolar Fluoride alters ameloblast lineage cells *in vitro*. J. Dent. Res. 86:336-340.

Yáñez L, García E, Rojas E, Carrizales L, Mejía J, Calderón J, Razo I, Díaz F (2003). DNA damage in blood cells from children exposed to arsenic and lead in a mining area. Environ. Res. 93:231-240.

Zeiger E, Shelby M, Witt KE (1993). Genetic toxicity of fluoride. Environ. Mol. Mutag. 21:309-318.

Zhang M, Wang A, Xia T, He P (2008). Effects of fluoride in DNA damage, S phase cell-cycle arrest and the expression of NF-KB in primary cultured rat hippocampal neurons. Toxicol. Lett. 179:1-5.

# 7

# Evaluation of polycyclic aromatic hydrocarbons (PAHs) content in foods sold in Abobo market, Abidjan, Côte d'Ivoire

Pierre MANDA[1]*, Djédjé Sébastien DANO[1], Ehouan Stephane-Joel EHILE[1,3], Mathias KOFFI[2], Ngeussan AMANI[3] and Yolande Aké ASSI[2]

[1]Laboratoire de Toxicologie et Hygiène Agro-industrielle, UFR des Sciences Pharmaceutiques et Biologiques, Université de Cocody, BPV 34 Abidjan, Côte d'Ivoire.
[2]Laboratoire Central pour l'Hygiène Alimentaire et l'agro-industrie, Laboratoire National pour le Développement Agricole, Abidjan Côte d'Ivoire.
[3]UFR des Sciences et Techniques des aliments, Université d'Abobo Adjamé, Abidjan Côte d'Ivoire.

This work was aimed to record the concentrations of eight polycyclic aromatic hydrocarbons (PAHs): (benzo[*a*]anthracene, benzo[b]fluoranthene, benzo[k]fluoranthene, benzo[a]pyrene, dibenzo[a,h]anthracene, indeno[1,2,3-c,d]pyrene, benzo[g,h,i]perylene and chrysene) in meats and fishes sold in Abobo market in Abidjan, Côte d'Ivoire. The amount of PAHs present in each sample was quantified using high-performance liquid chromatography (HPLC) equipped with ultraviolet (UV) detector. PAHs were present in all samples in variable quantity. More over benzo[a]pyrene (B[a]P) was present in majority of samples, in quantity above the limit fixed by European Union. With regard to cooking processes, smoking produce more PAHs compared to frying or grilled cooking. Concerning the nature of the matrices, no significant differences were found between meat and fish except benzo[g,h,i]perylene. The study declared that PAHs contamination in the tested foods exceeded the acceptable limit. Health risks linked with the consumption of these foods is a real danger that requires further study.

Key words: Polycyclic aromatic hydrocarbons (PAHs), fishes, meats, high-performance liquid chromatography (HPLC).

## INTRODUCTION

Polycyclic aromatic hydrocarbons (PAHs) are a group of environmental contaminants that originate from the pyrolysis or incomplete combustion of organic matter (Costes and Druelle, 1997). They are universal contaminants of our environment and of the human food chain (Lacoste et al., 2003). In food, PAHs are formed during processing and food preparation, either industrial or domestic, especially during the processes of smoking, drying and cooking (Moret et al., 1997; Bardolato et al.,

2006). Food contamination may also occur during periods of atmospheric pollution in which PAHs are deposited on seeds, fruits or vegetables, which are then consumed (Guillen et al., 1994; FAO/OMS 2008; Rey-Salgueiro et al., 2008). Experimental data related to PAHs in animals have shown that some of these compounds can induce many health effects such as systemic effects (hepatic, hematological and immunological effects and the development of arteriosclerosis) genotoxic and carcinogenic effects (Nisbet and Lagoy, 1992; Ramesh et al., 2004). Hence, special interest was given to studying the toxicity of PAHs by different international bodies: The Scientific Committee for Food (SCF) and Joint FAO/WHO Expert Committee on Food Additives (JECFA).

*Corresponding author. E-mail: mandapierre@yahoo.fr.

Based on reviews of profiles of PAHs in food and the results of the carcinogenicity study of two coal tar led by Culp et al. (1998), the SCF and JECFA suggested that B[a]P should be used as a marker for the occurrence and carcinogenic effect of PAHs in food. In 2005 and again in 2008, the European Commission has established maximum limits for PAHs in different foodstuffs (Regulation (EC) No 1881/2006). More recently, the Scientific Panel on Contaminants in the food chain (CONTAM Panel) of EFSA reviewed the available data on the occurrence and toxicity of PAHs. The CONTAM Panel concluded that B[a]P alone is not a valid indicator of the occurrence of PAHs in food. The group proposed four PAHs (benzo[a]anthracene, benzo[b]fluoranthene, benzo[a]pyrene, chrysene) or eight PAHs (benzo[a]anthracene, benzo[b]fluoranthene, benzo[k]fluoranthene, benzo[g,h,i]perylene, benzo[a]pyrene, chrysene, dibenzo[a,h]anthracene and indeno[1,2,3-cd]pyrene) as best indicators of PAHs in food to preserve the health of consumers better.

In fact, according to Kluska (2003), food contributes significantly to human exposure to PAHs. Many studies have shown that cereals, vegetables (COT, 2002), oil and fat (De Vos et al., 1990; Moret et al., 2002) are the main contributors to the ingestion of PAHs. However, grilled or smoked fishes and meats show a relatively low contribution, except in specific cases or due to socio-cultural reasons that cause these foods to occupy a prominent place in the diet (Kazerouni et al., 2001; Jira, 2004; FAO/WHO, 2008).

PAHs are most often identified and quantified using several analytical techniques: either gas chromatography with flame ionization detection (GC/FID) or coupled to mass spectrometry (GC/MS), high performance liquid chromatography with ultraviolet (HPLC/UV) or fluorescence (HPLC/FL) detection or coupled to mass spectrometry (HPLC/MS). The European Food Safety Authority (EFSA, 2007) in the report on PAH, revealed that concerning the analytical method used, 4% indicated GC-FID, 28% the HPLC-FL, 26% the HPLC-UV/FL and 43% the GC-MS. Nowadays GC/MS and HPLC/FLU techniques are the most sensitive and therefore more currently used for the analysis of PAH in foodstuff.

Presently, the two main analytical techniques used for determining PAHs in foods are high performance liquid chromatography (HPLC) coupled to a fluorescence detector (FLD) and gas chromatography-mass spectrometry (GC-MS). Both methods are sufficiently sensitive for determining PAH concentrations usually found in foods. Earlier, HPLC with an ultraviolet (UV) or a photo-diode array (PDA) detector and GC with a flame ionization detector (FID) were also methods often applied.

In Cote d'ivoire, meats and fishes are principal protein sources. However, the available literature relates very few national studies on food contamination linked to PAHs. It is also important to note the work of Yeboué-

Kouamé et al. (2003), who examined the risk involved in fish smoking in Abidjan through urinary 1- Hydroxypyrene (1-OHP) dosage. Their results show that the professional exhibition to the fish smokers was weak. This absence or insufficiency of basic data regarding PAHs in food in Cote D'ivoire led to the present study. Our objective has been to promote production of food free of contamination for human consumption. This study tried to evaluate the level of PAHs contamination in the two most consumed foodstuff, which are fish and meat sold in the markets in Abidjan.

## MATERIALS AND METHODS

### Standards and reagents

All solvents used were of HPLC quality. All of the individual standard solutions used were from Chiron (Denmark).The standards consisted of benzo[a]anthracene (B[a]A) (5 mg/ml), benzo[b]fluoranthene (B[b]F 1 mg/ml), benzo[k]fluoranthene (B[k]F; 0.2 mg/ml), benzo[a]pyrene (B[a]P; 1 mg/ml), benzo[g,h,i] perylene (B[g,h,i] P; 1 mg/ml), chrysene (CHR) and indenol(1,2,3,c,d)pyrene (IP; 0.2 mg/ml) packaged in 1 ml of toluene and dibenzo[a,h]anthracene (DB[a,h]A; 1 mg/ml) in solid form. From these standard stock solutions, solutions of individual PAHs at different concentrations up to 300 μg/l in an appropriate solvent (acetonitrile, HPLC grade, Scharlau) and 50, 100, 200 and 300 μg/l mixed standard solutions of the 8 PAHs were prepared. All of these solutions were prepared and stored in amber bottles in the dark at a temperature of 4°C.

### Materials used for solid phase extraction

Two types of cartridges were used, C18 grafted-phase cartridges (Mega Bond Elut SPE Be-C18, VARIAN) with a phase of 2 g and a capacity of 12 ml and transplanted Florisil phase cartridges (Mega Bond Elut SPE Be-Fl VARIAN) with a phase of 1 g and a capacity of 6 ml.

### Samples

The biological materials studied were made of two matrices (meat and fish) commonly consumed by the population. The sampling point of each matrix was its transformation by the processes of smoking, frying and broiling. Foods samples collected were from those sold to the population. Meat matrices were made of smoked pork samples, fried chicken samples and cooked or grilled mutton samples. Fish matrices were made of smoked sardine samples, fried tuna samples and carp samples cooked on the grill or broiled. Food matrices sampled were in three separate campaigns by applying the Directive 2005/10/EC. Indeed, the number of shelves on which food were sold was counted. A method applied to determine the number of tables representing each lot was stratified sampling method. The displays were by a chancy selection technique without repetition. Thus, for each sampling campaign, the lot defined the amount of ways present on the selected tables. The elementary sample constituted the quantity of removal materials at divers' corners on the lot. The total sample corresponded to the aggregation of all elementary samples; it was the sample made available to the laboratory. The stock of the three campaigns was 18 samples and 3 kg per sample. Samples were collected in aluminum foil, transported using coolers and then after grinding, mixing, weighing and labeling, kept in the freezer (-18°C) until analysis.

## Sample preparation

Jira (2004) has described the extraction method used with some modifications. It consisted of a liquid extraction followed by two extraction-purifications (solid phase extraction) on two different cartridges. 10 ml of the acetone-acetonitrile mixture (60:40; v/v) was added to the sample (2.5 g). The samples were then vortexed for 30 s, followed by an extraction for 5 min in an ultrasound bath and a centrifugation at 4000 rpm/min for 5 min. The upper phase extracted and collected were in a Teflon conical tube. This process was repeated three times and the extracts were concentrated using a rotary evaporator (Rotavap, Butchi) at 35°C. The residue was taken up in 2 ml of acetone-acetonitrile mixture (60:40; v/v), stirred for 10 s by vortexing, centrifuged at 4000 rpm/min for 5 min and the remains were collected. The process was repeated three times. This extraction step was followed by a double purification. The first purification performed was on a grafted C18 cartridge. The cartridge previously conditioned with 2 ml methanol and 24 ml acetonitrile. The three extracts were transferred into the cartridge, the tube was rinsed with 2 × 2 ml acetone-acetonitrile (60:40; v/v), which were subsequently transferred to the head cartridge. The elution was done with 5 ml acetonitrile-acetone (60:40; v/v) at atmospheric pressure. The eluate obtained was evaporated using a rotary evaporator at 35°C. The dry residue was taken up in 1 ml hexane.

The second purification was carried out on a bonded-phase Florisil cartridge. The cartridge was previously conditioned with 15 ml dichloromethane and 12 ml hexane. After vortexing for 15 s, the extract was transferred to the head cartridge; the conical tube was rinsed 2 times with 2 ml hexane-dichloromethane (75:25; v/v) and then transferred to the head cartridge. The elution was performed with 4 ml hexane-dichloromethane (75:25; v/v).The eluate evaporation occurred until 1 ml remained, and then 500 µl toluene (retainer) was added. Evaporation continued up to about 50 µl. The required volume of solvent (acetonitrile) was added to obtain 1500 µl and its volume was calculated according to Equation (1):

$$V \text{ added} = 1500\text{-}m/d \qquad (1)$$

where m = sample mass in mg and d = 0.8669 g/ml (density of toluene). Obtained samples were then transferred into vial for HPLC analysis.

## HPLC apparatus and conditions

The HPLC system was equipped with a Shimadzu SIL-20 AC automatic injector (Kyoto, Japan), a Shimadzu LC-20 AD pump (Kyoto, Japan), a Shimadzu DGU-20A degasser (Kyoto, Japan), a specified HPLC column: Prevail column C18, 150 × 4.6 mm, 5 µm particles, kept constant at 40°C by a Shimadzu CTO -20A oven (Kyoto, Japan) and a UV-visible SPD-20A Shimadzu detector (Kyoto, Japan) at 284 nm. The whole chain was controlled by a Shimadzu CBM-20A system controller.

The injection volume was 20 µl. The mobile phase consisted of solvent A (water) and solvent B (acetonitrile). The speed was 1.5 ml/min. The gradient was binary with 72% solvent B, 28% solvent A at 0 to 4 min, then 100% solvent B at 15 min. The calculation of PAHs was obtained by the method of external normalization following Equation (2), with Ci corresponding to the PAH content in a sample calculated in µg/kg fresh weight; Ai, the area of the peak (average of 2 injections) of PAHs in the sample solution; Air, the peak area (mean dose) of PAHs in the standard solution; Cir, PAH concentration in the standard solution in µg/l; V, the volume of the

final extract and m, the sample mass in grams.

$$Ci = \frac{Ai \times Cir \times V}{Air \times m} \qquad (2)$$

## Validation of the method

The validation of the analysis method was achieved according to NF V03-110 (1998) Standard. This procedure consists of the determination of detection and quantification limits, the calculation of the coefficient of variation of tests of repeatability and reproducibility and the determination of the linearity domain, the rate of recovery, the sensitivity and the specificity.

## Statistical analysis

Data were analyzed using SPAD version 4.01. The tests used for the comparison of the PAHs according to the nature of foods were the t-test for independent samples when variances were equal and the Mann-Whitney test for unequal variances. Concerning the comparison of cooking processes, ANOVAs (analysis of variance) were used when variances were equal. In contrast, the Kruskal-Wallis test was applied in cases of unequal variance. To detect levels of differences between cooking processes, the Kolmogorov-Smirnov test with two samples were used. Significance was accepted at a level of 5%.

# RESULTS AND DISCUSSION

## Validation of the method

In our study, we used HPLC/UV to quantify PAHs. The method by HPLC/UV also provides good results. The maximum rate of PAHs and particularly that of B[a]P is set at 5 µg/kg. This value is well above the HPLC/UV limit of detection. To reduce interference due to the low sensitivity of UV detector, a double purification was performed: on C18 bonded phase cartridge and the bonded phase florisil cartridge. The method used presented good linearity, established from 0 to 200 µg/kg with a coefficient of determination $r^2 = 0.9997$. The coefficients of variation of repeatability and reproducibility were respectively 1.34 (n = 10) and 4.74 (n = 5). Recovery rates were obtained from the standard addition method. PAHs standards were added to a quantity of 5 g of each matrix powder. The average recovery rate obtained for all eight PAHs was 91% (that is 80 to 101%). The detection limits ranged from 0.03 µg/kg for benzo(a)pyrene, benzo(k)fluoranthene to 0.36 µg/kg for chrysene and benzo(k)fluoranthene. As for the limits of quantification, they ranged from 0.06 to 1.2 µg/kg. In the EFSA (2007), the limits of detection varied from 0.0002 to 1 µg/kg with a mean of 0.12 µg/kg for B[a]P by the GC/MS technique.

## PAHs concentration in samples

Table 1 reports the average concentration of individual

**Table 1.** Average amount of PAHs (μg/kg) in samples.

| Sample | CHR | B[a]A | B[b]F | B[k]F | B[a]P | DB[a,h]A | IP | B[g,h,i]P |
|---|---|---|---|---|---|---|---|---|
| Smoked meats | 10.36 | 12.7 | 25.41 | ND | 8.76 | 0.97 | 64 | 0.11 |
| Fried meats | ND | 2.86 | 0.69 | 0.59 | 2.32 | 0.66 | 51 | 4.42 |
| Grilled meats | 5.04 | 12.6 | 19.22 | ND | 7.15 | 0.73 | 32.31 | ND |
| Smoked fishes | 13.67 | 45.07 | 66.67 | ND | 34.07 | 1.37 | 37.31 | 2.03 |
| Fried fishes | ND | 0.5 | 0.41 | 0.36 | 6.54 | 0.7 | 74.19 | 19.29 |
| Grilled fishes | 0.7 | 6.7 | 1.56 | 0.15 | 2.56 | 0.14 | 37.41 | 0.59 |

ND, Below detection limit; CHR, chrysene; B[a]A, benz[a]anthracene; B[b]F, benzo[b]fluoranthene; B[k]F, benzo[k]fluoranthene; B[a]P, benzo[a]pyrene; DB[a,h]A, dibenz[a,h]anthracene; IP, indeno[1,2,3-cd]pyrene; B[ghi]P, benzo[ghi]perylene.

PAHs in the various food samples. PAHs are present in all samples in varying concentrations. The highest mean levels of individual PAHs were observed in samples of smoked fish with B[b]F (66.67 μg/kg), B[a]A (45.07 μg/kg) and B[a]P (34.07 μg/kg). B[a]P and IP were present in all foods with concentration greater than 1 μg/kg. Concerning matrices of smoked foods, meats contained an average amount of 8.76 μg/kg B[a]P. This concentration is greater than the amount observed by Jira (2004), who measured an average concentration of 0.12 μg/kg B[a]P in samples of smoked ham and sausage, and is higher than the European Commission standard fixed at 5 μg/kg. Smoked fishes had 34.07 μg/kg B[a]P, which is much higher than the European Standard fixed at 5 μg/kg.

Concerning matrices of fried samples, fishes had 6.54 μg/kg B[a]P and fried meats had 2.32 μg/kg B[a]P. This concentration in fried meats is lower than the European Standard (5 μg/kg). However, it is higher than the average value of 0.05 μg/kg found in fried meats by Kazerouni et al. (2001). In broiled food matrices, fishes had an average of 2.56 μg/kg B[a]P and meats had an average of 7.15 μg/kg B[a]P. These amounts are higher than the European Standard fixed at 2 and 5 μg/kg respectively, for broiled fishes and meats. These values are also higher than the average concentration of 1.72 μg/kg found by Kazerouni et al. (2001). However, they are similar to those of Akpambang et al. (2009) who found that five of the six tested grilled food samples sold in Nigeria exceeded the limit of PAHs fixed by the European Union.

**Total sum of the amount of various contents of PAHs 4 and PAHs 8 in samples**

Table 2 shows the sum of the concentrations of different components of PAHs 4 and PAHs 8. The panel recommended the calculation of the sum of the contents of PAHs 4 or PAHs 8 during the last evaluation of the toxicity of PAHs in food. The CONTAM Panel proposed that the two sums are the best indicators of the occurrence of PAHs in food instead of B(a)P. Thus, in our

study, the sum of the contents of PAHs 4 was highest in samples of smoked fish (159.48 μg/kg) followed by samples of smoked meats (57.23 μg/kg). Samples of fried meats and fried fish presented the lowest amount 5.87 μg/kg and 7.45 μg/kg, respectively. About PAHs 8, also samples of smoked fish (200.19 μg/kg) and smoked meat (122.31 μg/kg) have the highest amounts. These values are higher than that observed by Akpanbang (2009) in samples of grilled fish in Nigeria. The proportion of B(a)P found in the amounts of the sum PAHs 4 and PAHs 8 are evaluated and presented in Table 2. The highest rates were found in samples of smoked fish (17%) and grilled meat (9.27%).

**Evaluation according to food nature**

Table 3 shows the studied PAHs profile according to the nature of the food matrix considered. The IP has the highest average levels but equal in the two matrices: 49.10 μg/kg for meat and 49.64 μg/kg for fish. The lowest levels are observed in B(k)F with 0.20 μg/kg for meat and 0.17 μg/kg for fish. The statistical analysis reveals that except for B[g,h,i]P (P=0.004), the amount of the other PAHs in meats and fishes was statistically equal (P>0.05). Thus, contamination by PAHs does not only depend on the nature of the food but also on the binary combination of the nature and cooking process applied to the food or on a likely environmental contaminant.

**Evaluation according to cooking process**

Table 4 shows the average amount of the individual PAHs in studied sample, according to the cooking process. PAHs appear in all samples because of the cooking processes. In the smoking process, the highest average amounts of individual PAHs were found in IP and B[b]F with 50.66 and 46.04 μg/kg, respectively. The amount of B[k]F found was inferior to the limit of detection . The B[a]P has a non negligible amount of 21.41 μg/kg. The lowest amounts of PAHs after smoking were 1.07 μg/kg B[g,h,i]P and 1.17 μg/kg DB[a,h]A. In the smoking

**Table 2.** Sum in µg/kg of PAHs 4 and PAHs 8 contents and its percentages in B[a]P.

| Parameter | Smoked meat | Fried meat | Grilled meat | Smoked fish | Fried fish | Grilled fish |
|---|---|---|---|---|---|---|
| PAHs 4 | 57.23 | 5.87 | 44.01 | 159.48 | 7.45 | 11.52 |
| PAHs 8 | 122. 31 | 62.54 | 77.13 | 200.19 | 101.99 | 49.81 |
| B[a]P / PAHs 4 (%) | 15.3 | 39.52 | 16.24 | 21.36 | 87.78 | 22.22 |
| B[a]P / PAHs 8 (%) | 7.16 | 3.70 | 9.27 | 17.01 | 6.41 | 5.13 |

**Table 3.** Average amount of PAHs (µg/kg) and comparisons according to the nature of matrix

| Nature | CHR | B[a]A | B[b]F | B[k]F | B[a]P | DB[a,h]A | IP | B[g,h,i]P |
|---|---|---|---|---|---|---|---|---|
| Meat | 5.13 | 9,39 | 15.11 | 0.20 | 6.08 | 0.79 | 49.10 | 1.54 |
| Fish | 4.79 | 17.42 | 22.88 | 0.17 | 14.39 | 0.74 | 49.64 | 7.30 |
| Significance | NS | NS | NS | NS | NS | NS | NS | S |

S, Significant comparatively; NS, not significant comparatively.

**Table 4.** Average amount of PAHs (µg/kg) according to cooking process and comparisons.

| Nature | CHR | B[a]A | B[b]F | B[k]F | B[a]P | DB[a,h]A | IP | B[g,h,i]P |
|---|---|---|---|---|---|---|---|---|
| Smoking | 12.01 | 28.88 | 46.04 | ND | 21.41 | 1.17 | 50.66 | 1.07 |
| Frying | ND | 1.68 | 0.55 | 0.47 | 4.43 | 0.68 | 62.59 | 11.85 |
| Broiling | 2.87 | 9.65 | 10.39 | ND | 4.85 | 0.43 | 34.86 | 0.33 |
| Significance | S | S | S | NS | S | NS | NS | S |

S, Comparatively significant; NS, comparatively not significant.

process, the use of traditional and rudimentary methods, such as dry wood and coconut wadding for fuel contribute to increased amounts of PAHs in foods (Lozada et al., 1998). In addition, the covering of foods with cardboards at the operation time could modify the photosensitive properties of the PAHs, causing their accumulation due to the presence or absence of light and/or oxygen modification of the PAHs content in foods (Simko, 2005). In addition, the long smoking times and uncontrolled temperatures would cause an increase in PAHs quantity. Indeed, according to Kazerouni et al. (2001), the quantity of B[a]P depends on the smoke temperature and the exposure time and has a tendency to concentrate PAHs in cooked products.

In the frying process, the highest amount of any individual PAH was IP (62.60 µg/kg), followed by B(g,h,i)P (11.85 µg/kg), B[a]P (4.43 µg/kg), DB[ah]A (0.68 µg/kg), B[b]F (0.55 µg/kg) and no traces of CHR was found. High amounts of PAHs could be due to the quality of the oil used. Indeed, oil manufactured under good conditions does not contain PAHs (Larson et al., 1987). The use of active coal in the refinement contributes to the total elimination of PAHs in oils during their manufacture (Lacoste et al., 2003). The repeated use of oil during frying would contribute to an increase of PAHs in the oils.

In the process of grilled cooking, IP presented the highest amount (34.86 µg/kg), followed by B[b]F (10.39 µg/kg), B[a]P (4.86 µg/kg), CHR (2,87 µg/kg), DB[a,h]A (0.44 µg/kg) and B[g,h,i]P (0.33 µg/kg). B[k]F was present with the lowest concentration (0.07 µg/kg). The application of sauce generally rich in fat could influence the amount of PAHs in foods cooked on a grill. Jägerstad and Skog (2005) pointed out that sauces smeared on meat increase the burnt surface of the meat and encourage the production of PAHs. In addition, during grilling, the greases melt and flow onto the heat source, causing their pyrolysis and the formation of PAHs.

According to Stolyhwo and Sikorski (2005), foods prepared under controlled conditions generally contain low amounts of PAHs. A lack of knowledge about PAH formation and the failure to employ restrictive parameters is the origin of the PAH contamination associated with the use of traditional methods could be incriminated in the formation of PAHs in all cooking processes.

## Detection limits of different cooking methods

To better identify the differences between the cooking processes in the genesis of PAHs, they were compared

**Table 5.** Differences in detection levels according to cooking process.

| PAH | Comparisons of the cooking method in pairs | | |
|-----|------------------------|------------------------|------------------------|
|     | Smoking/frying (Sm/Fr) | Smoking/grilled (Sm/Gr) | Frying/grilled (Fr/Gr) |
| CHR | Sm > Fr | Sm > Gr | NS |
| B[a]A | Sm > Fr | Sm > Gr | Fr < Gr |
| B[b]F | Sm > Fr | Sm > Gr | Fr < Gr |
| B[k]F | Sm < Fr | NS | NS |
| B[a]P | Sm > Fr | Sm > Gr | NS |
| DB[a,h]A | Sm < Fr | NS | NS |
| IP | NS | NS | NS |
| B[g,h,i]P | Sm < Fr | NS | Fr > Gr |

NS, Comparison not significant; Sm > Fr, smoking generate more PAHs than frying.

statistically in pairs (Table 5). This table shows the result of this comparison. The comparisons show that with the exception of B[g,h,i]P, smoking was the method that generated the most PAHs among the three practices, as measured by CHR, B[a]A, B[b]F, B[k]F and B[a]P. This result has been confirmed by Azeredo et al. (2006), who found that smoking generated significant amounts of B[a]P. Frying and broiling generates PAHs in equal proportions for CHR, B[k]F, B[a]P, DB[ah]A and IP.

## Conclusion

Smoked, grilled and fried meats and fishes sold in the Abobo market demonstrate contamination with different levels of PAHs. This contamination is marked by the presence of B[a]P in foods in quantity above European Standard. Smoking was revealed as the process that generated the most PAHs. Finally, according to the nature of the food, meats and fishes have high amounts of PAHs; however, there were no significant difference between the two types of matrices. Considering the potential carcinogenicity of PAH contamination and the importance of this food group in the Cote d'Ivoire food regime, the establishment of a regulation and surveillance plan should be considered as a high priority.

## REFERENCES

Akpambang VOE, Purcaro G, Lajide L, Amoo IA, Conte L, Moret S (2009). Determination of polycyclic aromatic hydrocarbons (PAHs) in commonly consumed Nigerian smoked/grilled fish and meat. Food Add Cont. Part A., 26(7):1096–1103.

Azeredo A, Toledo MCF, Camargo MCR (2006). Determination of benzo(a)pyrène in fish products. Ciênc Tecnol Aliment., 26:89-93

Bardolato E, Martins M, Aued-Pimentel S, Alaburda J, Kamagai J, Baptista G, Rosenthal A(2006). Systematic study of Benzo[a]pyrene in coffee samples. J Braz Chem Soc., 17:989–993.

Costes JM et Druelle V (1997). Les hydrocarbures aromatiques polycycliques dans l'environnement : la réhabilitation des anciens sites industriels. Rev de l'Institut Français du pétrole. 52:425-445.

COT. PAH in the UK diet: (2000). Total diet study samples. http://www.food.gov.uk/multimedia/pdfs/31pah.pdf.

De Vos R H, Van Dokkum W, Schouten A, De Jong-Berkhout P(1990).

Polycyclic aromatic hydrocarbons in Dutch total diet samples (1984–1986). Food Chem. Toxicol., 28: 263-268.

Findings of the EFSA Data Collection on Polycyclic Aromatic Hydrocarbons in Food, First issued on 29 June 2007. The EFSA Journal.

Food and Agricultural Organization / World Health Organization (2008). Report of a joint FAO/WHO expert consultation, March 31 – April 04, 2008; Food and Nutrition paper. Rome: FAO, 130p

Guillen MD. 1994. Polycyclic aromatic compounds: extraction and determination in food. Food Add. Cont., 11:669–684.

Jägerstad M, Skog K(2005). Genotoxicity of meat processed foods. Mutat. Res., 574:156-172.

Jira W(2004). A GC/MS method for the determination of carcinogenic polycyclic aromatic hydrocarbons (PAH) in smoked meat products and liquid smokes. Eur. Food Res. Technol., 218:208–212.

Kazerouni N, Sinha R, Hsu CH, Greenberg A, Rothman N(2001). Analysis of 200 food items for benzo(a)pyrene and estimation of its intake in an epidemiologic study. Food Chem. Toxicol., 39:423-436.

Kluska M(2003). Soil contamination with polycyclic aromatic hydrocarbons in the vicinity of the ring road in Siedle City. Polish J. of Environ. Studies, 12: 309–380.

Lacoste F, Raoux R, Dubois D, Soulet B(2003). Problématique des Hydrocarbures aromatiques polycycliques dans les corps gras. OCL, Oléagineux, Corps gras, Lipides. 10: 287–295

Lozada EP, Varca LM, Dimapilis AM(1998). Minimising polycyclic aromatic hydrocarbons (PAHs) levels in cooking oil. Paper presented at: National Academy of Science and Technology (NAST) 20th scientific meeting 1998.

Mc Grath T, Sharma R, Hajaligol M(2001). An experimental investigation into the formation of polycyclic aromatic hydrocarbons (PAH) from pyrolysis of biomass materials. Fuel, 80:1787-1797.

Moret S, Purcaro G, Conte SL( 2005). Polycyclic aromatic hydrocarbons in vegetable oils from canned foods. Eur. J. Lipid Sci. Technol., 107:488–496.

Moret S, Conte LS(2002). A rapid method for polycyclic aromatic hydrocarbons determination in vegetable oil. J. Separat. Sci., 25:96–100.

Moret S, Piani B, Bortolomeazzi R, Conte LS (1997). HPLC determination of polyaromatic hydrocarbons in olive oils. Zeitschrift für Lebensmittsmittel Untersuchung und Forshung A. 205:116–120.

Nisbet ICT and Lagoy PK(1992). Toxic equivalency factor (TEFs) for polycyclic aromatic Hydrocarbons (PAHs). Reg. Toxicol. pharmacol., 16:290–300.

Ramesh A, Walker SA, Hood DB, Guillen MD, Schneider K, Weyand EH (2004). Bioavailability and risk assessment of orally ingested polycyclic aromatic hydrocarbons. Int. J. Toxicol., 23:301–333.

Rey-Salgueiro L, Martinez-Caballo E, Garcia-Falcon MS, Simal-Gandara J(2008). Effects of chemical company fire on occurrence of polycyclic aromatic hydrocarbons in plants foods. Food Chem., 108:347–353.

Simko P(2005). Factor affecting elimination of polycyclic aromatic hydrocarbons from smoked meat foods and liquid smoke flavorings. Mol. Nutr. Food Res., 49:637–647.

Stolyhwo A, Sikorski ZE(2005). Polycyclic aromatic hydrocarbons in smoked fish – a critical review. Food Chem., 95:303–311.

Yeboué-Kouamé B, Bonny J, Assi A, Wognin S, Kouassi Y, Aka J,

Nisse C, Haguenoer J, Frimat P( 2003). Exposition aux Hydrocarbures Aromatiques Polycycliques : Dosage du 1-hydroxypyrène urinaire chez des fumeuses de poissons en zone urbaine et péri urbaine dans la région d'Abidjan. Arch. Mal. Prof. Med. Trav., 64:265-270

# 8

# Hot spot biomonitoring of marine pollution effects using cholinergic and immunity biomarkers of tropical green mussel (*Perna viridis*) of the Indonesian waters

**Khusnul Yaqin[1]\*, Bibiana Widiati Lay[2], Etty Riani[2], Zainal Alim Masud[2] and Peter-Diedrich Hansen[3]**

[1]Department of Fisheries, Faculty of Marine science and Fisheries, Hasanuddin University, Jalan Perintis Kemerdekaan Km 10, Makassar 90245, Indonesia.
[2]Environmental Science Study Programme, Bogor Agricultural University, Darmaga Campus, Bogor 16680, Indonesia.
[3]Department of Ecotoxicology, Technische Universitaet, Faculty VI, Franklin Strasse 29 (OE4), D-10587 Berlin, Germany.

Selected biomakers, Cholinesterase (ChE) and phagocytic activities have been investigated with the exposed green mussel *Perna viridis* in Indonesian coastal waters. An operative effect-based monitoring on two polluted sites and one reference area were investigated for aquaculture enterprises and human health aspects. Between two heavily polluted sites, green mussels from Cilincing indicated a lower level of the ChE activity than those from Kamal Muara. The phagocytic activity of green mussels from the polluted sites demonstrated significant higher activity than that of green mussels from the pristine site, Pangkep. However, there were no significant differences of phagocytic activities between the polluted sites. This might indicate that the existing pollutants in Jakarta Bay were more neurotoxic rather than immonotoxic substances. The results showed clearly that both selected biomarkers were potential valuable tools for effect-based monitoring and pollution impacts in coastal zones of Indonesia.

**Key words:** Green mussel, biomarkers, coastal zone management, Indonesia.

## INTRODUCTION

A biological approach has been used as a counterpart of a classic chemical approach for surveying marine pollution effects in many international programs (Cajaraville et al., 2000; Devier et al., 2005; Lehtonen et al., 2006; Orbea et al., 2006; Minier et al., 2006). A chemical analysis solely is considered as an invaluable analysis for interpretation of the pollutant impact in marine ecosystem since it does not illustrate the harmful effects (Walker, 1998; Damiens et al., 2004) and the fate of chemical compounds on living organism through biotransformation of xenobiotic substances within living organism body (Nicholson and Lam, 2005). In many cases, the biotransformation may increase xenobiotic substances toxicity on organism via producing reactive metabolite compounds that are more toxic than original

parent compounds (Belden and Lydy, 2000). Moreover, the chemical approach is costly, usable to only a small proportion of the xenobiotic compounds in the environment, produces a little biologically meaningful data, and consequently simplifies the complexity of the ecosystem under monitoring (Butterworth, 1995). For those reasons, the classic chemical analysis should be accompanied by the biological approach which is so called "biomarker" that elucidates biological responses of environmental pollution.

Biomarkers have been considered as sensitive and suitable tools for detecting either exposure, or effects of, pollutants (Hansen, 1995; Narbonne et al., 2001; Lagadic, 2002) since they can provide more comprehensive and biologically more relevant information on the potential impact of pollutants on the health status of organism (Van der Oost et al., 1996; Picado et al., 2007; Galloway, 2006). In respect to pollutants that has a lower stability in water such as organophosphate and

*Corresponding author. Email: khusnul@gmail.com.

carbamate pesticides, biomarkers are reliable tools for assessing the impacts of the pollutants on biota even if the existing of the pollutants in water cannot be detected (Sturm et al., 1999). It is because biomarkers can detect persistence responses and/or effects of the pollutants in such duration of biota lifetime (Depledge and Fossi, 1994). Therefore, they have been used enormously in biomonitoring to assess the risk of marine ecosystem pollution (Cajaraville et al., 2000; Martin-Diaz et al., 2004).

Mytilid mussels have received tremendous concerns as a sentinel organism when applying biomarkers in many pollution monitoring programmes (Cajaraville et al., 2000; Dizer et al., 2001a; Livingstone et al., 2000; Castro et al., 2004; Nesto et al., 2004; Leinio and Lethonen 2005; De Luca-Abbott et al., 2005; Halldórsson et al., 2007: Verlecar et al., 2008). As sedentary and filter-feeder animals, marine mussels do not escape from contaminated water where they are living and can accumulate many contaminants to the level higher than existing in water (Widdows and Donkin, 1992). Hence, the behaviors are providing realistic sentinel organisms that indicate the biologically available concentrations. The realistic bioavailability of contaminants in mussels is also demonstrated by the fact that they have inefficient detoxifying enzymes permitting small portion of contaminants that can be transformed within their body (Nicholson and Lam, 2005). Consequently, mussels have been considered as notable eco-sentinel organisms for effect-based monitoring activity and have represented the sensitivity of detection harmful effect of pollutions (Goldberg et al., 1978; Kim et al., 2008).

The extensive use of mussels and biomarkers for that purpose were carried out in temperate region by using blue mussels, Mytilus edulis (Halldórsson et al., 2007; Gagné et al., 2008; Tedesco et al., 2008; Yaqin and Hansen 2010). However, compare to the use of blue mussel in temperate regions there are few studies conducted concerning biomarkers in tropical region by using native species, green mussels, Perna viridis (Nicholson and Lam, 2005). It has been postulated that genetic and ecosystem differences of two marine mussels generated complicated inherent difficulties, when an extrapolation of M. edulis data to P. viridis was conducted. Therefore, a hot spot investigation of biomarkers in tropical regions by employing P. viridis is required to enhance the understanding of biological response of indigenous species toward contaminants to enforce biomonitoring of marine pollution effects programs in tropical region.

This study applied selected biomarkers which are cholinesterase (ChE) and phagocytic activities to monitor effects of pollution in coastal areas of Indonesia. ChE activity has been widely used as a biomarker (biochemical response) for neurotoxic effects of organophophorous and carbamate pesticides. There are some influences of ChE activity by several metals, PAH

and surfactants exposure (Tabche et al., 1997; Guilhermino et al., 1998; Akcha et al., 2000; Moreira et al., 2004). Moreover, the immune system is a vital part of the organism and associates intimately with the function of many organs and organ system (Fournier et al., 2000). In invertebrate, the phagocytic activity which is part of the immune system can be induced by wide range of xenobiotics. Hence, the phagocytic activity is considered as a less specific early indicator of immunotoxicity or as a biomarker (Oliver and Fisher, 1999; Blaise et al., 2002). The two selected biomarkers were employed in the current study based on microtiterplate techniques in order to provide a rapid, cost-effective, justifiable (Blaise et al., 2002), and well-adapted application in developing countries.

## MATERIALS AND METHODS

### Chemicals

Acetylthiocholine iodide, 5,5'-Dithio-bis-(-2-Nitrobenzoic acid) (DTNB), γ-globuline, Bovine Serum Albumin, Fluoresceinisothiocyanate were purchased from Sigma. Bradford reagent was purchased from Bio-Rad Laboratories GmBH, Germany. All others reagents used were analytical grade products.

### Study area

The study was conducted on three different areas of Indonesian coastal zone. A coastal area of Pangkajene Kepulauan (Pangkep) regency in South Sulawesi was chosen as reference site (station 1; Figure 1) because there are relatively minimal anthropogenic activities that were performed in this place such as traditional fisheries, which use static fishing equipment. On the other hand, two sites of Jakarta Bay, Kamal Muara and Cilincing (Figure 2) were chosen and considered as heavily anthropogenic polluted sites (station 2 and 3) since they received almost domestic and industrial wastes from Jakarta and neighboring cities of Jakarta. Moreover, some studies based on chemical analysis indicated that Jakarta Bay was under threatened by anthropogenic pollutants (Williams et al., 2000; Sudaryanto et al., 2002; Munawir, 2005). Whilst, many traditional fisheries activities such as green mussel aquacultures are situated along Jakarta Bay. Hence, Jakarta Bay is considered also as highly valued fisheries resources of coastal area, which plays indispensables role for preserving marine food resources and economic basis of small scales fishermen.

### Sample collection and preparation

Thirty two mussels (5 to 6 cm) were handpicked on traditional green mussel cultures along Jakarta Bay at Kamal Muara and Cilincing, and from the Pangkep wild reference population attached naturally on traditional static fishing equipments.

The collected living mussels were directly transferred to the laboratory using cool box under humid condition. Prior to dissecting out of the mussels, 1 ml of hemolymph was withdrawn from posterior adductor muscle (PAM) sinus using 1 ml syringe and 0.4 mm needle followed by phagocytosis assay as described thus. Gill, foot, mantle and PAM were cut off, blotted dry and weighted before placing them in 2 ml potassium phosphate buffer in Eppendorf tube (0.1 M/pH 8.0). Prior to transferring the tissues to Ecotoxicology Department Laboratory (Technische Universitaet of Berlin,

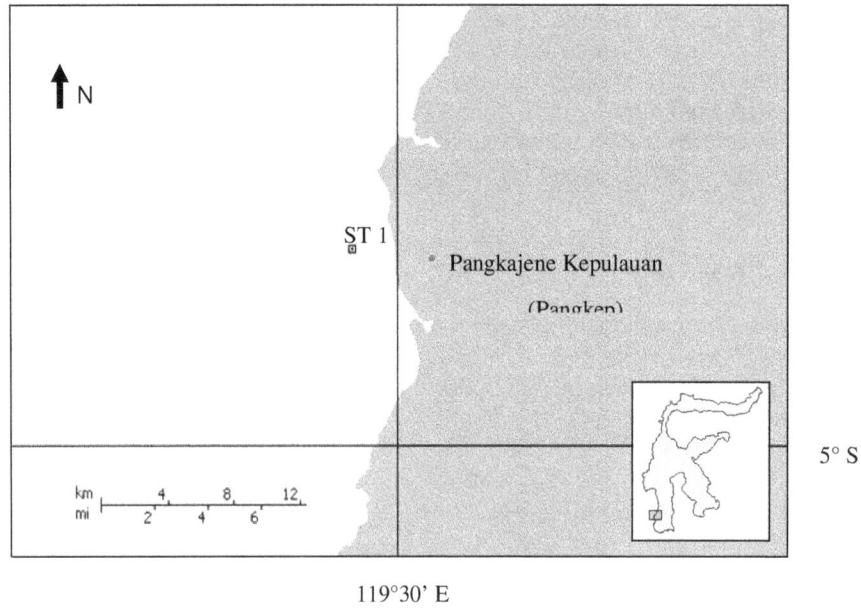

**Figure 1.** Sampling station (ST) in Pangkajene Kepulauan (Pangkep).

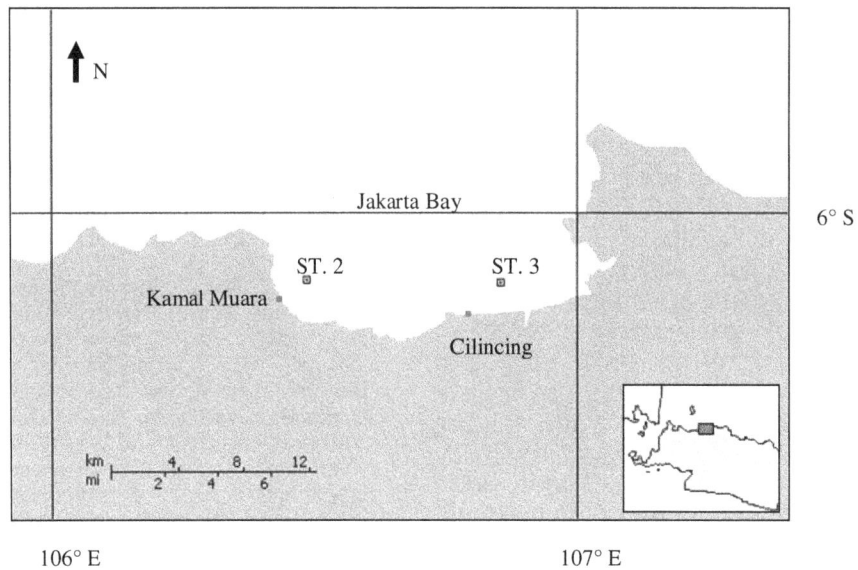

**Figure 2.** Sampling stations (ST) in Jakarta Bay.

Germany) using cool box that were filled by dry ice, the tissues were stored at -70 °C.

### Cholinesterase activity

The enzyme activity was measured following the modified Ellman method (Ellman et al., 1961) for a 96-well plate. A Dounce homogenizer was used to homogenize 0.3 g of each tissue in 2 ml potassium phosphate buffer (0.1 M/pH 8.0). The homogenate was centrifuged for 10 min at 10000 × g and the supernatant was harvested and stored at -80 °C prior to the analysis of ChE activity and protein content. The supernatant was diluted in 1:2 of

potassium phosphate buffer (0.1 M/pH 8.0) following the enzyme measurement.

The enzyme measurement was carried out by placing 50 μl of the diluted sample into each well of the microplate. A blank was made by putting 50 μl of potassium phosphate buffer into the blank section of the microplate wells. The plate was incubated for 5 min in 25°C with 200 μl of 0,75 mM 5,5'-Dithio-bis-(-2-Nitrobenzoic acid) prior to the reaction which started by addition of 50 μl of 3 mM Acethylthiocholine iodide. Accordingly, the plate was read by photometry for microtiter plate (Spectra Thermo TECAN) in an interval of 30 s for 5 min at 405 nm. Four independent measurements of the ChE activity were carried out for each individual of *P. viridis*, and the average activity were calculated.

## Protein |measurement for cholinesterase assay

Protein content measurement was carried out by diluting the gill extract 1:10 with distilled water. It was measured previously by placing 10 µl of the diluted extract and 10 µl of serial dilutions of $\gamma$-globuline protein standard into a separate well section of the microplate. A blank was made by placing 10 µl of distilled water into the blank section of the microplate. After the addition of 5% Bradford-reagent solutions (200 µl) into the microplate, the samples were left in room temperature for 20 minutes to allow color development. The absorbance was read at 620 nm using photometry (Spectra Thermo TECAN).

The ChE activity is expressed as nmoles of product developed per minute per mg of protein (nmol/min/mg protein). The ChE activity was measured on each tissue to recognize which tissue has the highest ChE activity.

## Phagocytic activity

Phagocytic activity of hemocytes was determined by a microplate-based fluorescence measurement method (Hansen, 1992; Anderson and Mora, 1995). This method is based on the number of fluorescence labeled yeast cells that were phagocytosed by mussel hemocytes. The yeast cells were treated and labeled by Fluoresceinisothiocyanate (FITC) (Anderson and Mora, 1995) and kept in aliquots at -80°C. After withdrawing hemolymph from the PAM sinus of the mussels, 100 µl of hemolymph was dropped into 96-microplate wells. Five replicates were used to analyze the phagocytic activity and 3 replicates were used for the protein analysis. The density of hemocytes from each mussel was calculated by using a hemocytometer under a light transmission microscope. After the incubation of the plate for 30 min to allow hemocytes deposition at the bottom of the microplate wells, 25 µl of the FITC-labeled yeast was added into each phagocytic activity section of the microplate wells. A standard was made by adding 100 µl of phosphate buffer saline (PBS) and 25 µl of the FITC-labeled yeast into the microplate wells. One column (8 wells) was used as a blank section by adding 125 µl of PBS. The plate was incubated for 90 min in 21°C at dark condition. At the end of the incubation, 50 µl of 1 % glutaradehyd was added into each phagocytosis section of microplate wells, while 50 µl of methanol was dropped into the protein section of microplate wells. Before transferring to the laboratory in Germany, the plates were covered by a film and wrapped in aluminum and stored at 5°C in darkness. Accordingly, the fixatives were removed carefully and replaced by 125 µl of PBS when samples have arrived in the laboratory. For quenching the fluorescence background of unphagocytosed cells, 25 µl of 0.6 mg/ml trypan blue dissolved in PBS was added to each well of the microplate. The plate was incubated for 20 min prior to removing of all supernatants. The fluorescence of the ingested FITC-labeled yeast cells were read at excitation of 485 nm and emission of 535 nm using a microtiter plate fluorometer (Dynatech, Fluorolite 1000).

## Protein measurement for phagocytic assay

A protein content measurement was carried out using hemocytes only. Prior to the measurement, the buffer was removed carefully and hemocytes were lysed with 50 µl of 0.1 N NaOH. After incubating the lysed hemocytes for 10 minutes in a shaking chamber, 10 µl of the lysed hemocytes and the serial dilutions of protein standard (Bovine Serum Albumin) were added to 96-microplate wells. Accordingly, 200 µl of 5% Bradford-reagent solution were added into the plate and incubated for 20 minutes to allow color development. The absorbance of protein was measured at 620 nm using photometer (Spectra Thermo TECAN).

Accordingly, the phagocytic activity was expressed as Relative Fluorescence Units (RFU) and finally calculated as a Phagocytic Index: RFU/mg hemocyte protein.

## Statistical analysis

The statistical analyses were performed using non-parametric test, Kruskall-Wallis to distinguish the differences of ChE and phagocytic activity among the sites. If there were differences among the sites ($p < 0.05$), the test was continued by Dunn's multiple comparison test to determine the different between two sites. The statistical analyses were conducted using GraphPad Prism trial version 5.0 for Windows, GraphPad Software, San Diego California USA.

## RESULTS

### Cholinesterase activity

It has been reported that the ChE activity level differed among organs in marine mussels (Bocquené et al., 1990; Brown et al., 2004). The current study was started by recognizing which organ of green mussel, *P. viridis* that posses the highest ChE activity. It has been performed using *P. viridis* tissues from expected clean area. The results presented in Figure 3 demonstrated the median of the ChE activity in the gill which had the significant highest ChE activity namely 83.56 ± 12.19 nmol/min/mg protein followed by the foot (46.16 ± 4.18 nmol/min/mg protein), the mantle (27.35± 2.50 nmol/min/mg protein) and the PAM (4.94 ± 4.08 nmol/min/mg protein). Accordingly, the gill was used as a tissue target for measuring the ChE activity since the highest ChE activity of such organ should be the most suitable for measurement of the ChE activity inhibition (Bocquené et al., 1990; Escartin and Porte 1997; Valbonesi et al., 2003; Lau et al., 2004; Brown et al., 2004; Damiens et al., 2007; Taleb et al., 2009; Yaqin 2010).

Statistical analysis showed the difference ChE activity in the gills of the samples ($p < 0.05$) (Figure 4). The animals collected from the reference site had the significant highest ChE activity (83.56 ± 12.19 nmol/min/mg protein) followed by the green mussel collected from heavily polluted areas, Kamal Muara (49.92 ± 3.29 nmol/min/mg protein) and Cilincing (27.20 ± 1.80 nmol/min/mg protein). Between two heavily polluted sites, the animals inhabited in Kamal Muara showed significant less inhibition of the ChE activity than those from Cilincing ($p < 0.05$).

### Phagocytic activity

In the present study, the phagocytic activity expressed as phagocytic index, and hemocytes numbers and total cell protein content were measured simultaneously. The results were represented in Figures 5 and 6. Statistical analysis of the median of circulating hemocytes numbers exhibited no difference numbers of hemocytes ranging

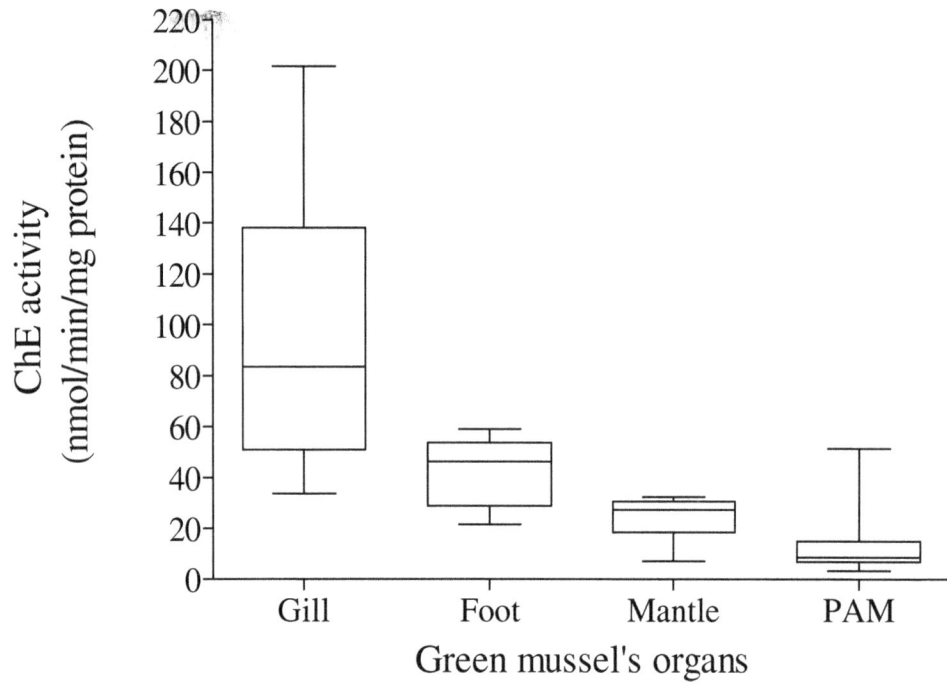

**Figure 3.** Cholinesterase activity of different organs of green mussel, *Perna viridis* from Pangkep Indonesia.  Data were expressed as median (25 and 75 % quartile, 5 and 95 % confidence interval).

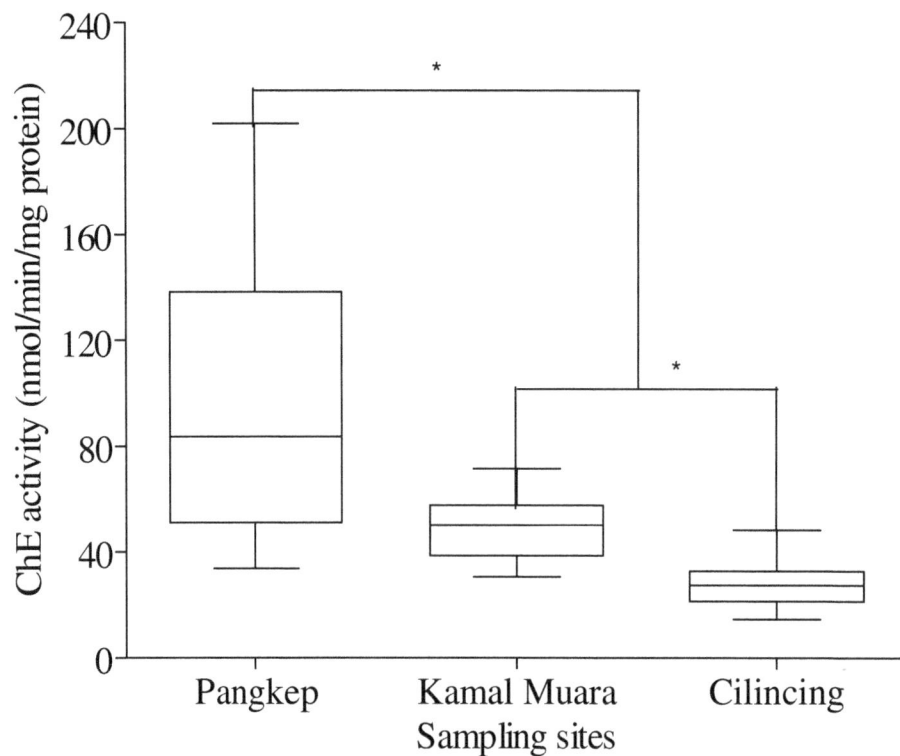

**Figure 4.** Gills ChE activity of green mussel, *Perna viridis* collected in the selected areas of Indonesian waters.   Data were expressed as median (25  and 75 % quartile, 5 and 95 % confidence interval). * indicate significant difference ($p < 0.05$) of the gills ChE activity.

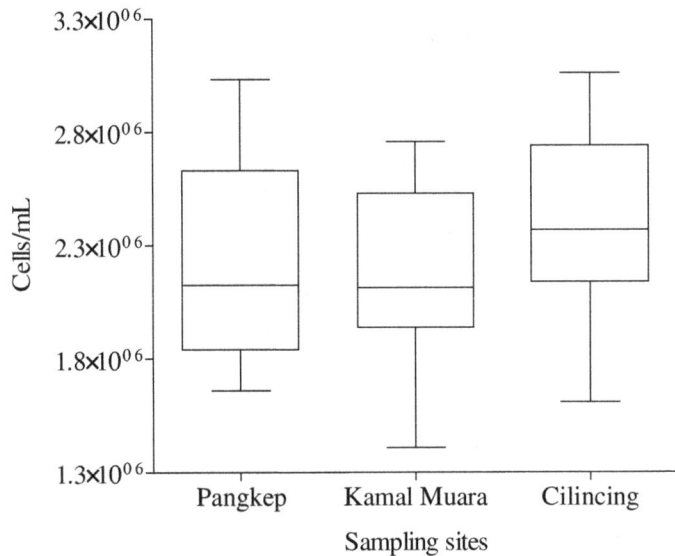

**Figure 5.** Circular hemocytes of green mussel *Perna viridis* collected in the selected areas of Indonesia waters. Data were expressed as median (25 and 75% quartile, 5 and 95% confidence interval).

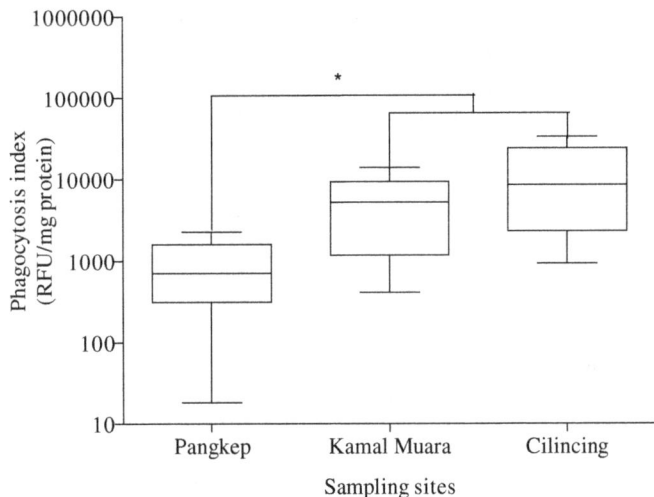

**Figure 6.** Phagocytotic Index of green mussel *Perna viridis* collected in the selected areas of Indonesia waters. Data were expressed as median (25 and 75% quartile, 5 and 95% confidence interval). *indicate significant difference (p < 0.05) of the hemolymph phagocytic activity. Y-axis is logarithmic scale.

from 2,1250,000 to 2,370,000 cells/ml. In contrast, the median of the phagocytotic index demonstrated significant different phagocytic activities of *P. viridis* collected from gradient pollutions of Indonesian coastal waters (p < 0.05). The animals collected from the two heavily polluted sites in Jakarta Bay showed significant higher phagocytic index than those collected from reference sites. Nevertheless, there was no significant

different phagocytic index within the polluted site (p = 0.118). The highest phagocytic index was demonstrated in hemocytes of *P. viridis* from Cilincing (23410.10 RFU/mg protein) which followed by Kamal Muara (7566.84 RFU/mg protein) and reference site, Pangkep (1714.19 RFU/mg protein).

## DISCUSSION

### Cholinesterase activity

Cholinesterases (ChEs) are enzymes that hydrolyses and inactivates neural transmitter acetylcholine (ACh) for regulating neural transmission impulse in the synaptic gap of cholinergic synapses and neuromuscular junctions (Soreq and Seidman, 2001). ACh play an important role both as excitatory and inhibitory transmitters of the gill muscle of bivalve (Gainey et al., 2003). In blue mussel, *Mytilus edulis*, ciliary movement of the gill is controlled by acetylcholine, dopamine and 5-hydrotryptamine (Aiello, 1990). Organophosphorous and carbamate pesticides inhibit ChE activity which may lead to severe physiological impairment of marine animals (Dauberschmidit et al., 1997) such as reduction in feeding efficiency of marine mussels (Donkin et al., 1997).

Since ChEs was purified by Wachtendonk and Neef (1979) in marine mussels hemolymph, a measurement of ChE activity in marine mussels has been used as a biomarker in laboratory test (Galloway et al., 2002; Rickwood and Galloway, 2004; Canty et al., 2007; Yaqin and Hansen 2010) and several international monitoring programs in the field (Narbonne et al., 1999; Cajaraville et al., 2000; Dizer et al., 2001a,b; Roméo et al., 2003; Bocquené et al., 2004; Gagné et al., 2008).

Characterization of ChEs in bivalve has been conducted in some bivalves e.g. in *M. galloprovincialis* the ChE specific activity was predominantly localized in the gills compare to others organs (Mora et al., 1999; Porte et al., 2001; Taleb et al., 2009). Moreover, the ChE activity from *M. galloprovincialis* gill was observed more sensitive to organophosphorous pesticides than that from the digestive gland (Escartin and Porte, 1997). In *M. edulis*, Bocquené et al. (1990) found that the highest ChE activity occurred in the gill compare to others organs such as the hepatopancreas, the mantle and the adducent muscle. By characterizing and comparing the ChEs in different organ of *M. edulis*, Brown et al. (2004) found that 'mitochondrial' fraction of foot had the highest ChE specific activity with very low recovery of activity.

Accordingly, the gill 'microsomal' activity had the second highest ChE specific activity with useful level of recovery and therefore was the most suitable fraction for biomarker application. The highest ChE activity in the gill compared to others organs such as the adducent muscle and the digestive gland were observed in the Antarctic scallop, *Adamussium colbecki* (Corsil et al., 2004).

Compared to the foot, the gill of the bivalve, *Scapharca inaequivalvis*, demonstrated the higher specific ChE activity level as well (Romani et al., 2005). Eventually, Bonacci et al. (2008) observed that the highest ChE activity also occurred in the gill of scallop, (*Pecten jacobaeus*) compared to others organs which were the adducent muscle and the digestive gland. Kopecka-Pilarczyk (2010) observed that the ChE activity from gill of *Mytilus trossulus* was the most sensitive enzyme activity compared to the activity from others organs and whole body tissue when exposed to carbaryl and metals.

The current study compared the ChE activity of green mussel, *P. viridis* in different organs such as the gill, the foot, the mantle and the PAM. The results demonstrated that the gill of *P. viridis* had significant higher of the ChE activity compared to the foot, the mantle and the PAM. Porte and Albaiges (2002) demonstrated that the ChE activity from the gill of blue mussels, *Mytilus galloprovincialis* was more sensitive enzyme activity than that of digestive gland and it revealed a certain correlation with the concentration of fenitrothion in whole mussels. It has been reported that the gill of *P. viridis* collected from Hong Kong waters had the higher ChE activity than that of the whole tissue and this ChE activity was not size-dependent (Lau and Wong, 2003). This is conceivable because mussels use their gills not only as a respiratory apparatus but also as filter feeder organ thereby ambient water filtered and managed for gaseous exchanges and sifting food (Bayne et al., 1976). Since the gill are the front line of contact with contaminants and the first line of defense (Lau and Wong, 2003), detoxification compounds such as ChEs are necessary to be produced to protect other organs. Consequently, the production of ChEs not only provides as the control of neurotransmission, but also serves as contaminants detoxification particularly for organophosphorus and carbamate pesticides (Soreq and Seidman, 2001). In addition, it has been reported that the protein level of *P. viridis* gill was not seasonal dependent which lead to reduce the intrinsic variability of the biochemical responses in different growth phase throughout the year (Lau et al., 2004). Those evidences set up the gill as a *par excellence* tissue for biomarkers application to minimize effects caused by the natural reproductive cycles and the dilution effect due to large variation in the total tissue protein (Lau et al., 2004). The selection of the gill as tissue target for conducting biomarkers were also shown by the nature of the gill, which comes into contacts with relatively large volumes of seawater compared to the rest of the animal so that conferring them with the potential for being a suitable target tissue for xenobiotic substance exposure.

The present study used the gill of *P. viridis* to investigate pollutants effect to ChE activity in some coastal areas of Indonesia. The results suggested that the ChE activity was a sensitive tool to detect neurotoxic effects of pollutants since it could discern different levels

of two heavily polluted areas. It was supported by the evident that the ChE activity of the gill of *P. viridis* from reference site was significantly higher than that from the gill of *P. viridis* which inhabit two polluted sites. The inhibition of the ChE activity from the gill of *P. viridis* collected from Kamal Muara was about 49.2%. Statistically, the greatest inhibition of the ChE activity was indicated in mussels from Cilincing, which was about 72.41%. By exposing brown mussels, (*Perna perna*) to furadan (carbamate pesticide), Alves et al. (2002) observed that the ChE activity of the gill was suppressed by 35%.

Ludke et al. (1975) classified the percentage of ChE activity inhibition based on comparison of the individual value with the activity of the normal population for providing the interpretation of the environmental risk. The following are the risk criteria of inhibition percentage of ChE activity that were proposed by Ludke et al. (1975):

0 to 20% = zone of normal variation
20 to 50% = presence of exposure or zone of reversible effects
50 to 100% = life-threatening situation or zone of irreversible effects

In respect to estuarine fishes, Coppage (1972) suggested that inhibition level of the ChE activity in the range of 20 to 70% could be classified as an indication of organophsophorous exposure. Subsequent studies observed that the inhibition of the ChE activity in the fish brain, which reached 70 to 90% indicated mortality (Coppage et al., 1975; Coppage and Matthews, 1975). Sandahl et al. (2005) observed that the inhibition of the ChE activity in brain and muscle from juvenile coho salmon (*Oncorhynchus kisutch)* was correlated well with the behavior disruption, that is, feeding and swimming ability when the fish were exposed by chlorphyrifos. At the lowest concentration (0.6 µg/l), chrlophyrifos caused 12% inhibition of the muscle ChE activity reducing 27 % of the swimming rate, while no mortality was observed when fish exposed by the high concentration (2.5 µg/l) inhibiting 67% of muscle's ChE activity. By conducting microcosm study using mixtures of selected organophosphorous pesticides, Sibley et al. (2000) observed that 10% mortality was correlated with approximately 50% inhibition of AChE activity, while 50% mortality was correlated with approximately 90% inhibition of AChE activity of fathead minnows. Fleming et al. (1995) found the die-off freshwater mussels (*Elliptio steinstansana*) from sites that were influenced by agricultural activities with the inhibition of ChE activity from 65 to 73% compared to the reference site. Based on the criteria and the results of those studies, it is suggested that discharged pollutants into coastal area of Jakarta Bay indicated neurotoxic compound causing from reversible to irreversible effects of the neurological activity of the green mussel population. By compiling the data of ChE activity from the research above which ranging from bird to freshwater mussel it is suggested that the green mussels which populated in Kamal Muara indicated reversible effects, while those from Cilincing showed irreversible conditions.

The link between the inhibition of ChE activity of sentinel organism and the discharged neurotoxic compounds from agricultural, urban and industrial activity to aquatic environment has been suggested by many studies (Fulton and Key 2001; Printes and Callaghan 2004; Galloway et al., 2002; Crane et al., 2002; Rickwood and Galloway, 2004; Canty et al., 2007; Warberg et al., 2007). However, the relationship between ChE activity and higher level biomarker such as feeding rate in green mussel has not been studied yet. Therefore, a chronic *in vivo* study on the response of ChE activity in green mussel and other behavioral biomarkers such as feeding rate to the serial concentrations of pollutants, which picturize suspected pollution area concentrations, is indispensable to translate the inhibition of ChE activity induced by pollutants into ecological perspective. The translatable of ecological consequence of the suppressed ChE activity is a vital consideration in ecological risk assessment in the coastal zone. It is because an appropriate ecological relevance of biomarkers can eliminate the primary source of uncertainty in application of ecological risk assessment (Sibley et al., 2000).

## Phagcytosis activity

Green mussel hemolymph contains both hemocyte and humoral defense factors which are responsible for the defense system. Hemocytes circulating in hemolymph are the principal cellular effectors of invertebrate immunity (Mitta et al., 1999) which have a capability to perform phagocytosis of foreign materials (Cheng, 1984; Carballal et al., 1997) and cytotoxicity via the production of radicals (Winston et al., 1996).

Phagocytosis of mussel hemocytes can be affected by various chemical stressors in the aquatic environment (Anderson and Mora, 1995). Biphasic patterns of mussel phagocytic responses induced by xenobiotic have been demonstrated in many laboratory studies (Cole et al., 1994; Pipe et al., 1999; Parry and Pipe, 2004). Theoretically, the phagocytic activity will be stimulated when mussels are exposed to low level of contaminants, while it will be suppressed when mussel are exposed to high level of contaminants. Consequently, measurement of the phagocytic activity, which is as part of immune system of mussel, has been used as a biomarker of xenobiotic substances effect (Anderson and Mora, 1995; Oliver and Fisher, 1999; Blaise et al., 2002; Gagné et al., 2002).

In spite of mussel hemocytes playing an important role in the phagocytic activity, it is difficult to depict the correlation pattern between circulated hemocytes number

and the phagocytic activity of mussel. The current study showed that there was no different numbers of circulating hemocytes of green mussel, which were collected from both polluted and clean sites. However, significant differences of the phagocytic activity between the collected green mussels from polluted sites and those from clean site were evident. The data showed that discharged pollutants in Jakarta Bay have stressed cultivated green mussels, which stimulated significantly their phagocytic activity compared to the phagocytic activity of the green mussels collected from the clean site. The modulation of mussel phagogytic activity was in accordance with Luengen et al. (2004) who observed the elevation of phagocytic activity of mussels that collected from polluted sites. The elevation of phagocytic activity induced by the pollutants may be a part of mussel's strategy to sequester the toxic materials from vulnerable organs (Oliver et al., 2001). Nevertheless, Dizer et al. (2001b) found that high number of circulating hemocytes of mussels collected from control site followed by relatively low phagocytic activity, while relatively low number of hemocytes from polluted sites had a high phagocytic activity. They could not depict clearly the relationship between hemocytes number and the phagocytic activity of mussels.

The complicated relationship between hemocytes number and the phagocytic activity of mussels may result from dynamic association/dissociation between hemocytes and bivalve tissues that enable to change the total size of the hemocytes population within bivalve body over short time (Ford et al., 1993). The population could not be simply depicted by circulating number of hemocytes, which were drained from the PAM sinus as the mussel has the open circulatory blood system, which circulate the blood to whole organs. In addition, commonly the mussel hemocytes are composed by phagocytotic and unphagocytotic hemocytes which can be altered by xenobiotic substances (Pipe et al., 1999). Unfortunately, most of the techniques to measure the phagocytic activity including the technique used in the present study were based on the mixture of hemocytes sub-population so that an estimation of capability level of each sub-population of hemocytes was not possible.

Although, the present study enabled to distinguish the phagocytic activity of green mussels dwelled in polluted and clean sites, the difference of the phagocytic activity within the polluted site could not be differentiated significantly. Having taking into account the data from the ChE activity, which enable to distinguish the magnitude effects of released pollutants within the polluted sites, it is tempting to suggest that released pollutants in Jakarta Bay seem to be ChEs inhibitors, which raised greater impact on the ChE activity rather than the phagocytosis activity. For that purpose, the chemical analysis of water/sediment samples and relevant pollutants within mussel's tissue should be taken into account. Regardless of the chemical analysis approach, the ChE activity indicated a more responsive tool compared to the phagocytic activity so that it could distinguish between two heavily polluted sites. However, it is hard to justify that the ChE activity is more sensitive compared to phagocytic activity as was observed by Perez et al. (2004) in ChE activity of invertebrates, *Scrobicularia plana* (clam) and *Nereis diversicolor* (marine worm). The authors delineated higher sensitivity of ChE activity compared to others biomarkers that were used in biomonitoring of Spain waters. Therefore, the useful results that recorded by the current study are the information on neurotoxicity and immunotoxicity compounds which were present in Jakarta Bay and the magnitude impact of neurotoxicity contaminants to induce an effect is greater than the immunotoxicity contaminants.

## Conclusion

Conclusively, the results suggested that the use of the selected biomarkers is a reliable and preferential strategy in the ecological risk assessment of released xenobiotic compounds in coastal waters due to their ability to elucidate bio-effects of neuro-immuno systems disruptors.

## ACKNOWLEDGEMENTS

The authors wish to thanks to DAAD (Deutscher Akademischer Austausch Dients/German Academic Exchange Service) for funding the research through the Special Programme for Young Indonesian Marine and Geoscience Researchers. The authors would like also to thank distinguish colleague Arifin from Marine Science and Fisheries Faculty, Hasanuddin University, Makassar for his invaluable assistence for collecting the green mussels in Pangkajene Kepulauan waters.

### REFERENCES

Aiello E (1990). Nervous control gill ciliary activity in *Mytilus edulis*. In: Stefano GB, editor. *Neurobiology of Mytilus edulis*. Manchester: Manchester University Press, pp. 189-208.

Akcha F, Izuel C, Venier P, Budzinski H, Burgeot T, Narbonne J-F (2000). Enzymatic biomarker measurement and study of DNA adduct formation in benzo[a]pyrene-contaminated mussels, *Mytilus galloprovincialis*. Aquat. Toxicol., 49: 269-287.

Alves SRC, Severino PC, Ibbotson DP, da Silva AZ, Lopes FRAS, Saenz LA, Bainy ACD (2002) Effects of furadan in the brown mussel *Perna perna* and in the mangrove oyster *Crassostrea rhizophorae*. Mar. Environ. Res., 54: 241-245.

Anderson RS, Mora LM (1995). Phagocytosis: A microtiter plate assay. *In* Stolen JS, Fletcher TC, Smith SA, Zelikoff JT, Kaattari SL, Anderson RS, Söderhäll K, Weeks-Perkins BA. (eds). Techniques in Fish immunology-4. Immunology and Pathology of Aquatic Invertebrates. Fair Haven, NJ, USA: SOS Publication, pp. 109-112.

Bayne BL, Thompson RJ, Widdows J (1976). Physiology: I. *In* B.L. Bayne (ed). Marine mussels: their ecology and physiology. London: Cambridge University Press, pp. 121-206.

Belden JB, Lydy MJ (2000). Impact of atrazine on organophosphate insecticide toxicity. Environ. Toxicol. Chem., 19: 2266-2274.

Blaise C, Trottier S, Gagné F, Lallement C, Hansen PD (2002). Immunocompetence of bivalve hemocytes as evaluated by a miniaturized phagocytosis assay. Environ. Toxicol., 17: 160-169.

Bocquené G, Chantereau S, Clérendeau C, Beausir, E, Ménard D, Raffin B, Minier C, Burgeot T, Leszkowicz AP, Narbonne JP (2004). Biological effects of the "Erika" oil spill on the common mussel (*Mytilus edulis*). Aquat. Living Resour., 17: 309–316.

Bocquené G, Gaglani F, Truquet P (1990). Characterization and assay condition for use of AChE activity from several marine species in pollution monitoring. Mar. Environ. Res., 30: 75-89.

Bonacci S, Corsil, Focardi S (2008). Cholinesterase activities in the scallop *Pecten jacobaeus*: Characterization and effects of exposure to aquatic contaminants. Sci. Total. Environ., 392: 99–109.

Brown M, Davies IM, Moffat CF, Redshaw J, Craft JA (2004). Characteristic of choline esterases and their tissue and subcellular distribution in mussel (*Mytilus edulis*). Mar. Environ. Res., 57: 155-169.

Butterworth FM (1995). Introduction to biomonitors and biomarkers as indicators of environmental change. *In* Butterworth FM, Corkum LD, Rincon JG (eds). Biomonitor and biomarkers as indicators of environmental change: A handbook. . New York: Plenum Press, pp. 1-8.

Cajaraville MP, Bebianno MJ, Blasco J, Porte C, Sarasquete C, Viarengo A (2000). The use of biomarkers to assess the impact of pollution in coastal environments of the Iberian Peninsula: A practical approach. Sci. Total. Environ., 247: 295-311.

Canty MN, JA Hagger, RTB More, L Cooper, TS Galloway (2007). Sublethal impact of short term exposure to the organophosphate pesticide azamethiphos in the marine Mollusc *Mytilus edulis*. Mar. Poll. Bull., 54: 396-402.

Carballal MJ, Lopez C, Azevedo C, Villalba A (1997). *In vitro* study of phagocytotic ability of *Mytilus galloprovincialis* Lmk. Haemocytes. Fish. Shellfish. Immunol., 7: 403–416.

Castro M, Santos MM, Monteiro NM, Vieira N (2004). Measuring lysosomal stability as an effective tool for marine coastal environmental monitoring. Mar. Environ. Res., 58: 741-745.

Cheng TC (1984). A classification of molluscan hemocytes based on functional envidence. *In* Cheng TC (ed). Comparative pathobiology vol. 6. *Inverterbrate blood cells and serum factors*. New York: Plenum Press, pp. 111-146.

Cole JA, Farley SR, Pipe RK (1994). Effects of fluranthene on the immunocompetence of the common marine mussel, *Mytilus edulis*. Aquat. Toxicol., 30: 367-379.

Coppage DL (1972). Organophospahte pesticides: Specific level of brain AChE inhibition related to death in sheeps head minnows. Transact. Am. Fish. Soc., 101: 534–536.

Coppage DL, Matthews E, Cook GH, Knight J (1975). Brain acetylcholinesterase inhibition in fish as a diagnosis of environmental poisoning by malathion, O,O-dimethyl S-(1,2 dicarbothoxyethyl) phosphorodiyhioate. Pest. Biochem. Physiol., 5: 536-542.

Coppage DL, Matthew E (1975). Brain acetylcholinesterase inhibition in marine teleost during lethal and sublethal exposures to 1,2-dibromo-2,2-dichloroethyl dimetyl phosphate (Naled) in sea water. Toxicol. Appl. Pharmacol., 31: 128 -133.

Corsil, Bonacci S, Santovito G, Chiantore M, Castagnolo L, Focardi S (2004). Cholinesterase activity in the Antarctic *scallop Adamussium colbecki*: Tissue expression and effect of $ZnCl_2$ exposure. Mar. Environ. Res., 58: 401-406.

Crane M, Sildanchandra W, Kheir R, Callaghan A (2002). Relationship between biomarker activity and developmental endpoints in *Chironomus riparius* Meigen exposed to an organophosphate insecticide. Ecotox. Environ. Safe, 53: 361-369.

Damiens G, Gnassia-Barelli EHM, Quiniou F, Roméo M (2004). Evaluation of biomarkers in oyster larvae in natural and polluted conditions. Comp Biochem. Physiol. C: Toxicol. Pharmacol., 138: 121-128.

Damiens G, Gnassia-Barelli M, Loques F, Romeo M, Salbert V (2007). Integrated biomarker response index as a useful tool for environmental assessment evaluated using transplanted mussels. Chemosphere, 66: 574–583.

Dauberschmidit C, Dietrich DR, Schlatter C (1997). Organophospates in the Zebra Mussel *Dreissena polymorpha*: subacute exposure, body burdens and organ concentrations. Arc. Environ. Contam. Toxicol., 33: 42-46.

De Luca-Abbott SB, Richardson BJ, McClellan KE, Zheng GJ, Martin M, Lam PKS (2005). Field validation of antioxidant enzyme biomarkers in mussels (*Perna viridis*) and clams (*Ruditapes philippinarum*) transplanted in Hong Kong coastal waters. Mar. Pol. Bul., 51: 694-707.

Depledge MH, Fossi MC (1994). The role of biomarkers in environmental assessment (2). Invertebrate. Ecotoxicol., 3: 161-172.

Devier MH, Augagneur S, Budzinski H, Le Menach K, Narbonne J-F, Garrigues P (2005). One-year monitoring survey of organic compounds (PAHs, PCBs, TBT), heavy metals and biomarkers in blue mussels from the Arcachon Bay, France. J. Environ. Monit., 7: 224-240.

Dizer H, de Assis HCS, Hansen PD (2001a). Cholinesterase activity as bioindicator for monitoring marine pollution in the Baltic Sea and the Mediterranean Sea. *In* Garrigues Ph, Barth H, Walker CH, Narbonne J-F (eds). Biomarkers in marine organisms a practical approach. Amsterdam: Elsevier Science BV, pp. 331-342.

Dizer H, Unruh E, Bissinger V, Hansen PD (2001b). Investigation of genotoxicity and immunotoxicity for monitoring marine pollution in the Baltic Sea and Mediterranian Sea. *In* Garrigues Ph, Barth H, Walker CH, Narbonne JF (eds). *Biomarkers in marine organisms a practical approach.* Amsterdam: Elsevier Science BV, pp. 237-257.

Donkin P, Widdows J, Evans SE, Staff FJ, Yan T (1997). Effect of neurotoxic pesticide on the feeding rate of marine mussels (*Mytilus edulis*). Pest. Sci., 49: 196-209.

Ellman GL, Courtney KD, Andres VJr, Featherstone RM (1961). A new and rapid colorimetric determination of acetylcholinesterase activity. Biochem. Pharmocol., 7: 88-95.

Escartin E, Porte C (1997). The use of cholinesterase and carboxylesterase activities from *Mytilus Galloprovincialis* in pollution monitoring. Environ. Toxicol. Chem., 16: 2090-2095.

Fleming WJ, Augspurger TP, Alderman JA (1995). Freshwater mussel die-off attributed to anticholinesterase poisoning. Environ. Toxicol. Chem., 14: 877–879.

Ford SE, Kanaley SS, Littlewood DTJ (1993). Celluar response of oyster infected with *Haplosporidium nelsoni*, change in circulating and tissue-infiltrating hemocytes. J. Invert. Phatol., 61: 49-57.

Fournier M, Cyr D, Blakley B, Boermans H, Brousseau P (2000). Phagocytosis as a biomarker of immunotocity in wildlife species exposed to environmental xenobiotics. Am. Zoo., 40: 412-420.

Fulton MH, Key PB (2001). Acetylcholinesterase inhibition in estuarine fish and invertebrate as an indicator of organophosphorus insecticide exposure and effects. Environ. Toxicol. Chem., 20: 37-45.

Gagné F, Blaise C, Aoyama I, Luo R, Gagnon C, Couillard Y, Campbell C, Salazar M (2002). Biomarker study of a Municipal effluent dispersion plume in two species of freshwater mussels. Environ. Toxicol., 17: 149-159.

Gagné F, Burgeot T, Hellou J, St-Jean S, Farcy E, Blaise C (2008). Spatial variations in biomarkers of *Mytilus edulis* mussels at four polluted regions spanning the Northern Hemisphere. Environ. Res., 107: 201-217.

Gainey LF, Walton JC, Greenberg MJ (2003). Branchial musculature of a venerid clam: pharmacology, distribution, and innervation. Biol. Bull., 204: 81-95.

Galloway TS (2006). Biomarkers in environmental and human health risk assessment. Mar. Pol. Bull., 53: 606–613.

Galloway TS, Millward N, Browne MA, Depledge MH (2002). Rapid assessment of organophosphorous/carbamate exposure in the bivalve mollusc *Mytilus edulis* using combined esterase activities as biomarkers. Aquat. Toxicol., 61: 169-180.

Goldberg E D, Bowen VT, Farrington JW, Harvey G, Martin JH, Parker PL, Risebrough RW, Robertson W, Schneider E, Gamble E (1978). The mussel watch. Environ. Conserv., 5: 101–125.

Guilhermino L, Barros P, Silva MC, Soares AMVM (1998). Should the use of inhibition of cholinesterases as a specific biomarker for organophosphate and carbamate pesticides be questioned? Biomarker, 3: 157–163.

Halldórsson HP, De Pirro M, Romano C, Svavarsson J, Sarà G (2007).

Immediate biomarker responses to benzo[a]pyrene in polluted and unpolluted populations of the blue mussel (*Mytilus edulis* L.) at high-latitudes. Environ. Inter., 34: 483-489.

Hansen PD (1992). Phagocytosis in *Mytilus edulis*, a system for understanding the sublethal effects of anthrophogenic pollutants (xenobiotic) and the use of AOX as an integrating parameter for the study of the equilibrium between chlorinated hydrocarbons in *Dreissena polymorpha* following long term exposures. Limnol. Aktuell, 4: 171-184.

Hansen PD (1995). The pontential and limitation of new technical approaches to ecotoxicology monitoring. *In* Richardson M (ed). Environmental Toxicology Assessment. Taylor & Francis Inc. London, pp. 13-28.

Kim Y, Powell EN, Wade TL, Presley BJ (2008). Relationship of Parasites and Pathologies to Contaminant Body Burden in Sentinel Bivalves: NOAA Status and Trends 'Mussel Watch' Program. Mar. Environ. Res., 65: 101-127.

Kopecka-Pilarczyk J (2010). The effect of pesticides and metals on acetylcholinesterase (AChE) in various tissues of blue mussel (*Mytilus trossulus* L.) in short-term *in vivo* exposures at different temperatures. J. Environ. Sci. Health Part B, 45: 336 – 346.

Lagadic L (2002). Biomarkers: Useful tools for the monitoring of aquatic environments. Revue Méd. Vét., 153: 581-588.

Lau PS, Wong HL (2003). Effect of size, tissue parts and location on six biochemical markers in the green-lipped mussel, *Perna viridis*. Mar. Poll. Bull., 46: 1563-1572.

Lau PS, Wong HL, Garrigues P (2004). Seasonal variation in antioxidative responses and acetylcholinesterase activity in *Perna viridis* in eastern oceanic and western estuarine of Hong Kong. Continent. Shelf. Res., 24: 1969-1987.

Lehtonen KK, Schiedek D, Köhler A, Lang T, Vuorinen PJ, Förlin L, Baršienė, J, Pempkowiak J, Gercken J (2006). The BEEP project in the Baltic Sea: Overview of results and outline for a regional biological effects monitoring strategy. Mar. Pollut. Bull., 53: 523-537.

Leinio S, Lehtonen K (2005). Seasonal variability in biomarkers in the bivalves *Mytilus edulis* and *Macoma balthica* from the northern Baltic Sea. Comp. Biochem. Physiol. C, 140: 408-421.

Livingstone DR, Chipman JK, Lowe DM, Minier C, Mitchelmore CL, More MN, Peters LD, Pipe RK (2000). Development of biomarkers to detect the effects of organic pollution on aquatic invertebrate: recent molecular, genotoxic, cellular and immunological studies on the common mussel (*Mytilus edulis* L.) and other mytilids. Inter. J. Environ. Poll., 13: 56-91.

Ludke JL, Hill EF, Dieter MP (1975). Cholinesterase (ChE) response and related mortality among birds fed ChE inhibitors. Arc. Environ. Contam. Toxicol., 3: 1-21.

Luengen AC, Friedman CS, Raimondi PT, Flegal AR (2004). Evaluation of mussel immune responses as indicators of conta-mination in San Francisco Bay. Mar. Environ. Res., 57: 197–212.

Martın-Dıaz ML, Blasco J, Sales D, DelValls TA (2004). Biomarkers as tools to assess sediment quality: Laboratory and field surveys. Trends. Analyt. Chem., 23: 807-818.

Minier C, Abarnou A, Madoulet AJ, Le Guellec, AM, Tutundjian R, Bocquené G, Leboulenger F (2006). A pollution-monitoring pilot study involving contaminant and biomarker measurements in the seine estuary, france, using zebra mussels (*Dreissena polymorpha*). Environ. Toxicol. Chem., 25: 112–119.

Mitta G, Vandenbulcke F, Hubert F, Roch P (1999). Mussel defensins are synthesised and processed in granulocytes then released into the plasma after bacterial challenge. J. Cell. Sci., 112: 4233-4242.

Mora P, Fournier D, Narbonne J-F (1999). Cholinesterases from the marine mussels *Mytilus galloprovincialis* Lmk. and *M. edulis* L. and from the freshwater bivalve *Corbicula fluminea* Müller. Comp. Biochem. Physiol. C., 122: 353–361.

Moreira SM, Santos MM, Ribeiro R, Guilhermino L (2004). The 'Coral Bulker' Oil Spil on the North Coast of Portucal: Spatial and temporal biomarker responses in *Mytilus galloprovincialis*. Ecotoxicol.,13: 619–630.

Munawir K (2005). Pemantauan Kadar Pestisida Organoklorin Di Beberapa Muara Sungai Di Perairan Teluk Jakarta. Oseanol. Limnol. Indonesia, 37: 13-23.

Narbonne J-F, Garrigues MP, Michel X, Lafaurie M, Salaun JP, Den

Blasten P, Pagano G, Porte C, Livingstone D, Hensen P-D, Herbert A (1999). Biological markers of environmental contamination in marine ecosystems: Biomar project. J. Toxicol. Toxin Rev., 18: 205-220.

Narbonne JF, Duabeze M, Baumard P, Budzinski H, Clerandeau C, Akca F, Mora P, Garrigues P (2001). Biochemical markers in mussel, *Mytilus* sp., and pollution monitoring in European Coasts: Data analysis. *In*: Garrigues Ph., Barth H, Walker CH, Narbonne J-F (eds). Biomarker in Marine Organisms: A Practical Approach. Amsterdams: Elsevier Sci., pp. 215-236.

Nesto N, Bertoldo M, Nasci C, Da Ros L (2004). Spatial and temporal variation of biomarkers in mussels (Mytilus galloprovincialis) from the Lagoon of Venice, Italy. Mar. Environ. Res., 58: 287-291.

Nicholson S, Lam PKS (2005). Pollution monitoring in Southeast Asia using biomarkers in the mytilid mussel *Perna viridis* (Mytilidae: Bivalvia). Environ. Inter., 31: 121-132.

Oliver LM, Fisher W (1999). Appraisal of prospective bivalve immunomarkers. Biomarkers, 4: 510-530.

Oliver LM, Fisher WS, Winstead JT, Hemmer BL, Long ER (2001). Relationships between tissue contaminants and defense-related characteristics of oysters (*Crassostrea virginica*) from five Florida bays. Aquat. Toxicol., 55: 203–222.

Orbea A, Garmendia L, Marigómez I, Cajaraville MP (2006). Effects of the 'Prestige' oil spill on cellular biomarkers in intertidal mussels: results of the first year of studies. Mar Ecol Prog Ser., 306: 177-189.

Parry HE, Pipe RK (2004). Interactive effects of temperature and copper on immunocompetence and disease susceptibility in mussels (*Mytilus edulis*). Aquat. Toxicol., 69: 311-325.

Perez E, Blasco J, Sole M (2004). Biomarker responses to pollution in two invertebrate species: *Scorbicularia plana* and *Nereis diversicolor* from the Cadiz Bay. Mar. Environ. Res., 58: 275-279.

Picado A, Bebianno MJ, Costa MH, Ferreira A, Vale C (2007). Biomarkers: a strategic tool in the assessment of environmental quality of coastal waters. Hydrobiologia, 597: 79-87.

Pipe RK, Coles JA, Carrisan FMM, Ramanathan K (1999). Copper induced immunomodulation in the marine mussel, *Mytilus edulis*. Aquat. Toxicol., 46: 43-54.

Porte C, Albaiges J (2002). Residues of pesticides in aquatic organisms. Revue. Méd. Vét., 153: 345-350.

Porte C, Escartin E, Borghi V (2001). Biochemical tools for the assessment of pesticide exposure in a deltaic environment: The use of Cholinesterase and Carbaoxylesterases. *In* Ph. Garrigues, H Barth, CH Walker, JF Narbonne [Editors]. Biomarkers in marine organisms a practical approach. Elsevier Science B.V. Amsterdam, pp. 259-278.

Printes LB, Callaghan A (2004). A comparative study on the relationship between acetylcholinesterase activity and acute toxicity in *Daphnia magna* exposed to anticholinesterase insecticides, 23: 1241–1247.

Rickwood CJ, Galloway TS (2004). Acetylcholinesterase inhibition as a biomarker of adverse effect: A study of *Mytilus edulis* exposed to the priority pollutant chlorfenvinphos. Aquat. Toxicol., 67: 45-56.

Romani R, Isani G, De Santis L, Giovannini E, Rosi G (2005). Effects of chlorpyrifos on the catalytic efficiency and expression level of acetylcholinesterases in the bivalve mollusk *Scapharca inaequivalvis*. Environ. Toxicol. Chem., 24: 2879-2886.

Roméo M, Mourgaud Y, Geffard Y, Gnassia-Barelli M, Amiard JC, Budzinski H (2003). Multimarker approach in transplanted mussels for evaluating water quality in Charentes, France, coast areas exposed to different anthropogenic conditions. Environ. Toxicol., 18: 295-305.

Sandahl JF, Baldwin DH, Jenkins JJ, Scholz NL (2005). Comparative thresholds for acetylcholinesterase and behaviour impairment in coho salmon exposed to chlorphyrifos. Environ. Toxicol. Chem., 24: 136 – 145.

Sibley PK, Chappel MJ, George TK, Solomon KR, Liber K (2000). Integration effects of stressors across levels of biological organization:examples using organophosphorus insecticides mixtures in field-level exposure. J. Aquat. Ecosyst. Stress. Recov., 7: 117-130.

Soreq H, Seidman S (2001). Acetylcholinesterase – new roles for an old actor. Nature Rev. Neurosci., 2: 8 – 16.

Sturm A, de Assis HCS, Hansen P-D (1999). Cholinesterase of marine teleost fish: Enzymological characterization and potential use in the monitoring of neurotoxic contamination. Mar. Environ. Res., 47: 389 – 398.

Sudaryanto A, Takahashi S, Monirith I, Ismail A, Muchtar M, Zheng J, Richardson BJ, Subramanian A, Prudente M, Hue ND, Tanabe S (2002). Asia-Pacific Mussel Watch: Monitoring of Butyltin Contamination in Coastal Waters of Asian Developing Countries. Exp. Toxicol. Chem., 21: 2119-2130.

Tabche LM, Mora BR, Faz CG, Castelan IG, Ortiz MM, Gonzalez VU, Flores MO (1997). Toxic efect of sodium dodecylbenzenesulfonate, lead, petroleum, and their mixtures on the activity of acetylcholinesterase of *Moina macropa in vitro*. Environ. Toxicol. Water Quality, 12: 21-215.

Taleb ZM, Benali I, Ykhlef-Allal A, Bachir-Bouiadjra B, Amiard J-D, Boutiba Z (2009). Biomonitoring of environmental pollutionon the Algerian west coast using caged mussels *Mytilus galloprovincialis*. Oceanologia, 51: 63–84.

Tedesco S, Doyle H, Redmond G, Sheehan D (2008). Gold nanoparticles and oxidative stress in *Mytilus edulis*. Mar. Environ. Res., 66: 3-131.

Valbonesi P, Sartor G, Fabbri E (2003). Characterization of cholinesterase activity in three bivalves inhabitating the North Ardriatic sea and their possible use as sentinel organisms for biosurveillance programmes. Sci. Total. Environ., 312: 79– 88.

Van der Oost R, Goksøyr A, Celander M, Heida H, Vermeulen NPE (1996). Biomonitoring of aquatic pollution with feral eel (*Anguilla anguilla*): II. Biomarkers: Pollution-induced biochemical responses. Aquat. Toxicol., 36: 189-222.

Verlecar XN, Jena KB, Chainy GBN (2008). Seasonal variation of oxidative biomarkers in gills and digestive gland of green-lipped mussel *Perna viridis* from Arabian Sea. Estuar. Coast. Shelf Sci., 76: 745-752.

Wachtendonk Von D, Neef J (1979). Isolation, perufication and molecular properties of an acetylcholinesterase (E.C. 3.1.1.7) from the haemolymph of the sea mussel *Mytilus edulis*. Comp. Biochem. Physiol., 63C: 279-286.

Walker CH (1998). Biomarker strategies to evaluate the environmental effects of chemicals. Environ. Health. Perspect Suppl., 106: 613-620.

Warberg MB, Coen LD, John E, Weinstein JE (2007). Acute Toxicity and Acetylcholinesterase Inhibition in Grass Shrimp (Palaemonetes pugio) and Oysters (*Crassostrea virginica*) Exposed to the Organophosphate Dichlorvos: Laboratory and Field Studies. Arch. Environ. Contam. Toxicol., 52: 207–216.

Widdows J, Donkin P (1992). Mussels and environmental contaminants: Bioaccumulation and physiological aspects. *In*: Gosling E (ed). The mussel Mytilus: Ecology, physiology, genetics and culture. Amsterdam: Elsevier Science Publishers BV, pp. 383-424.

Williams TM, Rees JG, Setiapermana D (2000). Metals and trace organic compounds in sediments and waters of Jakarta bay and the pulau seribu complex, Indonesia. Mar. Poll. Bull., 40: 277-285.

Winston GW, More MN, Kirchin MA, Soverchia C (1996). Production of reactive oxygen species by hemocytes from the marine mussels, *Mytilus edulis*: Lysosomal localization and effect of xenobiotics. Comp. Biochem. Physiol., 113C: 221-229.

Yaqin K (2010). Potential use of cholinesterase activity from tropical green mussel, *Perna viridis* as a biomarker in effect-based marine monitoring in Indonesia. Coast. Mar. Sci., 34: 156–164.

Yaqin K, Hansen PD (2010). The use of cholinergic biomarker, cholinesterase activity of blue mussel *Mytilus edulis* to detect the effects of organophosphorous pesticides. Afri. J. Biochem. Res., 4: 265-272.

# 9

# Acute haematological response of a cichlid fish *Sarotherodon melanotheron* exposed to crude oil

Oriakpono Obemeata[1*], Hart Aduabobo[1] and Ekanem Wokoma[2]

[1]Department of Animal and Environmental Biology, Faculty of Science, University of Port Harcourt, P. M. B. 5323 Rivers State, Nigeria.
[2]Department of Crop and Soil Science, Faculty of Agriculture, University of Port Harcourt, P. M. B. 5323 Rivers State, Nigeria.

The acute haematological response of a brackish water cichlid fish *Sarotherodon melanotheron* exposed to crude oil was evaluated. They were exposed for 96 h to crude oil concentrations of 0, 50, 125, 250, 375, and 500 mg/L of water obtained from the fish source. Haematological analyses were carried out at 12, 24, 48, 72, and 96 h, respectively. Haematological analysis revealed that the red blood cells (RBC), haemoglobin (Hb), packed cell volume (PCV), thrombocytes, and lymphocytes of the control group were significantly higher ($P \leq 0.05$) than the crude oil treated groups while the white blood cells (WBC), neutrophils, leucocrit (Lct) and monocytes of the crude oil treated groups were significantly higher than the control group, indicating an immune response to the toxicant. These parameters can be standardised and used as biomarkers in biomonitoring programs.

**Key words:** Haematological response, biomarkers, biomonitoring, *Sarotherodon melanotheron*, crude oil.

## INTRODUCTION

Haematological indices have been employed in effectively monitoring the responses of organisms to stressors and thus its health status under such adverse conditions. Generally, haematological tests are used to establish normal health status and to diagnose diseases caused by various factors namely heavy metals, environmental stress, parasitic infections, genotoxic effect of pollutants, nutrition, and pollution in human and veterinary science (Fedato et al., 2010). Haematological parameters act as physiological indicators to changing external environments (Caruso et al., 2005) as a result of their relationship with energetic (metabolic levels), respiration (haemoglobin) and defence mechanisms (leukocyte levels). Haematological parameters also provide an integrated measure of the health status of an organism, which over time manifests in changes in weight (Yaji and Auta, 2007).

The assessment of haematological values of fishes are carried out to ascertain the effect of certain chemical pollutants such as insecticides or heavy metals and the variation with age, sex and season (Van Vuren and Hattingh, 1978; Clarke et al., 1979), to determine the effect of disease condition or parasite on the blood values (Barham et al., 1980), and to establish a normal range of blood parameters (Siddiqui and Naseen, 1979). Haematological parameters have been recognised as valuable tools for the monitoring of fish health (Bhaskar and Rao, 1984; Schuett et al., 1997). However, the standardization of haematological parameters is difficult in fish because these parameters can be influenced by deficient diets, diseases and environmental stress situations (Silveira and Rigores, 1989). Nevertheless, the analysis of these parameters may improve the diagnosis of fish health (Blaxhall and Daisley, 1973; Anderson, 1974; Aldrin et al., 1982). This study provides standard haematological values for *Sarotherodon melanotheron*, a brackish water fish, as a way of establishing fish in healthy, disease and various stress conditions.

*Corresponding author. E-mail: obytrees@yahoo.com.

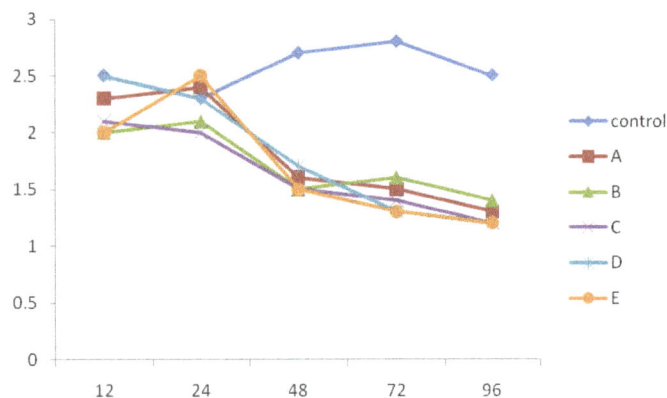

**Figure 1.** Red Blood Cells – RBC (cells × 106/L) of *Sarotherodon melanotheron* during the 96-h period.

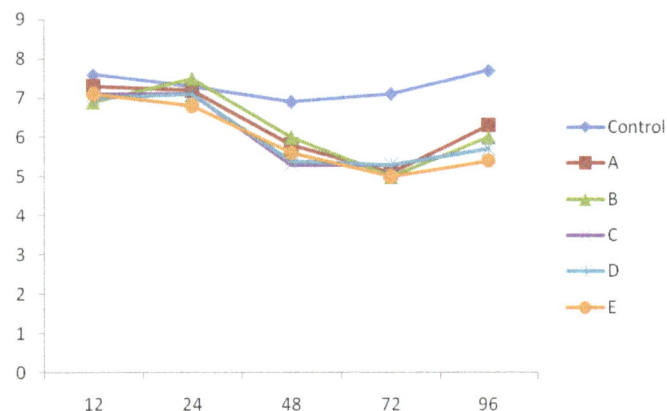

**Figure 2.** Haemoglobin -Hb (g/L) of *Sarotherodon melanotheron* during the 96-h period.

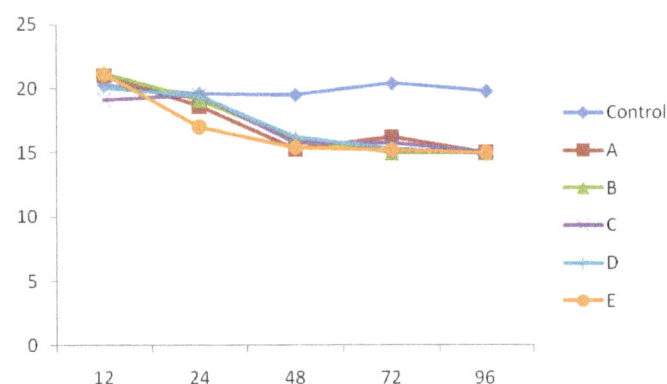

**Figure 3.** Packed cell volume -PCV (%) of *Sarotherodon melanotheron* during the 96-h period.

## MATERIALS AND METHODS

Three hundred and eighty adult male *S. melanotheron* (mean weight 372.56 ± 9.27 g; mean length 19.32 ± 4.48 cm) were purchased from Brackish Water fish farms, in Buguma and sexed.

They were allowed to acclimatize for seven days, after which they were divided into six vessels in triplicates based on body weight and labelled Control, A, B, C, D, and E, representing concentrations of 0, 50l, 125, 250, 375, and 500 mg/L crude oil exposure. The concentrations were chosen after preliminary studies were conducted with varying concentrations of the test solution. Blood sampling was conducted at the expiration of 12, 24, 48, 72 and 96 h. Blood samples were collected from 90 male *S. melanotheron* with heparinized plastic syringe, fitted with 21 gauge hypodermic needle and preserved in disodium salt of ethylene-diaminetetraacetic acid (EDTA) bottles for analysis. The Blaxhall and Daisley (1973), Brown (1980) and Wedemeyer et al. (1983) haematological methods were adopted for this study. The cyano-haemoglobin method was used to determine haemoglobin (Hb) using diagnostic kits from Sigma diagnostics USA, and packed cell volume (PCV) was determined by the microhaematocrit method. Red blood cell (RBC), leucocrit (LCT) and thrombocyte count were determined with the improved Neubauer haemocytometer according to (Dacie and Lewis, 1991). White blood cells (WBC) was determined with the improved Neubauer counter, while differential counts such as neutrophils, lymphocytes and monocytes were determined on blood film stained with May-Grunwald-Giemsa stain (Mirale, 1982). The completely randomised design was used and analysis of variance was conducted using the SAS software and differences among means were separated.

## RESULTS

Figures 1 to 9 show the results obtained for haematological indices of *S. melanotheron* during the 96-h assay. Statistical analysis conducted with the recorded values of RBC, lymphocytes, thrombocytes, haemoglobin and packed cell volume (PCV) of *S. melanotheron* indicate that the control groups were significantly higher $P < 0.05$ than the crude oil treated groups A, B, C, D, and E. There were no significant differences between the values of the crude oil treated groups, except for haemoglobin where treatments A and B were significantly higher than treatments C, D, and E, respectively.

Furthermore, the white blood cells (WBC), leucocrit, neutrophils and monocytes of *S. melanotheron* revealed that the crude oil treated groups were significantly higher $P < 0.05$ than the control. The crude oil treated groups showed an increase in WBC, leucocrit, neutrophils and monocytes. However, there were no significant differences between the crude oil treated groups except for the monocytes that had significant differences as treatments A, C, D and E were significantly higher than treatment B.

## DISCUSSION

During this study, water quality parameters were maintained within recommended limits. The haematological response of *S. melanotheron* following exposure to different crude oil concentrations revealed a crisis situation indicating they can serve as biomarkers in fish under stress, or when faced with the challenge of a pollutant.

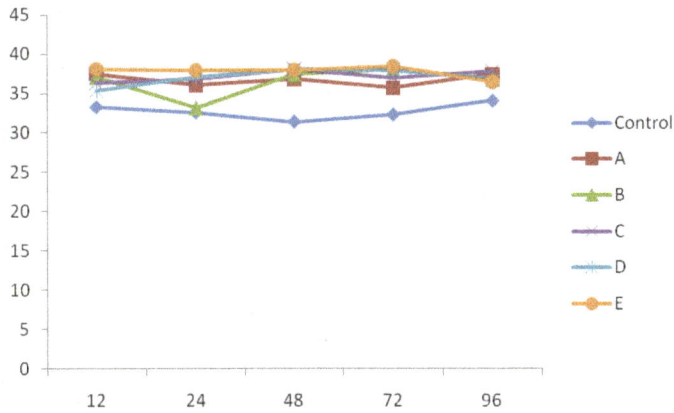

**Figure 4.** White blood cells -WBC (cells x 109/L) of *Sarotherodon melanotheron* during the 96-h period.

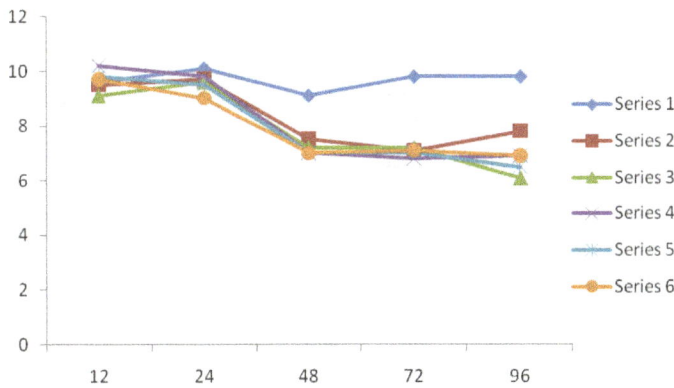

**Figure 5.** Leucocrit (cells × 1012/L) of *Sarotherodon melanotheron* during the 96-h period.

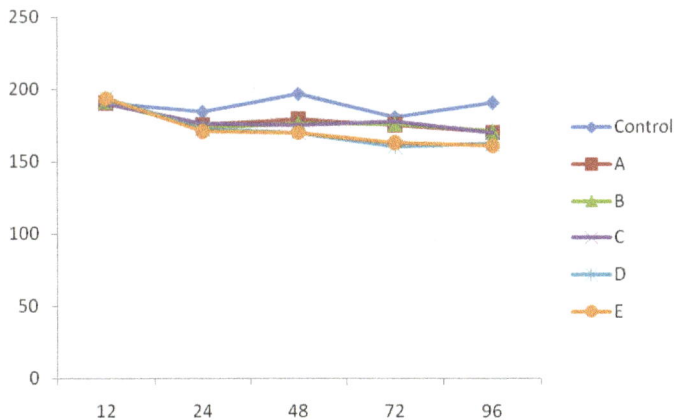

**Figure 6.** Thrombocytes (%) of *Sarotherodon melanotheron* during the 96-h period.

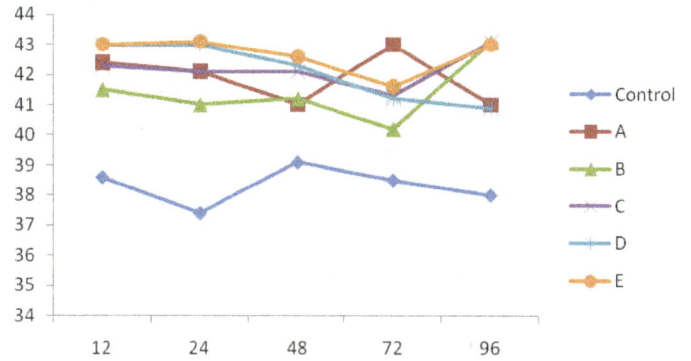

**Figure 7.** Neutrophils (%) of *Sarotherodon melanotheron* during the 96-h period.

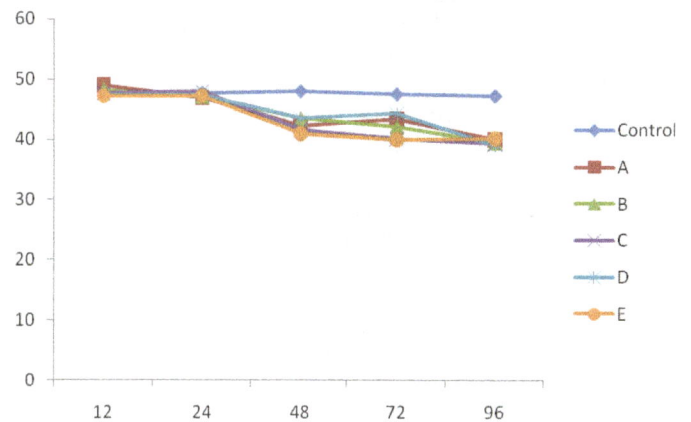

**Figure 8.** Lymphocytes (%) of *Sarotherodon melanotheron* during the 96-h period.

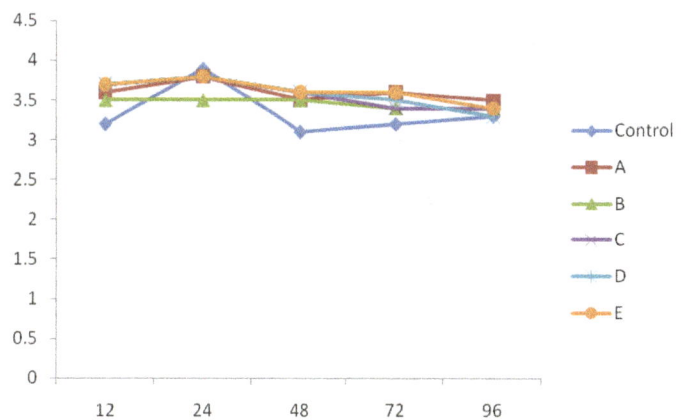

**Figure 9.** Monocytes (%) of *Sarotherodon melanotheron* during the 96-h period.

The variations found in haematological indices when exposed to crude oil are a defensive mechanism against crude oil toxicity through stimulation of erythropoiesis, which corresponds with studies on *Tilapia guineensis* and eels (Kirsch and Mayer, 1973; Hwang et al., 1989). Moreover, there was a significant reduction in the red blood cell and haemoglobin of the fish species. This reduction similar to that recorded on juvenile cobia (*Rachycentron canadium*) exposed to various degrees of salinity (Denson et al., 2003). The result of this study also corroborates the report of Munkittrick and Leatherland (1983) who stated that a change in water quality characteristic specific to an area inhabited by a fish population could affect their haematological indices. The changes in these parameters may be attributed to osmoregulatory dysfunction induced by changes in the total hardness, total alkalinity and salinity of the water body (Weirich and Tomasso, 1991).

Putman and Freel (1978) stated that different rates of fish activity demand different levels of metabolic activity; such activity requires several physiological adjustments including adjustments in the haematological parameters. The reduction in the RBC and Hb may be due to the presence of stressors which manifest in form of a change in the environment resulting to haemagglutination due to impaired osmoregulation (Rottman et al., 1992) or erythropoiesis in the organs responsible for the production of RBC. Packed cell volume is a major haematological parameter that changes with fish activity and environ-mental stress. In the course of this study, the PCV value of the fish species was observed to reduce with increasing concentration of crude oil and exposure time of the fish species. This may be attributed to the changes in water balance, which could cause a decrease in blood volume and an increase in the white blood cells resulting in reduced PCV (Cameron, 1970). On the other hand, the white blood cells of the fish species were observed to increase considerably with increase in experimental time. This increase was also observed in the highest crude oil concentration. The result agrees with the finding of Davids et al. (2002) who reported increase in size and monocytes of *Tilapia guineensis* and *S. melanotheron* after exposure to industrial effluents. The increase in WBC may be due to recruitment of more cells to combat the stressor (Ajani et al., 2007). This increase may also be attributed to non specific immune response to stress as a result of interaction of prolactin and cortisol hormones to restore ion balance in isosmotic salinity (Anyanwu et al., 2007), and a stimulation of the immune system in response to toxicity of crude oil.

The reduction observed in the leucocrit value may be due to the reaction of fish to the effect of the stress induced by the new environment. Dick and Dixon (1985) reported a significant reduction in leukocyte and lymphocyte of rainbow trout (*Salmo gaidneri*) after acute exposure to copper for 24 h. This was attributed to a generalized stress response resulting from increased pituitary-interrenal activity. Alkahem (1994) also observed a decrease in total leucocrit of *O. niloticus* exposed to sub-lethal levels of nickel. This was attributed to a reduction in the number of circulating thrombocytes and lymphocytes due to a reduction in lymphocytes delivery to the circulatory system and a rapid destruction of cells

which leads to an increased rate of peripheral removal of lymphocytes. Moroad and Houston (1988) attributed such lymphopenia to the lysis of lymphocytes after exposure to stressors in the environment. In our study, thrombocytes and lymphocytes of the fish species reduced in values during the experimental period. Thrombocytes were observed to drop sharply at 24 h, while lymphocytes dropped gradually then sharply at 48 h. This decrease may be attributed to lysis of the lymphocytes (lymphopenia) after exposure of the fish species to crude oil which altered the physicochemical characteristics of the water body. The decrease in thrombocytes and lymphocytes in crude oil exposed *S. melanotheron* is similar to that recorded in Atlantic *S. gairdneri* and *O. niloticus* by Matushima and Mariano (1996) who suggested a suppression of production from haematopoietic organs. The reduced lymphocytes and thrombocytes indicate a weakened defence and delay clotting in the event of an injury to the fish in the new environment.

Neutrophils and monocytes were observed to increase steadily during the experimental period and with increasing concentration, indicating a response of the two fish species to the crude oil concentrations particularly the two highest concentrations used. This increase is due to a non-specific immune response to stress, and the recruitment of more cells to combat the stressor (Ajani, et al., 2007). This study recommends further studies on haematological indices to ensure their appropriate use as index (biomarkers) in fish to monitor changes in environmental conditions, and organisms in healthy, disease and those undergoing stress conditions. This study recommends the integration of genetic toxicology and genetic ecotoxicology studies in Nigeria as being critical to the development of standardised biomonitoring programs.

## REFERENCES

Ajani F, Olukunle OA, Agbede SA (2007). Hormonal and Hematological Responses of *Clarias gariepinus* (Burchell 1822) to Nitrite Toxicity. J. Fish. Int. 2:48-53.

Aldrin JF, Messager JL, Laurencin FB (1982). La Biochemie Clinique en Aquaculture. Interet et perspective cnexo. Actes Colloq. 14:291-326.

Alkahem HF, Ahmed Z, Al-Akel AS, Shamusi MJK (1994). Toxicity bioassay and changes in haematological parameters of *oreochromis niloticus* induced by trichloroform. Arab Gulf J. Sci. Res., 16: 581-593.

Anderson DP (1974). Fish Immunology. Neptune TFH Publications, New Jersey. p 239.

Anyanwu PE, Gabriel UU, Anyanwu AO, Akinrotimi AO (2007). Effect of salinity changes on Haematological parameters of *Sarotherodon melanotheron* from Buguma Creek, Niger Delta. J. Anim. Veter. Adv. 6(5): 658-662.

Barham WT, Smit GL, Schonbec HJ (1980). The haematological assessment of bacteria infection in rainbow trout *Salmon gairdneri* (Richardson). J. Fish Biol. 17:275-281.

Bhaskar BR, Rao KS (1984). Influence of environmental variables on haematological ranges of milk fish, *Chanos chanos* (forskal), in brackish-water culture. Aquaculture 83:123-136.

Blaxhall PC, Daisley KW (1973). Routine haematological methods for use with fish blood. J. Fish Biol. 5:771-781.

Brown BA (1980). Haematology, Principles and Procedure, 3rd Edition. Lea and Fabiger, Philadelphia. P 356.

Cameron JN (1970). The influence of environment variables on the hematology of pinfish, *Lagodon Rhomboids* and Stryred mullet *Musil Cephalus* Comp. Biochem. Physiol. 32:175-192.

Caruso G, Genovese L, Maricchiolo G, Modica A (2005). Haematological, biochemical and immunological parameters as stress indicators in *Dicentrarchus labrax* and *Sparus aurata* farmed in off-shore cages. Aquacult. Int. 13:67-73

Clarke S, White more DH, McMahon RF (1979). Consideration of Blood Parameters of Largemouth Bass, (*Micropterus salmonides*). J. Fish Biol. 14:147-154.

Dacie JV, Lewis SN (1991). Practical Haematology, Fifth edition. Churchhill Livingstone, Edinburgh. p.390.

Davids CBB, Ekweozor EAS, Daka ER, Dambo WB, Bartimacus EAS (2002). Effects of Industrial Effluents on some Hematological Parameters of *Sarotherodon melenothero* and *Tilapia guineensis*. Global J. Pure Appl. Sci. 8: 305-310.

Denson MR, Stuart KR, Smith TIJ (2003). Effects of salinity on growth, survival and selected Haematological parameters of Juvenile cobi-*Rachycentron canadum*. J. World. Aqucult. Soc. 34: 496-503.

Dick PT, Dixon DG (1985). Changes in Circulating Blood Levels of Rainbow Trout, *Salmo gaindineri* following acute and Chronic Exposure to Copper. J. Fish. Biol. 26:475-481.

Fedato RP, Simonato JD, Martinez CBR, Sofiaa SH (2010). Genetic damage in the bivalve mollusk *Corbicula fluminea* induced by the water-soluble fraction of gasoline. Mutat. Res. 700:80-85.

Hwang PP, Sun CM, Wus M (1989). Changes of plasma osmolality, Chloride concentration and gill Na-K-ATPase activity in tilapia *Oreochromis mossambicus* during sea-water acclimation. Mar. Biol. 100:295-299.

Kirsch R, Mayer GN (1973). Kinetics of Water and Chloride exchanges during adaptation of the European eel to sea water. J. Fish Biol. 27:259-263.

Matushima ER, Mariano M (1996). Kinetics of the inflammatory reaction induced by genin in the Swim Bladder of *Oreochromis niloticus* (Ivile Tilapia) Braz. J. Vet. Res. Anim. Sci. 33:5-10.

Mirale JB (1982). Laboratory Medicine haematology, (6th Edn.). CV Mosby Co., London. p 883.

Moroad A, Houston AH (1988). Leucocytes and Leucopoiesis Capacity in Gold Fish, *Cerassius averatus* exposed to sub-lather levels of Cadmium. Aquacult. Toxicol. 13:141-154.

Munkittrick KR, Leatherland JF (1983). Haematocrit values in the feral goldfish, *Crassus curatus* as indicators of the health of the population. J. Fish Biol. 23:153-161.

Putman W, Freel RW (1978). Hematological Parameters of five Species of Marine Fishes. Comp. Biochem. Physiol. 61:585-588.

Rottman RW Francis-Floyd R, Durborow H (1992). The role of stress in fish disease. Southern Regional Aquaculture Centre (SRAC) Publication. p 474.

Schuett DA, Lehmann J, Guerlich R, Hamers R (1997). Haematology of *Swordtail xiphiphorus helleri*. I: Blood parameters and light microscopy of blood cells. J. Appl. Icthiyol. 13:83-89.

Siddiqui AK, Naseen S (1979). Blood Dyscrasia in a teloest *Colisa fasciatus* after acute exposure to sublethal concentrations of head. J. Fish Biol. 14:199-204.

Silveira R, Rigores C (1989). Caracteristics hemalogicas normales de *Oreochronis aureus* em cultivo. Rev. Latinoam Acuic. 39: 54-56.

Van V, Hattingh J (1978). The Effects of Toxicants on the Hematology of *Labeo umbratus* (Teleostei: Cyprindae). Com. Biochem. Physiol. C 83:155-159.

Wedemeyer GA, Gould RW, Yasutake WT (1983). Some haematological potentials and assessment methods. J. Fish Biol. 23:711-716.

Weirich CR, Tomasso JR (1991) Confinement- and transport-induced stress on red drum juveniles: effects of salinity. Progres. Fish Culturist 53:146–149.

Yaji AJ, Auta J (2007). Sub-lethal effects of monocrotophos on some haematological indices of African catfish *Clarias gariepinus* (Teugels). J. Fish. Int. 2(1):115-117.

# 10

# Public health risks associated with apples and carrots sold in major markets in Osogbo, Southwest Nigeria

M. A. Adeleke[1]*, A. O. Hassan[2], T. T. Ayepola[1], T. M. Famodimu[1], W. O. Adebimpe[3] and G. O. Olatunde[1]

[1]Public Health Entomology and Parasitology Unit, Department of Biological Sciences, Osun State University, P. M. B 4429, Osogbo, Nigeria.
[2]Microbiology Unit, Ladoke Akintola University Teaching Hospital, Osogbo, Nigeria.
[3]Department of Community Medicine, College of Health Sciences, Osun State University, Osogbo, Nigeria.

This study investigated the public health risks associated with the consumption of carrots and apples sold in major markets in Osogbo metropolis, Osun State Nigeria. Hundred samples of the fruits (49 apples and 51 carrots) were obtained from five randomly selected spots in the four major markets, namely, Igbonna, Oke-fia, Alekuwodo, and Orisumbare in Osogbo metropolis. The samples were screened for microbial and parasitic contaminants using standard procedures. Seven microbial isolates, *Pseudomonas aeruginosa*, *Staphylococcus aeureus*, *Enterococcus faecalis*, *Bacillius cereus*, *Listeri monocytogenes*, *Citrobacter species*, and *Candida* species; and two parasitic organisms, cysts of *Entamoeba coli* and ova of *Ascaris lumbricoides* were isolated from the fruits. The frequency of contaminants and the microbial load were higher in carrots than apples, though the variations were not statistically significant (P>0.05). There were significant variations in the level of parasitic contaminants of the fruits between the markets (Apple, P=0.035; Carrot, P=0.007). The results therefore demonstrated that carrots and apples sold in the major markets in Osogbo metropolis are contaminated with microbial pathogens and parasites that are capable of causing food-borne disorders to consumers. The vendors and the residents need to be educated on the public health risks inherent in unwholesome hygienic practices and its attendants effects in causing food-borne illnesses in the study area.

Key words: Fruits, carrots, apple, contamination, parasites, microorganisms, Nigeria.

## INTRODUCTION

Fruits constitute the natural sources of vitamins and mineral nutrients to the body. They are used as nutritional remedies for many patients suffering from different ailments such as diabetes, constipations, and stroke (Obeta et al., 2011). Carrot and apple to be specific, have been known to contain antioxidants, vitamin A, vitamin C, fiber, and carotene which help in boosting immunity of the system and protect the cellular anatomical composition of the body (Eni et al., 2010; Whitney-Chavex, 2011). The two fruits also help in boosting insulin in the body, therefore targeted as part of nutritional composition of diabetic patients (Whitney-Chavex, 2011). However, despite the health benefits associated with these two fruits, the risks of contamination with parasitic and pathogenic microorganisms cannot be under evaluated. These fruits are irrigated with water that may probably be contaminated with pathogenic organisms. Parasitological and microbial studies on the soil and rivers in Nigeria have shown heavy contamination with many pathogenic organisms (Sam-Wobo and Mafiana, 2005; Olayemi, 1994; Kanu and Achi, 2011; Agbabiaka and Oyeyiola, 2012). The pre-harvesting and harvesting processes of the two fruits make them to be prone to parasitic and microbial contaminants in the soil and water (Eni et al., 2010).

While it is acknowledged that some authors have made attempts to document the microbial and parasitic qualities of fruits and vegetables sold in different parts of Nigeria (Chukwu et al., 2010; Eni et al., 2010; Alli et al., 2011; Obeta et al., 2011; Oranusi and Braide, 2012), a review

---

*Corresponding author. E-mail: healthbayom@yahoo.com.

of their results showed significant variations in the prevalence and the species composition of the pathogenic organisms isolated from these food items. These variations, which could be traced to differences in environmental pollution, possibly point to the fact that non-uniform epidemiological patterns of the pathogenic organisms associated with the fruits and vegetables should be expected in different parts of Nigeria. This observation, therefore calls for nationwide surveillance to better our understanding on the epidemiological pattern of extrinsic food borne diseases and recommend the appropriate measures to circumvent the risks. Carrot and apples constitute the predominant fruits in Osogbo metropolis and they are available all year round. These carrots are sold by various vendors, and in most cases, are eaten either without being washed or washed with water whose quality is unknown. Only few people take proper care in washing them thoroughly before eating. To the best of our knowledge and available literature search, no previous record exist on the parasitic and microbial contaminants of the fruits sold in Osogbo metropolis, an urban area in Southwestern Nigeria.

The objective of this study was to investigate the parasitic and microbial contaminants associated with apples and carrots sold in major markets in Osogbo metropolis, Southwestern Nigeria with the underline aims of understanding the epidemiology of food-borne diseases and recommend appropriate measures in eliminating the risks towards attaining healthy living of the people in the study area.

## MATERİALS AND METHODS

### Study area

This study was conducted in Osogbo metropolis which lies on latitude 7°49' N and a longitude 4°37' E in Southwestern Nigeria. Osogbo is the state capital of Osun State and the city is occupied by both elites and indigenous residents.

### Samples collection

Both carrots and apples are not grown in the study area. The fruits are normally transported in cartons by the vendors far away from Southwestern Nigeria. A total of 100 samples of the two fruits were obtained from five randomly selected spots in the four major markets, namely Igbonna, Oke-fia, Alekuwodo, and Orisumbare in Osogbo metropolis. Five vendors were visited per market to obtain five samples of the fruits (a minimum of two of each fruit per vendor). The samples were collected into sterile containers and transported to the laboratory for analysis.

### Isolation of microorganisms

The samples were washed with normal saline in 100 ml round bottom clean plastic containers. 0.01 ml of the sample was then taken from each container and cultured on MacConkey, Sabouraud's dextrose agar, and chocolate agar plate and were incubated overnight at 37°C.

### Bacteriological analysis of the samples

The colonies were identified by standard bacteriological procedures as described by Cowan and Steel (1975). Gram's staining was performed to determine if the organism is gram negative or gram positive. A smear of the test organism was made on a clean slide, dried and covered with crystal violet for 30 to 60 s. It was washed off with clean water and was covered with Lugol's iodine for 30 to 60 s and later washed off with clean water. The slide was decolorized with acetone-alcohol, and was washed immediately with clean water and was covered again with neutral red stain for 2 min, and was washed off with clean water. The back of the slide was wiped clean and placed in a draining rack for the smear to air dry. The slide was examined microscopically with the oil immersion lens after the application of the oil on the slide. Gram-positive bacteria gave a dark purple colour while Gram-negatives give a red colour.

### Biochemical tests

Series of biochemical tests such as catalase, citrate, coagulase, oxidase, and urease were performed on the bacteria isolates in accordance with procedures highlighted by Baron and Finegold (1990) and Chaichanawongsaroj et al. (2004).

### Mycological analysis of the samples

The fungi isolates were identified by microscopic examination of the actively growing mould using morphological characters such as the absence or presence rhizoid, colour, and micro-morphology of their sporulating structures and conida (Evans and Richrdson, 1989; Onions et al., 1981).

### Microbial load determination

Enumeration of microorganisms present in each sample was done by 10-fold serial dilutions using the pour plate method. Counts were made on plates showing discrete colonies. The overall load of the organisms was counted and expressed as colony forming unit (c.f.u) together.

### Parasitological analysis of the samples

The samples's aliquots in 100 ml round bottom clean plastic container were allowed to stand on the bench for few hours to allow proper sedimentation in accordance with Alli et al. (2011). The supernatant was discarded with a Pasteur pipette leaving about 15 ml at the bottom. 10 ml of the deposit was transferred into a centrifuge tube and was spun for 5 min at 3,000 rpm. The supernatant was decanted while the deposit was re-suspended with 10% formal saline and was centrifuged. The supernatant was decanted and the deposit was then transferred into a clean glass slide. A drop of iodine was added as stain and covered with a cover slip. The slides were examined under ×400 microscope for parasite ova and cysts as previously described by Alli et al. (2011).

### Statistical analysis

The student t-test and Chi-square were used to determine the significant difference in the prevalence of the organisms and the microbial load between the markets using Statistical Package for Social Sciences (SPSS) version 16.0.

**Table 1.** The frequency of contamination of fruits by microbial agents.

| Market | No. of apple screened | No. of carrots screened | No. of apple contaminated | No. of carrot contaminated |
|---|---|---|---|---|
| Igbona | 12 | 13 | 6 | 9 |
| Alekuwodo | 13 | 12 | 3 | 6 |
| Okefia | 12 | 13 | 0 | 5 |
| Orisumbare | 12 | 13 | 3 | 6 |
| Total | 49 | 51 | 12 (24.0%) | 26 (50.9%) |

**Table 2.** Microbial load and diversity of the microorganisms isolated from the fruits.

| Micro-organism | No. of fruits screened | No. of isolate for carrot | Mean load (CFU) | No. of isolate for apple | Mean load (CFU) |
|---|---|---|---|---|---|
| *B. cereus* | 100 | 2 | $1.0 \times 10^4$ | 6 | $1.0 \times 10^3$ |
| *S. aereus* | 100 | 2 | $1.0 \times 10^2$ | 1 | $1.0 \times 10^1$ |
| *E. faecalis* | 100 | 1 | $1.0 \times 10^1$ | Not found | Nil |
| *L. monocytogenes* | 100 | 1 | $1.0 \times 10^4$ | Not found | Nil |
| *P. aeruginosa* | 100 | Not found | Nil | 1 | $1.0 \times 10^3$ |
| *C. species* | 100 | 3 | $1.0 \times 10^3$ | Not found | Nil |
| *C. species* | 100 | 3 | $1.0 \times 10^4$ | Not found | Nil |

**Table 3.** The frequency of contamination of the fruits by the parasites.

| Market | No. of fruits screened for apple | No. of fruits screened for carrot | No. of apple contaminated | No. of carrot contaminated |
|---|---|---|---|---|
| Igbona | 12 | 13 | 3 | 4 |
| Alekuwodo | 13 | 12 | 1 | 4 |
| Okefia | 12 | 13 | 1 | 2 |
| Orisumbare | 12 | 13 | 2 | 3 |
| Total | 49 | 51 | 7 (14.3%) | 25 (%) |

## RESULTS

### Microbial studies

The results of the bacteriological and mycological identifications revealed seven isolates namely, *Pseudomonas aeruginosa*, *Staphylococcus aeureus*, *Enterococcus faecalis*, *Bacillius cereus*, *Listeri monocytogenes*, *Citrobacter* sp. and *Candida* sp. were isolated from the fruits. However, more carrots (50.9%) were contaminated than apples (24%) (Table 1), the result of pair wise statistical analysis showed that there was no significant difference in the frequency of contamination of the two fruits (P=0.06; P>0.05). While the frequency of contamination of carrots varied significantly between the markets as revealed by *t*-test (P=0.005; P<0.05), there was no significant difference in the contamination of apples between the markets (P=0.092; P>0.05). All the isolates encountered were found as contaminants of carrots with the exception of *P.*

*aeruginosa*, while three out of the seven isolates were found as contaminants of apples with *B. cereus* constituting the most frequent contaminant of apple (Table 2). The microbial load of the isolates ranged between $1.0 \times 10^1$ and $1.0 \times 10^4$ in carrots and $1.0 \times 10^1$ to $1.0 \times 10^3$ in apples.

### Parasitological studies

The parasitological analysis of the samples showed that more carrots (25%) were also contaminated with parasites than apples (14.3%) (Table 3), but the difference in the level of contamination was not significant (P=0.058; P>0.05). There were significant variation in the level of contamination of the fruits between the markets (Apple, P=0.035; Carrot, P=0.007). Only the cysts of *Entamoeba coli* (2 cysts per fruit) and ova of *Ascaris lumbricoides* (1 ova per fruit) were found on the fruits.

## DISCUSSION

The observations from the present study revealed the microbial and parasitic contamination of the apples and carrots available for consumption in Osogbo Metropolis. The contamination, which could be a reflection of poor sanitary conditions of the environment where the fruits were cultivated or handled by the vendors. Studies have shown that most of the soil and rivers in Nigeria are heavily contaminated with pathogens (Sam-Wobo and Mafiana, 2005; Olayemi, 1994; Kanu and Achi, 2011; Agbabiaka and Oyeyiola, 2012). These fruits would have been contaminated from the field and the pathogens were subsequently transported to the store. The isolation of faecal microbial isolates and the recovery of the cysts of *E. coli* and ova of *A. lumbricoides* attest to the faecal contamination of the fruits either by water used for the irrigation or the soil of the cultivating area. On the other hand, poor sanitary conditions have also been reported among Nigerian after defecation (Sam-Wobo and Mafiana, 2005) and the tendency of contamination of the fruit samples by the vendors could not be ignored. This may also be the reason to justify the variation in the level of contamination of the fruits between the markets used for the study or the differences in the source of cultivation of the fruits. The public health implications of these findings are enormous. Prevalence of diarrhoeal diseases may be on the increase among subjects who consumed these fruits without proper washing. In Nigeria, carrots are important component of salad and similar foods, which are eaten raw or half cooked. Incidence of food poisoning cannot be ruled out in societies with poor personal hygiene and where food vendors are not routinely screened or registered.

Though, the frequency and microbial load of the pathogens were low when compared with similar studies in other parts of Nigeria (Eni et al., 2010; Alli et al., 2011; Kanu and Achi, 2011; Oranusi and Braide, 2012), the isolation of these organisms, no matter the level, raises considerable public health concern when considering their pathogenicity. Most of the isolates including the parasites (*A. lumbricoides* and *E. coli*) found on the fruits have been reported to cause gastrointestinal disorders and chronic diarrhorea (Alli et al., 2011; Oranusi and Braide, 2012). Therefore, consumption of these fruits without proper wash signifies the risks of gastrointestinal disorders among the residents.

The higher risk could be found among children consuming improperly washed fruits, including infestation by worms observed in this study. Worms infestation is a common cause of failure to thrive, growth retardation, and impaired cognitive functions among children (Sam-Wobo and Mafiana, 2005). It is thus important that these fruits pass the food hygiene test right from harvesting to consumption, while all stakeholders obeys existing public health laws towards morbidity reduction most especially among children consuming these fruits from time to time. The higher contamination of the carrots with parasites and microorganisms poses more public health challenge, in that the fruit is cheap when compared with other fruits and it is the most hawked fruit in Osogbo metropolis. Carrot is consumed by both economically low and affluent classes. While the later group may take precautionary measures to wash the fruit thoroughly before eating (possibly due to their educational status), the former group most of the time, eat the fruit as consumed from the vendors (Eni et al., 2010).

## Conclusion

The results of this study have demonstrated that carrots and apples sold in the major markets in Osogbo metropolis are contaminated with pathogens and parasites that are capable of causing food-borne disorders to the consumers. There is need for proper washing of the fruits with clean source of water before consumption. The vendors and the residents also need to be educated on the public health risks inherent in unwholesome hygienic practices and its attendants effects in causing food borne illness in the study area.

## ACKNOWLEDGEMENTS

The authors acknowledge the support of the Management and staff of Group Diagnostica, Oke-fia Osogbo to the study.

### REFERENCES

Agbabiaka TO, Oyeyiola G (2012). Microbial and physic-chemical assessment of Foma river, Itanmo, Ilorin Nigeria: An important source of domestic water in Ilorin metropolis. Int. J. Plant Anim. Environ. Sci. 1:209-216.

Alli JA, Abolade GO, Kolade AF, Salako AO, Mgbakor CJ, Ogundele MT, Oyewo AJ, Agboola MO (2011). Prevalence of intestinal parasites on fruits available in Ibadan Markets, Oyo State. *Acta Parasitological Globalis* 2(1):6-10

Cowan ST, Steel JL (1975). In: Manual for the identification of medical Bacteria, 2nd Ed. Cambridge University Press 1975. pp. 45-114.

Baron EJ, Finegold SM (1990). Bailey & Scott's diagnostic microbiology, VIII edn. St. Louis: Mosby Co. pp. 323-861.

Chaichanawongsaroj N, Vanichayatanarak K, Pipatkullachat T, Polrojpanya M, Somkiatcharoen S (2004). Isolation of Gram-negative bacteria from cockroaches trapped from urban environment. Southeast Asian J. Trop. Med. Public Health 35:681-684.

Chukwu CO, Chukwu ID, Onyimba IA, Umeh EG, Olarubofin F, Olabode AO (2010). Microbiological quality of pre-cut fruits on sale in retail outlets in Nigeria. Afr. J. Agric. Res. 5(17):2272-2275.

Eni AO, Oluwawemitan IA, Solomon US (2010). Microbial quality of fruits and vegetables sold in Sango Ota, Nigeria. Afr. J. Food Sci. 4(5):291-296.

Evans EGV, Richrdson MD(1989). Medical mycology: a practical approach. Oxford: Oxford University Press 1989, pp154.

Kanu I, Achi OK (2011). Industrial effluents and the impact on water quality of receiving rivers in Nigeria. J. Appl. Technol. Environ. Sanita. 3:75-86

Obeta SE, Nwakonobi TU, Adikwu OA (2011). Microbial effect of selected stored fruits and vegetables under ambient conditions in Markudi, Benue State, Nigeria. Res. J. Appl. Sci. Engine. Technol.

3(5):393-398

Olayemi AB (1994). Bacteriological water assessment of an urban river in Nigeria. Int. J. Environ. Health Res. 4:165-164.

Onions AHS, Allsopp A, Eggins HOW (1981). Smiths Introduction to Industrial Mycology, Edward Arnold, London. p. 389

Oranusi US, Braide W (2012). A study of microbial safety of ready-t-eat foods vended on highways, Onitsha-Owerri, Southeast Nigeria. Int. Res. J. Microbiol. 3(2):66-71

Sam-Wobo SO, Mafiana CF (2005). The effects of surface soil physio-chemical properties on the prevalence of helminths in Ogun State, Nigeria. Univers. Zambia J. Sci. Technol. 9(2):13-20.

Whitney-Chanex E (2011). The nutritional value of juices, carrots, beets, apple and celery. Food and drink nutrition. *www.livinstrong.com*

# 11

# The socioeconomic impact of Arsenic poisoning in Bangladesh

## Shakeel Ahmed Ibne Mahmood* and Amal Krishna Halder

Bangladesh Arsenic Control Society, BACS, C/o, 2/24 Babar Road, Mohammadpur, Dhaka 1207, Bangladesh.

**The purpose of this paper is to propose a methodology to analyze the health effects, how people cope with the socioeconomic consequences of the disease and to predict the beneficial effects of various alternative mitigation methods and recommends governmental measures for prevention of Arsenic poisoning. This research has been evaluated by the Bangladeshi Health Care system for its ability to recognize, isolate, report and control cases of Arsenic. The statistics is provided through the latest Internet publications, literature on global and regional information on environment, and using the database of Bangladesh Demographic and Health Survey 2004. This research attempted to analyze how other international agencies are trying to prevent Arsenic in their countries, where the people are affected or infected by Arsenic. As preventive measures, surface water treatment including drinking or taking water from the pond, various educational program, support of Government and NGOs, using media materials and Pan American Center for Sanitary Engineering and Environmental Sciences (CEPIS) in Peru, called ALUFLOC and Danida and Water Aid developed technologies to remove Arsenic were mentioned.**

**Key words:** Arsenic poisoning, alternative mitigation methods, governmental measures.

## INTRODUCTION

The Arsenic poisoning, termed the biggest environmental disaster and a major public health issue in recent times in Bangladesh (Mahmood and Ball, 2004). The Bangladesh government recognizes Arsenic contamination of ground water as a serious health problem, and aims to mitigate the situation with international cooperation (New Nation, 2006).

Contamination of groundwater by Arsenic has been reported in many countries including Australia, Chile, China, Hungary, Mexico, Peru, Thailand, Vietnam, and the US, but the most seriously hit areas are Bangladesh and West Bengal (India). The contamination of ground water by Arsenic in the deltaic region, particularly in the Gangetic alluvium of Bangladesh, is said to be one of the most important natural misfortunes (Daily Star, 2006). The World Bank has offered $52 million in assistance to help the country find alternative drinking water sources, while the United Nations. Foundation has contributed

$2.5 million (UNU, 2002)

### Arsenic

Arsenic (As) is a naturally occurring element in the earth's crust, and traces of Arsenic can be found throughout the environment. Arsenic is the chemical element that has the symbol As, atomic number 33 and relative atomic mass 74.92. Arsenic is a metalloid. It can exist in various allotropes, although only the grey form is industrially important. The main use of metallic Arsenic is for strengthening alloys of copper and especially lead. Arsenic is a common n-type dopant in semiconductor electronic devices, and the optoelectronic compound gallium arsenide is the most common semiconductor in use after doped silicon (Wikipedia, 2010).

Arsenic belongs to group V of the periodic table (Tutor, 2010). High concentrations of Arsenic have also been found in groundwater from areas of bedrock and placer mineralization, which are often the sites of mining activities. Arsenic concentrations of up to 5000 µg/L have been

*Corresponding author. E-mail: shakeel.mahmood@gmail.com

found in groundwater associated with the former tin-mining activity in the Ron Phibun area of Peninsular Thailand, the source most likely being oxidized arsenopyrite (FeAsS) (IGAIC, 2010).

Arsenic in soil can originate naturally, and past human activities may have added to these levels in some areas. Historically, the heaviest use of Arsenic in this country has been as a pesticide. The current predominant use of Arsenic is as a wood preservative. In ground water, Arsenic occurs primarily in two forms, $As^{+3}$ (arsenite) and $As^{+5}$ (arsenate). Organic Arsenicals are not known to occur at significant levels in ground water. Arsenic may change chemical form in the environment, but it does not degrade (Department of Environmental Protection, 2006).

## METHODOLOGY

A systematic review was conducted from 1999 to 2010. Information was retrieved from documents available mainly in electronic database and on the websites of specialized agencies, using the terms Arsenic and impact on social economy in rural Bangladesh with other researchers work was undertaken, including 1 leading Bangladesh daily newspapers also analyzed. 47 research papers were retrieved from the database (websites) of several national and international agencies were browsed. The most important, being online collection from different journals on Arsenic related issues. These sites housed a number of reports on quantitative and qualitative studies, estimates of Arsenic cases, policy analysis of the existing Arsenic situation in Bangladesh, and government strategies. Histological observations were carried out and a cross-sectional prevalence study of Arsenic and socio economic impact in rural Bangladesh was also held. A scrutiny of the abstract revealed that, some presentation posted on the websites, which was presented in international conferences and few other presentations were published in journals. Collected documents were skim read to cases, whether they contained information on Arsenic in conjunction with socio economic impact. Data accruing from the study analyzed Bangladesh Demographic and Health Survey 2004 database downloaded from the internet (www.measuresdhs.com) using SPSS Version 11.5. Wealth quintiles are prepared applying principal component analysis (PCA) as the PCA technique describes better measurement of wealth category compared to any other wealth measurement techniques (Vyas and Kumaranayake, 2006).

## Socio-economic background of rural Bangladesh

Access to a safe water supply is one of the most important determinants of health and socioeconomic development (Cvjetanovic, 1986). The recognition of the importance of a safe water supply led to an emphasis on the provision of appropriate facilities in developing countries (UNICEF, 1998). The socio-economic status of Bangladesh is similar to that of other developing countries. Most of the people are ignorant about various aspects of public health, a wide range of superstitions and false beliefs being prevalent in these communities. These low levels of socio-economic discrimination (Barkat et al., 2002). However, in one study in Laximipur, Bangladesh, identified that better economic conditions leading to maintaining better nutritional status of population may have a role to prevent arsenocosis, at least partly (Dhar et al., 1998). One study says that, the Arsenic problem will have major economic impacts throughout Bangladesh, affecting agriculture and other related industries including water management and public

health, as well as the overall national economy (UNU, 2001). According to a report in the Bulletin of the WHO [2000; 78(9): 1093-1103] Bangladesh has the largest Arsenic contamination in ground water in history. The potential severity of the problem is estimated to be greater than that from the chemical accident at Bhopal, India in 1984 and the nuclear accident at Chernobyl, Ukraine in 1996 (Glimpse, 2002). The Public Health Department of Bangladesh (DPHE), a national agency under the Ministry of Local Government Rural Development and Co-operatives, is entrusted to provide a safe water supply, sanitation and drainage facilities throughout the country (except for three major cities). DPHE is the focal point for initiating a national policy framework and development plan for water supply and sanitation under the guidelines of the Ministry of Local Government and Planning Commission of the Government of Bangladesh.

DPHE is responsible for planning, designing, implementing and monitoring water supply and sanitation projects in both rural and urban areas of the country. It provides necessary technical support to local governments, selecting locations of water sources, infrastructure, operational training for maintenance of water supply, and sanitation facilities. It assists councils in the promotion of health and hygiene through education programs conducted by union WATSANs water supply and sanitation committees (DPHE website). Although switching to tubewells helped control waterborne diseases, this success has come at a price. The shift to groundwater created another problem, millions of people have been exposed to water naturally contaminated by Arsenic-rich rocks (Frisbee and Richard, 2002)

## The dilemma

Cases of Arsenic poisoning began surfacing in the 1980s (WHO, 2000; British Geological Survey/DPHE, 2000; Caldwell, 2003, Adeel, 2001a, b). The first Arsenicosis patient was diagnosed in West Bengal, India in 1983; the first Bangladeshi patient was diagnosed in 1987. The government probably knew as early as 1985 when Bangladeshis started going to India for skin complaints, diagnosed as Arsenic poisoning. By 1993, the DPHE confirmed Arsenic in a tubewell in Chapai Nawabgang, a village in the Rajshahi Division (similar to a province) and found eight Arsenicosis patients. At about the same time (1995), Arsenic in drinking water was internationally recognized at the first International Conference on Arsenic (Frisbee et al., 2002). In recent days, the awareness about Arsenic poisoning has risen in Bangladesh to a significant extent (Aziz et al., 2006). Now, an estimated 45 to 49% of Bangladesh's 21 million tubewells are thought to be pumping Arsenic contaminated water (Frisbee et al., 2002). More than 14,000 people have been identified as suffering from Arsenic-related diseases and an estimated 20 to 35 million (Caldwell, 2003; WHO, 2000; Adeel, 2001a, b) are exposed to drinking water in which the concentration of Arsenic exceeds the Bangladeshi drinking water standard. Bangladeshis are also at risk due to irrigation of farmland with Arsenic contaminated water. Arsenic has been found to be above permissible limits in vegetables, fruits and cereal soils irrigated with crops grown in contaminated water, raising fears that the entire population is at risk. Researchers have found that food, including rice (a major staple of the Bangladeshi diet) cooked with Arsenic-contaminated water contains high levels of Arsenic (Dhaka University and Commonwealth Scientific and Industrial Research Organization as cited in Mortoza, 2002). Researchers are examining whether Arsenic in food will affect individuals who consume it. The developmental effects on children resulting from the ingestion of Arsenic have also not been extensively investigated but it is probable that children are more sensitive to Arsenic induced toxicity than adults (Mahmood and Ball, 2004). Furthermore, malnutrition, which is a common in the poor, shows a strong association with Arsenicosis. This is because malnutrition increases the susceptibility to Arsenicosis (Ahmad et al., 2007; Sarkar and Mehrotra, 2005). The UN estimate that upward of 57 million people in Bangladesh, around 50% of the country's population, is at risk for cancer because of contaminated well water (Adeel, 2001a, b). Officials noted that the number of people at risk exceeds the number of people infected with HIV worldwide. Within a decade, one-tenth of the deaths in much of the south of the country could be caused by Arsenic; because Arsenic contamination has been found in all 64 districts, the number of proportion of deaths attributable to Arsenic could be higher. Although, Bangladesh is ranked second in terms of the availability of ground water, the Arsenic problem will affect agriculture, water management, and the economy as a whole, as well as the health of individuals and families (WHO, 2000). This paper analyzes susceptibility and the economic impacts of the urban and rural populations of Bangladesh to Arsenic, assesses the risky behavior of contracting Arsenic and assesses how people cope with the socioeconomic consequences of the disease and to forecast the positive effects of various alternative mitigation techniques with special emphasis on the current social status of the people of Bangladesh. The available options for safe water can be classified by source: groundwater, surface water and rainwater. Recent years have seen increasing acceptance of strategies for incremental improvement in the environment and health in general and of demand-driven approaches to water supply and sanitation in particular.

## Groundwater

The simplest and most immediately achievable option is the sharing of tubewells that are currently low or free from Arsenic. Arsenic-containing wells may still be used safely for laundry-washing for example, and a simple color coding (like that used for "traffic lights") may have a significant impact on community Arsenic exposure if carefully and continuously backed up by awareness raising and educational activities (WHO, 2006).

### The Shapla Arsenic

The filter technology invented by Prof. Islam operates based on the absorption of Arsenic through the reaction with activated iron oxides impregnated in crushed brick particles and simultaneous filtration. Arsenic absorbing media production involves the activation of crushed brick particles of proper size by iron salts and heating the solution at a specific temperature for a specified time, resulting in the formation of dispersed activated iron oxides in and on the pores and edges of brick particles (BSS, 2005).

### Alcan filter

Arsenic mitigation options were chosen in consultation with BRAC and based on their wide experience providing alternative water options to Arsenic exposed populations in the country. This system runs water through an activated alumina medium which efficiently removes Arsenic (ICDDR,B, 2004).

## Danida research project

Danida has conducted research on Arsenic removal in Noakhali, Bangladesh (similar to a state in the US) since November 1998. The research looks at the use of a mix of alum and $KMnO_4$. These ingredients are introduced into a large bucket, which is drained off after 1 to 1.5 h into another bucket. The cost of chemicals for an average family is Tk. 10/US $0.2 per month (WHO, 2006).

## CEPIS/PAHO ALUFLOC

The WHO PAHO Pan American Center for Sanitary Engineering and Environmental Sciences (CEPIS) in Peru has developed a technology called ALUFLOC for Arsenic removal at the household level. It has been tested in Argentina. ALUFLOC is a sachet containing chemicals that are added to a bucket of Arsenic contaminated tubewell water. After about 1 h of treatment, the water is safe for consumption. Preliminary field results suggest that, ALUFLOC is effective in reducing Arsenic content to safe levels (WHO, 2006).

## Surface water

Surface waters (rainwater, rivers, lakes, etc) are typically low in Arsenic and therefore, potentially attractive drinking water sources in Arsenic-rich areas. However surface waters are frequently contaminated with human and animal faecal matter and other materials that are unsafe. Such contamination originally led to the preference for groundwater sources in Bangladesh (WHO, 2006).

## Surface water treatment

Treatment of surface water can be achieved by several means. Slow sand filtration is a typical method for treatment in rural areas and small towns. Water passes slowly through a large tank filled with sand and gravel. Fine particles are filtered out and microorganisms are inactivated by a thin layer formed on the surface of the bed (Schmutzdecke) (WHO, 2006).

### Rainwater

Rainwater harvesting is a recognized water technology in use in many developing countries around the world (WHO/IRC 1997). UNICEF has promoted dispersion of the technology since 1994 in Bangladesh. The rainwater is collected using either a sheet material rooftop and guttering or a plastic sheet and is then diverted to a storage container. Water is not collected during the first few minutes of a rainstorm to avoid contamination by dust, insects, bird dropping, etc (WHO, 2006). Rainwater harvesting is largely capital intensive and is dependent on the availability of suitable roofing materials for guttering and storage tanks. Rainwater harvesting has proven to be successful in places including China (Taiwan), Sri Lanka and Thailand (WHO, 2006).

### Dug well

Before the (cheap and simple) tube well was introduced, many villages dug wells to obtain surface water. However, many, if not most, were NOT carefully installed, dug, covered, and chemically treated, and they were full of bacteria. Successful projects involving sanitary dugwells exist in other countries, and some that may be successful have been reported but not yet in the detail to convince the skeptics that no coliform bacteria exist. Wells must be remote from latrines, old or new.

Grameen Bank now has a similar list. They must be covered to prevent entry of animals and refuse but is aerated to allow oxidization. Water is taken out by a connected tube well and pump, making them as easy to use as a tube well. They are located at a distance from latrines including past latrines with their buried organic waste if possible. In many wells, the water is pumped by electricity to an overhead tank for which PVC pipe leads to half a dozen taps in individual houses or between houses. This system of running water is very popular (Harvard, 2010).

### A field-kit by the National Institute of Preventive and Social Medicine

BRAC (Bangladesh Rural Advancement Committee), an international non-governmental organization, has tested a simple and low-cost procedure of tubewell water, which can be implemented at the community level. This study demonstrated that a change in the water source ensuring safe water by introducing an effective, affordable, and simple procedure can be helpful to overcome the Arsenic contamination problem (Chowdhury et al., 1999).

### Technology choice

The choice between technologies should take into account their cost effectiveness in providing Arsenic free and microbiologically safe drinking water. Different options may have very different balances of cost between, for example, capital and recurrent costs and may impact differently on the household costs of water management. However, the criteria of sustainability and acceptance by rural users must be incorporated in the calculation of cost effectiveness. This should aid the decision making process regarding which mitigation method(s) to implement (WHO, 2006).

### DISCUSSION

#### Economic and social conditions of the affected families

Like any other disease, loss of productive hours has a negative impact on family income and on the economic conditions of poor families. As Arsenicosis is a 'new disease', patients move from doctor to doctor for diagnosis despite the lack of proven treatment; patients usually spend a lot of money on treatment even before diagnosis. Such substantial expenditures over a prolonged period of time worsen the economic conditions of poor families (APSU, 2006). As a result of misconceptions about Arsenicosis, it is often identified with contagious diseases. As a result, patients are ostracized which adversely affects their livelihood. For example, job loss as a result of arcenosis symptoms has been reported as has, decline in business (for example of a shopkeeper or a peddler). Women with Arsenicosis usually suffer the most. There are reports of broken marriages and problems in getting married. Thus, the adverse attitude towards Arsenicosis patients contributes to worsening economic conditions of families with Arsenicosis patients (APSU, 2006).

#### Economic differentials in social and economic impacts of Arsenicosis

Apart from the health effects, Arsenic poisoning also causes a wide range of social problems and economic loss. One study suggests a significant relationship between socio-economic status of a household and social and economic problems caused by Arsenicosis of a family member. The lower income group (Tk. 2500) group was significantly more likely to report facing social problems than higher income groups (APSU, 2006).

#### The role of socioeconomic status in Arsenic poisoning

In Bangladesh, it is likely that access to tubewell drinking water will be at least partially determined by social status. Therefore, the observed relationship between Arsenicosis prevalence and household income could be due to social barriers to access to

Arsenic-free water for poor households (WHO, 2006). There are also significant socio-economic barriers to switching wells as most wells are privately owned and there may be reluctance to sharing a water source. Privacy in another issue, because tubewells are usually installed near household latrines. Moreover, women, who traditionally collect water, are not usually allowed to leave their immediate household unaccompanied. Another point to consider is that, if the density of users at each well increases, this may affect the aquifer and the water source may, in turn, become Arsenic contaminated (Willingness to pay for Arsenic-free, 2003). Economic status was found to be a key factor determining acceptability and the price that households were willing to pay for Arsenic removal technology. Although most households showed a preference for the features of the more expensive technology, namely Alcan, very few could realistically afford this technology individually (Willingness to pay for Arsenic-free, 2003).

### FINDINGS

### Arsenicosis and economic status: The poor suffer most

As a result of the high coverage and low cost of the technology, the poor used to pay little for water in the rural areas Table 1. Those who did not own any private or community tubewells managed to collect water from the nearby government tubewell or private tubewells that were usually plentiful in the neighborhood. Although due to the prevailing social barriers, access to Arsenic-free water always may not be possible but because of social and religious reasons, drinking water is not usually denied to anybody in rural Bangladesh (Chowdhury, 2002). Thus, the 'tubewell revolution' ensured provision of one of the basic human needs-safe drinking water-for the rural poor in Bangladesh. The high service level of tubewells requires little time for collecting water, which has indirect but positive impacts on the economic conditions of poor families (Chowdhury, 2002).

Another study showed that the majority of Arsenicosis patients (71%) belong to the low-income group, while 29% are middle class. No patients were from the high-income group. All of these patients were from rural areas of the country and the majorities were related with the traditional occupations of the country, like cultivation (53%). In addition to lower levels of educational (81.5 %), most of the patients with chronic Arsenicosis were suffering from malnutrition (91%) (Sikder et al., 2002).

### Knowledge level regarding Arsenic issue

Data from the 2004 Bangladesh Demographic and Health Survey, a nationally representative survey, were used in this analysis; the survey methodology is described elsewhere. Overall, 84% of households had heard about Arsenic. The richest group (quintile) of people is significantly more aware of Arsenic knowledge (97%) than the poorest quintile (69%) Table 2. However, in terms of divisional coverage, Khulna division (95%) people have

**Table 1.** The costs of different alternative technologies in comparison with shallow tubewells.

| Alternative technological options | Unit Cost Taka | No. of family/Unit (Family size = 5 | Installation cost/person Taka |
|---|---|---|---|
| Shallow Hand Tubewell | 5 000 | 50 | 20 |
| Rainwater Harvesting | 6 200 | 1 | 1240 |
| Dug/Ring Well | 35 000 | 25 | 280 |
| Deep Tubewell | 45 000 | 50 | 180 |
| Pond Sand Filters | 35 000 | 50 | 140 |
| Arsenic Removal Filters | | | |
| --Community type | 75 000 | 25 | 600 |
| --Household type | 450-2 500 | 1 | 90-500 |

(Chowdhury, 2002).

**Table 2.** Heard of Arsenic by wealth quintiles and regions.

| Heard of Arsenic by wealth quintiles | | | | Heard of Arsenic by region | | | |
|---|---|---|---|---|---|---|---|
| Wealth quintile | Number | Row (%) | Column (%) | Region/ Division | Number | Row (%) | Column (%) |
| Poorest(n=2364) | 1622 | 18 | 69 | Barisal (n=632) | 494 | 6 | 78 |
| Poorer(n=2197) | 1750 | 20 | 80 | Chittagong (n=1816) | 1502 | 17 | 83 |
| Middle(n=2026) | 1730 | 20 | 85 | Dhaka (n=3364) | 3053 | 35 | 91 |
| Richer(n=1959) | 1828 | 21 | 93 | Khulna (n=1289) | 1220 | 14 | 95 |
| Richest(n=1933) | 1879 | 21 | 97 | Rajshahi (n=2753) | 2051 | 23 | 74 |
| | | | | Sylhet (n=624) | 488 | 6 | 78. |
| Total(n=10479) | 8809 | 100.0 | 84 | Total (n=10478) | 8808 | 100.0 | 84. |
| Chi-square trend: p=0.000 | | | | Chi-square trend: p=0.000 | | | |

Sources: BDHS 2004 database.

more knowledge than any other divisions and the people in Rajshahi division have the lowest level (78%), Tabel 1. Overall, it can be concluded herewith that, the poverty prone people are more vulnerable in terms of Arsenic knowledge compared to others, because it is shown that, the range of deference between inter-wealth index category of people is much more higher thatthe range of inter-divisional coverage. Among 84% of people those who heard about Arsenic, Table 3 exploring the statistics about, if they collect drinking water from any tubewell, whether the tubewell has been tested for Arsenic poisoning. Table 2 shows that, 90% of the population depends on tubewell water, as their safest drinking water source. However, almost half (48.67%) of the tubewells remained unmarked (in other words untested for Arsenic poisoning) and this is an alarming situation for the nation and in this regard; either Government and/or other associate agencies can have their instant attention in terms of testing these tubewells for Arsenic poisoning. This would help the nation to know that, the water they are collecting from the tubewells, whether it is Arsenic

free or not. However, there is a study where it was found that, the study population's exposure to sources of Arsenic related information did not result in adoption of avoidance measures (Aziz et al., 2006).

## Intensity of untested tubewells by wealth quintiles by regions

Table 4 make highlights that, the unmarked/untested tubewells are more or less equally distributed among the four lowest wealth quintiles, but this could be noted that the rate of unmarked/ untested tubewells are lowest in richest quintiles and moreover the difference between richest and any other quintile is statistically significant. On the other hand, some divisions are more vulnerable (Table 3).

For instance, untested/unmarked tubewells are highest in Rajshahi and Dhaka division. Sylhet, Barisal and Khulna divisions are comparatively in safer side in terms of testing and marking of tubewells for Arsenic

**Table 3.** Arsenic marking on tubewells.

| Marking information | # of HHs | % of HHs |
|---------------------|----------|----------|
| Red marked | 599 | 6.80 |
| Green marked | 2852 | 32.38 |
| Unmarked | 4287 | 48.67 |
| Do not Know | 170 | 1.93 |
| Do not collect from tubewell | 901 | 10.23 |
| Total | 8809 | 100.00 |

Sources: BDHS 2004 database

**Table 4.** Intensity of unmarked/untested tubewells.

| Unmarked/ untested tubewells by Socio-economic status | | | Unmarked/ untested tubewells by region | | |
|---|---|---|---|---|---|
| (Wealth quintiles) | Households | % households | Region/ Division | Households | % households |
| Poorest | 877 | 20.46 | Barisal | 291 | 6.8 |
| Poorer | 900 | 20.99 | Chittagong | 768 | 17.9 |
| Middle | 852 | 19.87 | Dhaka | 1180 | 27.5 |
| Richer | 922 | 21.51 | Khulna | 328 | 7.6 |
| Richest | 736 | 17.17 | Rajshahi | 1474 | 34.4 |
| Total | 4287 | 100.00 | Sylhet | 246 | 5.7 |
| | | | Total | 4287 | 100.0 |

Sources: BDHS 2004 database.

poisoning.

## Red marking tubewells (risk area/group) by socio-economic status and divisions

The red marked tubewells are found among the poorest and middle category of people highest and lowest in the richest category of people (Table 5) and in terms of divisional analysis, Dhaka and Chittagong is the highest and Barisal and Rajshahi is the lowest (Table 4). One possible reason of becoming lowest Arsenic intensity in Barisal region is because of river based area. Overall, people in the poorest category are deprived in terms of both knowledge and achievement in getting drinking water from tested tubewells.

## Coping with Arsenicosis

Once a family member becomes sick, a variety of coping methods come into play, depending upon the status of the afflicted person. A major amount of attention is paid to the effects of illness of the breadwinner, usually the father. Coping with the burden of treatment costs constitutes a first important issue for the family. However, adoption of methods to obtain Arsenic free water for the

bread-winner would also allow the other household members the opportunity to safe drinking water.

In Bangladesh, Pryer (1989) found that "large" medical expenditures are paid out of the sale of assets. It would need to be ascertained whether these assets are factors of production, such as land, that affect future income, or are smaller assets like beds, tables, chairs, fan or radio. It has also been found that, intra-household labor substitution takes place in case of chronic illness to preserve income. For example, family members (the breadwinner's wife and mother) could work additional hours. The children in the family may sell goods or foodstuff at the market. Another managing mechanism may be decreasing food consumption or other consumption of other basic needs of items such as clothing, education and housing. Pryer (1989) also established that some households accumulate large loans to finance lost income due to the breadwinner's illness.

The economic burden of Arsenicosis was revealed in a study, which found that the total of 7930 "years lived with disability" resulted in 1908 "disability-adjusted life years "in case of Arsenicosis (Molla et al., 2004). Coping is likely to differ between rural and urban areas. For example, the economy of poor households in urban areas is likely to be connected much more closely with the urban manufacturing sector, offering in principle a wider variety of coping mechanisms. Evidence on coping with illness

**Table 5.** Distribution of red marking tubewells.

| Identified red marked tubewells by Socio-economic status | | | Identified red marked tubewells by region | | |
|---|---|---|---|---|---|
| Wealth quintiles | Households | % households | Region/ Division | Households | % households |
| Poorest | 137 | 22.8 | Barisal | 8 | 1.3 |
| Poorer | 120 | 20.0 | Chittagong | 217 | 36.3 |
| Middle | 141 | 23.6 | Dhaka | 239 | 39.9 |
| Richer | 128 | 21.4 | Khulna | 95 | 15.8 |
| Richest | 73 | 12.3 | Rajshahi | 10 | 1.7 |
| Total | 599 | 100.0 | Sylhet | 30 | 5.0 |
| | | | Total | 599 | 100.0 |

Sources: BDHS 2004 database.

among urban slum residents in Dhaka City is available from Desmet et al. (1998). For daily wagers they found that "sacrificing holidays" is the first coping strategy following loss of income due to illness, followed by intra-household labor substitution, and foregoing consumption of commodities. Taking loans and using cash savings rank fourth and fifth in the list of coping strategies. On the whole, expenditure for basic need items such as staple foods, education, clothing and education does not seem to be affected (WHO, 2006).

It is important to prevent the increase of Arsenicosis for both health and economic reasons. However, drug treatment to eliminate Arsenic from the body is a bit time consuming and may be associated with some expense, leading to the conclusion that palliative care, including application of ointment in the case of keratosis, may be the only affordable treatment in rural areas of Bangladesh. Bangladeshi villagers affected by Arsenicosis are likely to lose a significant amount of productive time. In addition, the disease may become a burden on villagers' overall financial and time resources (WHO, 2006). In this context, it can also be noted that till date no chelating agent has shown definite success in removing Arsenic from the body (Das and Sengupta, 2008).

### Governments and international NGO's commitment

The former Bangladesh State Minister for Youth and Sports, Mohammad Fazlur Rahman, said that the present government has been working relentlessly to make Arsenic-free water available throughout the country. He said: "This is a part of the government's scores of epoch-making programs for the speedy progress and prosperity in national economy to bring about a positive change in rural life" Daily Star (2006). Here, this can be noted that Governmental strategies already adopted to combat the disease in Bangladesh and West Bengal (Ghosh et al., 2008). Today international agencies such as the World Bank, UNICEF, UNDP, WHO, BRAC, ICDDR,B and the Rotary have accepted that this is a problem of national

and international importance. UNICEF has funded projects to examine filters (Swash, 2003). Four technologies have thus far been approved by the government (Daily Star, 2004).

### Conclusion

From the review, it is evident that poor families are suffering more from the Arsenic problem in Bangladesh. They are being exposed to Arsenic through drinking water and they have less access to alternative safe drinking water sources. Poverty related malnutrition also appears to be making them more vulnerable to Arsenicosis. Once sick, the loss of productive hours and expenditures on 'treatment' contribute to the worsening of their economic condition (Chowdhury, 2002). A graver impact of Arsenicosis for the poor may be their vulnerability to social taboos because of their condition. Mitigation methods need to be implemented urgently, not least for reasons of poverty alleviation. It is important to assess the future health and socioeconomic impacts of alternative mitigation methods, so that policy-makers in Bangladesh can take informed decisions (WHO, 2006). In particular, important discussion and analysis will be needed on alternative financing strategies for hot-spot areas, thereby specifying the roles of donors, local and central Government, and households (WHO, 2006).

Finally, it is clear that the technologies introduced to supply Arsenic free safe drinking water are only short-term emergency solutions for areas severely affected by Arsenic contamination. The longer-term solutions may include the provision of a piped water supply to the population and the optimal use of surface water (Jakariya et al., 2005). This is most unfortunate for the victims; as like HIV/AIDS, it has the potential to affect almost all families in Bangladesh. This is creating and will create a loss of productivity damaging the economy (Mahmood and Ball, 2004). A bigger tragedy is that, despite the passage of time, there has been no fresh thinking on what promises to be a massive social and environmental problem. "There were many occasions "when a stitch in time could have saved us nine" be it in politics or in economics." (Daily Star, 2006).

Given the millions of people in Bangladesh who are currently suffering from Arsenic poisoning, health policy-makers (WHO, 2006) with the help from researchers need to devise policies capable of counteracting this threat. They need to show that, an increasing number of people will suffer from Arsenicosis, if mitigation methods are not implemented rapidly. It is also important to demonstrate the social and economic effects on households with Arsenicosis patients, and how mitigation methods can reduce the burden of those effects. Above all, Bangladesh should immediately implement appro-priate Arsenic and water supply and sanitation policies to benefit the people of this country. In addition to that, Bangladesh government should include in the national policies about implementing the beneficial effects of various alternative mitigation methods, as well as govern-mental measures for prevention of Arsenic poisoning. Any action taken must address the needs throughout the country because Arsenic contamination is not limited to particular geographic areas. Only by addressing the problem at a national level can the negative effect on economic growth be diminished.

## ACKNOWLEDGEMENT

The authors acknowledge with gratitude the cooperation of Dr. Elizabeth Oliveras, Sc.D. (Harvard University), Operations Research Scientist, Health System Economics Unit of Health Systems and Infectious Diseases Division, at ICDDR, B, Bangladesh.

### REFERENCES

Adeel Z (2001a). Policy Dimensions of the Arsenic Pollution Problem in Bangladesh. United Nations University. Environment and Sustainable Development Programme. Available at http://www.unu.edu/env/ArsenicArsenic/Adeel.pdf.

Adeel Z (2001b). Arsenic Crisis Today Strategy for Tomorrow. Policy Brief. UN University and the Earth Identity Project. Available at http://www.unu.edu/env/water/arsenicArsenic/policy-brief.html.

Ahmad SA, Sayed MH, Khan MH, Karim MN, Haque MA, Bhuiyan MS (2007). Sociocultural aspects of arsenicArsenicosis in Bangladesh: Community perspective. J. Environ. Sci. Health A Toxicol. Hazard Subst. Environ. Eng., 42: 1945-58.

APSU (2006). Ministry of Local Government Rural development and Cooperative, Government of People's Republic of Bangladesh. Social Aspects of Access to Healthcare for ArsenicArsenicosis Patients, Feb, p. 27.

Aziz SN, Boyle KJ, Rahman M (2006). Knowledge of arsenicArsenic in drinking-water: Risks and avoidance in Matlab, Bangladesh. Health Popul. Nutr.. 24: 327

Barkat, Abul, AKM Maksud, KS Anwar, AKM Munir (2002). Social and economic consequences of ArsenicArsenicosis in Bangladesh. Paper presented at the Bangladesh Environmental, BAPA., 1: 216- 230.

BDHS (2004). National Institute of Population Research and Training(NIRPORT), Dhaka, Bangladesh, Mitra and Associates, Dhaka, Bangladesh and ORC Macro, Calveton, Maryland, USA, May, 2005. Also available at: http://www.measuredhs.comhttp://www.measuredhs.com. p. 1

BSS (2005). Novel filter invented to free arsenic from water.

Available at:_http://www.bssnews.net/index.php?genID=BSS-02-2002-09-21&id=7http://www.bssnewsnet/index.php?genID=BSS-02-2002-09-21&id=7. p. 1

Caldwell BK, Caldwell JC, Caldwell, Mitra SN, Mitra, Smith W. Smith (2003). Searching for an Optimum Solution to the Bangladesh ArsenicArsenic Crisis. @ Soc. Sci. Med., 56(10): 2089-2096.

Choudhury AMR, Mohammed Jakariya, Ashiqul H Tareq, Jalaluddin Ahmed (1999). Village Health Workers Can Test Tubewell water for ArsenicArsenic. Paper presented at the Eight ASCON, June, p. 48.

Choudhury MA (2002). Impact of ArsenicArsenic on the rural poor in Bangladesh. Paper presented at the Bangladesh Environ., 1: 154.

Cvjetanovic B (1986): "Health effects and impact of water supply and sanitation." World Health Stat. Q., 39(1): 105-117.

Daily S (2004). Govt okays Marketing of Anti-arsenicArsenic Technologies. Available at http://www.thedailystar.net/2004/02/26/d40226100273.htm. 4:266

Daily S (2006). Govt. trying to ensure supply of arsenicArsenic-free water. Available at : http://www.thedailystar.net/2006/08/19/d60819060382.htm. 5:810

Das NK, Sengupta SR (2008). Arsenicosis Diagnosis and treatment.Department of Environmental Protection (2006). A Homeowner's Guide to ArsenicArsenic in Drinking Water. Available at : http://www.state.nj.us/dep/dsr/arsenic/guide.htmhttp://www.state.nj.us /dep/dsr/arsenicArsenic/guide.htm, p.1

Desmet M, Bashir I, Sohel N (1998). "Direct and indirect health-care user expenditure by slum residents in Dhaka-City, Bangladesh." Center for Health and Policy Research, ICDDR -B, working paper no, 5: 98.

Dhaka Community Hospital Trust. Bangladesh (2002). Causes, effects and remedies. Dhaka and School of Environmental Studies, Jadavpur University, Calcutta, India. 8-12. Feb. LGED Auditorium, Dhaka, Bangladesh, p. 51.

Dhaka University and Commonwealth Scientific and Industrial Research Organization (2002). ArsenicArsenic in Groundwater of Bangladesh: Contamination in the Food Chain. Cited in Sylvia Mortoza (11 Jun). ArsenicArsenic in the Food Chain. News from Bangladesh. Available at: http://bicn.com/acic/resources/infobank/nfb/2002-06-11-d11062002.htm. p. 1

Dhar RK, Biswas KK,Samanta G (1997). Ground water arsenic calamity in Bangtladesh, Curr. Sci., 73. 1(10): 1

Frisbee Seth, Richard Ortega (2002). The Concentrations of ArsenicArsenic and Other Toxic Elements in Bangladesh's Drinking Water @ Environmental Health Perspectives 110 (November): pp. 1057 -1152.

Ghosh P, Roy C, Das NK, Sengupta SR(2008). Epidemiology and prevention of chronic arsenicArsenicosis: An Indian perspective. Indian J. Dermatol Venereol Leprol, 74: 582-593

Glimpse, ICDDR, B ( 2002). ASMAT New Project for Arsenic Research in Matlab. 24:4

Harvard University (2010). Chronic Arsenic Poisoning:History, Study and, Dec, 2002 http://www.physics.harvard.edu/~wilson/arsenicArsenic/remediation/a rsenicArsenic_project_remediation_technology.html p.1

ICDDR, B (2004). ArsenicArsenic contamination in Matlab, Bangladesh. Available at: http://www.icddrb.org/pub/publication.jsp?classificationID=56&pubID= 5761

IGAIC (2010). Arsenic Arsenic in Groundwater Worldwide. Available at: http://www.igrac.net/publications/143 p. 1

Das NK, Sengupta SR (2008). Arsenicosis: Diagnosis and treatment. Indian J Dermatol Venereol Leprol. 74: 571-581

Jakariya Md, Mizanur Rahman, AMR Chowdhury, Mahfuzur Rahman, Yunus Md, Abbas Bhuiya, MA Wahed MA, Prosun Bhattacharya,Gunnar Jacks, Marie Vahter, Lars-Ake Persson (2005). Sustaibale safe water options in Bangladesh: experiences from the Arsenic Project at Matlab (Asmat), Natural Arsenic in Ground water: Occurrence, Remediation and Management- Bundschuh, pp. 319-330

Mahmood, Shakeel, Ahmed Ibne, Carolyn Ball (2004). Defining and implementing Arsenic policies in Bangladesh: Possible roles for public and private sector actors. J. Health Human Services Admin.,

(JHHSA), 27: 2.

Molla AA, Anwar KS, Hamid SA, Hoque ME, Haq AK (2004). Analysis of disability adjusted life years (dalys) among Arsenic victims:A cross-sectional study on health economics perspective. Bangladesh Med. Res. Counc. Bull., 30: 43-50

Mortoza Sylvia (27 June 20032). Arsenic: A False Sense of Security. @ Bangladesh Observer Magazine. p. 1

Pryer J (1989): "When breadwinners fall ill: preliminary findings from a case study in Bangladesh." Inst. Dev. Stud. Bull., April, 20(2): 49-57.

Harvard (2010). Chronic Arsenic Poisoning:History, Study and Remediation. Available at

Sarkar A, Mehrotra R (2005). Social dimensions of chronic arsenic Arsenicosis in West Bengal (India). Epidemiology,16: S68.

Sikder MS, AZM Maidul AZM, M Ali M(2002). Study on socio-economic status of chronic ArsenicArsenicosis patients in Bangladesh.1st National Conference on Environmental Health, p. 1

Tutor Vista (2010). Poistion in the Periodic Period. Available at: http://www.tutorvista.com/topic/arsenic-periodic-table-of-elementshttp://www.tutorvista.com/topic/Arsenic-periodic-table-of-elements.

UNICEF (1998): "State of the Children's Report". UNICEF, New York. p 54.

UNU (2001). Bangladesh water: Emergency measures urged to prevent "catastrophic" arsenic poisoning. Available at: http://unu.edu/media/archives/2001/pre19.01.html. p. 1

Vyas S, Kumaranayake L(2006).: Constructing socio-economic status indices: how to use principal components analysis. Health Policy Plann., 21:459-468

WHO (2006). Towards an assessment of the socioeconomic impact of Arsenic poisoning in Bangladesh. Available at: http://www.who.int/water_sanitation_health/dwq/Arsenic2/en/index.html

Wikipedia (2010). Arsenic. Available at http://en.wikipedia.org/wiki/Arsenic.

Willingness to pay for Arsenic-free (2003). Safe Drinking water in Bangladesh, PS Press Services Pvt. Ltd. August, ( 2003), p 9, p-11.

World Health Organization (WHO) (2000). Researchers Warn of Impending Disaster from Mass Arsenic Poisoning. Press Release. No. 55.

# 12

# A spectrophotometric method for quantification of sulphite ions in environmental samples

Peter Musagala[1], Henry Ssekaalo[2], Jolocam Mbabazi[2] and Muhammad Ntale[2]

[1]Department of Chemistry, Busitema University, P. O. Box 236, Tororo, Uganda.
[2]Department of Chemistry, Makerere University, P. O. Box 7062, Kampala, Uganda.

**This study described an alternative method developed for the quantification of sulphite ions in environmental samples. The method was based on results of an investigation of the reaction of excess pentacyanidonitrosylferrate(II) popularly known as nitroprusside (NP) and the sulphite anion. NP-SO$_3^{2-}$ reaction product by use of zinc-ethylenediamine complex cation(s) was stabilized. The NP-SO$_3^{2-}$ reaction product was stabilized for 30 minutes by use of zinc ethylenediamine complex cation(s) and the absorbance was enhanced, making determination of sulphite possible. The method has a limit of quantification of 2.321 µg SO$_3^{2-}$ mL$^{-1}$. Good accuracy was achieved for samples spiked with SO$_3^{2-}$ in the range from 1 to 10 µg SO$_3^{2-}$ mL$^{-1}$, which demonstrated the validity of the proposed procedure. The repeatability (CV) was not more than 2.37% and the limit of detection was estimated at 0.99 µg SO$_3^{2-}$ mL$^{-1}$. The method was applied to determine the concentration of sulphite ions in sugar and wine brands sold in local market. Comparable results were obtained between this method and an iodometric procedure for determination of sulphite in environmental samples.**

**Key words:** Determination of sulphite ions, wine, sugar, environmental samples.

## INTRODUCTION

Sulphite anion (SO$_3^{2-}$) is a major species of sulphur in which oxidation state IV is expressed. The detection of the anion has long held the interest of the analytical community, because of the large number of roles that it can play within environmental and physiological systems (Isaac et al., 2006). It may occur in boilers and boiler feed waters treated with it for dissolved oxygen control, in natural waters or wastewaters as a result of industrial pollution, and in treatment plant effluents dechlorinated with sulphur dioxide. Sulphites or sulphiting agents are the most common preservatives used in winemaking (Koch et al., 2010), and are also important additives in many food products, because they inhibit development of both enzymatic and non-enzymatic browning in a variety of processing and storage situations (American Public Health Association, 1998; Claudia and Francisco, 2009). The sulphite ion is a very effective microbial inhibitor in acid or acidified foods.

However, excess sulphite in boiler waters is deleterious, because it lowers the pH and promotes corrosion. Control of SO$_3^{2-}$ in wastewater treatment and discharge may be important environmentally, principally because of its toxicity to fish and other aquatic life and its rapid oxygen demand (American Public Health Association, 1998). Sulphites as additives can cause an asthmatic reaction; presence of excessive amounts of sulphites is respon-sible for off flavour in food products (McFeeters and Barish, 2003; Machado et al., 2008). Some of it added to foods often disappears as a result of reversible and irreversible chemical reactions. Thus, it is often important to measure both free and bound forms of sulphite that are present in foods.

Several techniques have been developed to quantify the sulphite anion alone or in combination with other sulphur species like sulphate, thiosulphate and dithionate. These include: titrimetric methods with potassium iodide-potassium iodate (American Public Health Association, 1998), copper sulphate (Shahine and Ismael, 1979), cerium(IV), mercury(II) (Crompton, 1996), spectro-photometric methods with 1,10-phenanthroline (American Public Health Association, 1998), Fuchsin N solution (Badri, 1988), mercuric chloranilate (Humphrey and Hinze, 1971), mercuric thiocyanate and $Fe^{3+}$, electro-chemical methods with mercury(I) chloride-mercury(II) sulphite electrodes (Marshall and Midgley, 1983) and the sulphite oxidase enzyme electrode (Smith, 1987). Other methods include molecular emission spectrometry (Schubert et al., 1979), high performance liquid chromatography (HPLC) with ultraviolet (UV) detection (McFeeters and Barish, 2003), ion-exchange chroma-tography (Edmond et al., 2003), chemiluminescence methods (Al-Tamrah, 1987; Koukli et al., 1988) and flow injection analysis techniques (Thanh et al., 1994; Araujo et al., 1998; Xiaoli and Wei, 1998; Atanassov et al., 2000; Hasson and Spohn, 2001; Claudia and Francisco, 2009).

The phenanthroline method is currently adopted as the standard one for determination of sulphite in water and wastewaters, while the sulphite in foods and beverages is determined by the traditional AOAC Official Method 990.28 (AOAC, 1995; Cunniff, 1995). The phenanthroline method requires elaborate technical specification and user expertise and as such, can incur substantial running costs. It involves purging of an acidified sample with nitrogen gas and trapping the liberated sulphur dioxide gas ($SO_2$) in an absorbing solution containing $Fe^{3+}$ ion and 1,10-phenanthroline. The $Fe^{3+}$ ion is reduced to $Fe^{2+}$ by $SO_2$, producing the orange tris(1,10-phenanthroline)iron(II) complex as illustrated in the following reaction schemes:

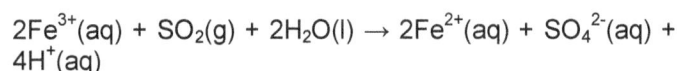

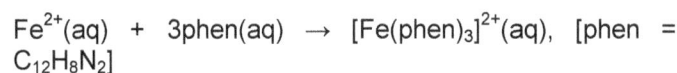

$$2Fe^{3+}(aq) + SO_2(g) + 2H_2O(l) \rightarrow 2Fe^{2+}(aq) + SO_4^{2-}(aq) + 4H^+(aq)$$

$$Fe^{2+}(aq) + 3phen(aq) \rightarrow [Fe(phen)_3]^{2+}(aq), [phen = C_{12}H_8N_2]$$

In solution, the sulphite anion reacts with pentacyanidonitrosylferrate(II) ion to form an unstable, red sulphite-nitroprusside ion ($[Fe(CN)_5(NOSO_3)]^{4-}$) (Fogg et al., 1966; Andrade and Swinehart, 1972; Leeuwenkamp et al., 1984; Araujo et al., 2005). In our laboratory, we succeeded in stabilizing both the sulphate (Mbabazi et al., 2011) and the sulphite anions through a series of spectrophotometric tests. The red sulphite-nitroprusside reaction product was stabilized by ethylenediamine complexes of zinc; the general reaction scheme for this behavior being;

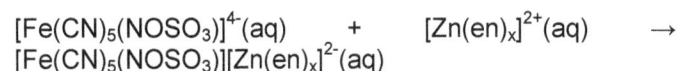

$$[Fe(CN)_5(NOSO_3)]^{4-}(aq) + [Zn(en)_x]^{2+}(aq) \rightarrow [Fe(CN)_5(NOSO_3)][Zn(en)_x]^{2-}(aq)$$

The main objective of this study therefore was to utilize the sulphite-nitroprusside stabilized product and describe an alternative spectrophotometric procedure for the quantification of the sulphite ion in solution. Our method has been compared with an iodometric titration method for determining the concentration of sulphite ions in wine and other beverages such as sugar that are commonly sold in local markets.

## MATERIALS AND METHODS

### Apparatus

Weighing was done on an AAA 160DL dual range balance (Adam Equipment Co. Ltd UK). The absorption spectra were recorded on a UV-VIS Shimadzu UV-1700 CE double beam spectrophotometer (Shimadzu Corporation, Japan) and absorbance measurements at a fixed wavelength were made with the same instrument in the photometric mode. pH measurements were done with a Corning Pinnacle 555 pH/ion meter (Corning Incorporated Life Sciences Corning, New York, 14831 USA). The addition of aqueous nitroprusside (NP) to aqueous $SO_3^{2-}$ was carried out using Transferpette micro pipettes (BRAND GMBH + CO KG Postfach, 11 55 97877 Germany).

All chemicals used were of analytical grade. The solutions used were prepared as subsequently described.

### Starch indicator solution

The starch indicator solution was prepared by dissolving analytical grade soluble starch (2 g) and salicylic acid (0.2 g) in hot deionized water (100 mL).

### Standard potassium iodate solution

Potassium iodate (0.0021 mol/L) solution was prepared by dissolving the solid (812.4 mg) in a minimum amount of deionized water and diluted to 1000 mL.

### Standard sodium thiosulphate

Sodium thiosulphate pentahydrate ($Na_2S_2O_3.5H_2O$, 6.204 g) was placed in a volumetric flask (1000 mL) and dissolved in deionized water (50 mL). Sodium hydroxide (0.4 g) was added and the solution made to the mark with deionized water. The solution was then standardized using a standard potassium hydrogen iodate solution as follows. Potassium iodide (2 g) was dissolved in an Erlenmeyer flask with deionized water (150 mL). Concentrated sulphuric acid (3 drops) and starch solution (20 mL, 0.0021 mol/L) were added, respectively. The solution was diluted to 200 mL and the liberated iodine was titrated against the sodium thiosulphate titrant with starch as the indicator. The concentration of sodium thiosulphate was found to be 0.025 mol/L.

### Standard iodine solution

Potassium iodide (25 g) was dissolved in a minimum amount of water and iodine (3.2 g) was added. After dissolution of the iodine, the solution was diluted to 1000 mL and standardized against $Na_2S_2O_3.5H_2O$ (0.025 mol/L) using starch solution as indicator.

### Sodium sulphite stock solution

Fresh sodium sulphite stock solutions were prepared by weighing sodium sulphite (0.5 g) and dissolving it in deionized water to make

100 mL of aqueous solution. Determination of the sulphite ion concentration was carried out iodometrically.

### Aqueous sodium nitroprusside (SNP)

A fresh stock solution of SNP (0.02 mol/L) was used. This was prepared by dissolving SNP crystals (6 g) in deionized water and the volume was made up to 1000 mL.

### Aqueous zinc acetate solution

Zinc acetate (220 g) was weighed into a volumetric flask and dissolved in deionized water. The solution was made up to 1000 mL.

### Aqueous ethylenediaminetetraacetic acid (EDTA) solution

The sodium salt of EDTA (0.0372 g) was dissolved in deionized water and the solution was made up to 10 mL.

### Spectrophotometric procedure for determination of sulphite ion

This was done as per our validated stabilization procedure. SNP (4.5 mL, 0.02 mol/L) was pipetted into in a 50 mL volumetric flask; EDTA (100 µl, 0.01 mol/L) was added followed by ethylenediamine (30 µl, 13.66 mol/L) and ethanol (10 mL). A solution of 0.025 g of gelatin in 5 mL of deionized water was added followed by perchloric acid to lower the pH to 6.5. An aliquot of intermediate standard aqueous $SO_3^{2-}$ followed by zinc acetate solution (0.5 mL, 1 mol/L) was added; the mixture was diluted to the mark by addition of an appropriate amount of deionized water and uniformly mixed for 5 s. The resultant red solution was immediately transferred to cuvettes which were inserted in the spectrophotometer for scanning its spectrum. A stable absorbance at $\lambda_{max}$ 482 nm was read off and then using a pre-prepared absorbance versus concentration calibration curve, the concentration of the sulphite ion was determined.

### Spectrophotometric determination of total sulphite in wine

Wine was treated to release matrix bound $SO_3^{2-}$ as follows: sodium hydroxide solution (1.6 mL, 4.0 mol/L) was added to 10 mL of the sample to release the bound $SO_3^{2-}$. The mixture was left to stand for 5 min and the pH of the solution was adjusted to 8.5 using sulphuric acid. SNP solution (4.5 mL, 0.02 mol/L) containing EDTA (4.5 µl, 0.01 mol/L) was pipetted into four different 50 mL volumetric flasks. Ethylenediamine (30 µl, 13.66 mol/L) was added to each flask followed by ethanol (10 mL) to stabilize $SO_3^{2-}$ in aqueous solution. A solution of 0.025 g of gelatin in 5 mL of deionized water was added (to avoid bubbling in the ethanol stabilized solutions) followed by treated wine (200 µl). Three of the flasks were then spiked with 6, 8 and 10 µg $SO_3^{2-}$ mL$^{-1}$. Zinc acetate solution (0.5 mL, 1 mol/L) was added; the mixture was diluted to the mark with deionized water and was uniformly mixed for 5 s. The resultant red solution was immediately transferred to a cuvette which was inserted in the spectrophotometer to read the absorbance. The concentration of the sulphite anion in the sample that was not spiked was obtained from absorbance versus concentration curve.

### Spectrophotometric determination of sulphite ion in sugar

SNP solution (4.5 mL, 0.02 mol/L) containing EDTA (4.5 µl, 0.01

mol/L) was pipetted into four different 50 mL volumetric flasks; ethylenediamine (30 µl, 13.66 mol/L) was added followed by ethanol (10 mL) to stabilize $SO_3^{2-}$ in aqueous solution. A solution of 0.025 g of gelatin in 5 mL of deionized water was added followed by a sugar solution (200 µl). Three of the flasks were then spiked with 6, 8 and 10 µg $SO_3^{2-}$ mL$^{-1}$. Zinc acetate solution (0.5 mL, 1 mol/l) was added; the mixture was diluted to the mark by addition of an appropriate amount of deionized water and was uniformly mixed for 5 s. The resultant red solution was immediately transferred to cuvettes which were inserted in the spectrophotometer to read the absorbance. The concentration of the sulphite anion in the sample that was not spiked with $SO_3^{2-}$ was obtained from absorbance versus concentration curve.

### Iodometric determination of sulphite ion

Standard iodine (10 mL, 0.0125 mol/L) was measured from a burette into a 250 mL conical flask. Sodium sulphite solution (2 mL) was added, the excess iodine was back titrated against standard $Na_2S_2O_3.5H_2O$ (0.025 mol/L) using starch as the indicator.

### Iodometric determination of sulphite ion in sugar

Sugar (1.0 g) was dissolved in deionized water to make 10 mL of solution. Starch (0.5 mL) was added and the resultant mixture was titrated against standard iodine solution.

### Iodometric determination of sulphite ion in wine

Wine (10 mL) was pipetted into a conical flask. Sodium hydroxide solution (1.6 mL, 4 mol/L) was added to release the bound $SO_3^{2-}$. The mixture was left to stand for 5 min. Sulphuric acid (1.7 mL, 10% v/v) was added followed by starch solution (0.5 mL). The resultant mixture was titrated against standard iodine solution.

### Method validation

A validation was carried out on the developed method and the following characteristics were evaluated; working range and linearity, accuracy, precision, selectivity and detection limits.

### Working range and linearity

The linearity of the method was evaluated by using calibrators in the entire working range of 1 to 10 µg mL$^{-1}$ for the analyte. The curve was a plot of the absorbance of the sulphite-nitroprusside reaction product against concentration. The regression equation with the slope, intercept and correlation coefficient ($r^2$) was generated using Microsoft Excel. Limit of quantification formed the lower end of the working range. The linearity was established using the square of correlation coefficient value ($r^2$) of the line of best fit.

### Accuracy and precision

Accuracy was determined by calculating the mean recovery of the seven portions spiked with standard $SO_3^{2-}$ solutions at three concentration levels, all within the working range. The accuracy was then expressed as;

$$\text{Accuracy} = \frac{y}{z} \times 100\%$$

**Table 1.** The accuracy and precision of the developed nitroprusside method at three concentration levels.

| Nominal concentration (µg $SO_3^{2-}$ $mL^{-1}$) | Concentration measured (µg $SO_3^{2-}$ $mL^{-1}$)*( mean ± SD) | CV (%; n = 7) |
|---|---|---|
| 1.521 | 1.466 ± 0.035 | 2.368 |
| 2.500 | 2.457 ± 0.046 | 1.888 |
| 5.031 | 4.846 ± 0.098 | 2.028 |

*Mean ± SD, at 95% confidence interval.

where y was the mean value of the concentration of the seven replicates and x was the spiked (nominal) concentration. The intra-batch variability in the measurement of the sulphite ion concentration (precision) was calculated from seven repeat determinations of spiked samples and was expressed as a coefficient of variation (CV).

### Selectivity

Several cationic and anionic species such as sulphide, thiosulphate, hydrogen phosphate, chloride, copper (II), iron (II) potassium and sodium ions were tested for interference. Different amounts of the ionic species were added to a solution containing sulphite ions of known concentration and the extent of interference caused by each species was investigated.

### Detection and quantification limits

The limit of detection (LoD) was defined as the lowest concentration of sulphite ion in a sample that could be detected but not necessarily quantified under stated conditions of the developed method. In this work, LoD was determined by analyzing 10 independent sample blanks for sulphite ion amount and the standard deviation (SD) among the values determined was calculated. LoD was then expressed as the sulphite ion concentration corresponding to the mean sample blank value + 3SD

The limit of quantification (LoQ) was defined as the lowest concentration of sulphite ion that could be determined with an acceptable level of repeatability, precision and trueness. In this work, LoQ was determined by analyzing 10 independent sample blanks for sulphite ion amount and SD was calculated. LoQ was then expressed as the sulphite ion concentration corresponding to the mean sample blank value + 10SD.

## RESULTS AND DISCUSSION

A comparison of sulphite content in wine and sugar using both NP and an iodometric titration method was made and the results are shown in Tables 3 and 4.

A lot of attention has been drawn to the concentration of sulphite ions in environmental systems because of their associated health effects. A method that is simple but accurate is therefore essential for monitoring the levels of the sulphite ions in such systems.

### Method development and validation

The optimum conditions chosen for this work were those that had been previously validated in our laboratory (Musagala, Busitema University, unpublished); in which ethylenediamine in the presence of zinc was found to stabilize the sulphite-nitroprusside reaction product for a period of 30 min. The range of ethylenediamine concentration used was 5.5 to 22 mmol/L, above this concentration the absorbance was further enhanced but the product was found to be very unstable. The spectrophotometric determination of the $SO_3^{2-}$ from its reaction with NP was carried out in aqueous slightly acidic media by using ethylenediamine and zinc concentrations of 11 and 10 mmol/L, respectively. This was done without alkali metal cations. Absorbance readings were made after a period of 5 to 10 min.

Mbabazi et al. (2011) noted that an excess of NP not greater than ten-fold is sufficient for analysis since the absorbance increases with NP concentration. A similar observation was also noted in this study. Again, it was noted that temperature, pH and the ratio of excess NP to $SO_3^{2-}$ exerted a significant effect on stability of the product of the NP-$SO_3^{2-}$ reaction. An optimum pH of 6.5 was found to be satisfactory for our purpose.

### Working range and linearity

The working range was found to lie between 1 and 10 µg $SO_3^{2-}$ $mL^{-1}$; above the upper limit, the decomposition of the product was relatively fast and often characterized by precipitation with time. The linearity of the method as measured by the correlation coefficient of inter-assay linear regression curves ($r^2$) was better than 0.99 in all cases in the measured range of 1-10 µg $SO_3^{2-}$ $mL^{-1}$ which was indicative of a good linear relationship between $SO_3^{2-}$ concentration and absorbance.

### Accuracy and precision

A comparison of the nominal concentration of the spiked solutions for seven replicates at three concentration levels with the concentration measured at each of the levels showed very good accuracy at all the three concentration levels (Table 1), an indication that the developed method was fit for the intended purpose. The precision of the method was also very good (CV not more than 2.368) in all cases, again suggesting that the method was precise in the concentration levels considered.

### Selectivity

The effect of some cations and anions on the method

**Table 2.** The effect of a number of anions and cations on the nitroprusside method for $SO_3^{2-}$ in the presence of ethylenediamine, using 3.10 µg $SO_3^{2-}$ $mL^{-1}$ for testing.

| Foreign ion added | Interference concentration (µg/mL) | $SO_3^{2-}$ recovered (µg/mL)* |
|---|---|---|
| $S^{2-}$ | 1000 | ND |
| $S_2O_3^{2-}$ | 1000 | ND |
| $HPO_4^{2-}$ | 1000 | ND |
| $Cl^-$ | 1000 | 3.05 ±0.02 |
| $Cu^{2+}$ | 100 | ND |
| $Fe^{2+}$ | 100 | ND |
| $K^+$ | 1000 | 3.05 ±0.02 |
| $Na^+$ | 1000 | 3.05 ±0.02 |

*Mean ± SD, at 95% confidence interval; ND = Not detected.

**Table 3.** Comparison of the developed nitroprusside method to iodometric titration method for the determination of sulphite in wine.

| Type of wine | Concentration of $SO_3^{2-}$ $µg^{-1}$ $mL^{-1}$ | |
|---|---|---|
| | Nitroprusside method | Iodometric titration |
| | 65.0 ± 6.6 | 70.0 ± 8.4 |
| | 68.0 ± 6.6 | 70.0 ± 8.4 |
| Baron de Vignon semi-dry white wine | 58.0 ± 6.6 | 60.0 ± 8.4 |
| | 60.0 ± 6.6 | 60.0 ± 8.4 |
| | 50.0 ± 6.6 | 50.0 ± 8.4 |
| | 47.0 ± 8.4 | 50.0 ± 8.4 |
| | 68.0 ± 8.4 | 70.0 ± 8.4 |
| Bellingham dry white wine | 58.0 ± 8.4 | 60.0 ± 8.4 |
| | 60.0 ± 8.4 | 60.0 ± 8.4 |
| | 50.0 ± 8.4 | 50.0 ± 8.4 |
| | 69.0 ± 6.9 | 70.0 ± 7.4 |
| | 50.0 ± 6.9 | 50.0 ± 7.4 |
| King fisher strawberry fruit red wine | 63.0 ± 6.9 | 65.0 ± 7.4 |
| | 60.0 ± 6.9 | 60.0 ± 7.4 |
| | 59.0 ± 6.9 | 60.0 ± 7.4 |

*Mean ± SD, at 95% confidence interval.

was studied in detail by adding different amounts of ionic species as shown in Table 2. The greatest anionic interference to this method would be expected to come from $S^{2-}$ that reacts with NP in a similar manner to produce a red product ($[Fe(CN)_5(NOS)]^{4-}$) with $\lambda_{max}$ 538 nm at pH above 10. Sulphide also forms a sparingly soluble precipitate with $Zn^{2+}$ in the reaction mixture. Other ionic species were also found to interfere during the determination of sulphite as indicated in Table 2. The presence of oxidizable species such as sulphide, thiosulphate, phosphate, and iron may lead to high levels of sulphite. Copper is known to catalyse oxidation of sulphite to sulphate when the sample is exposed to air leading to low results.

However, the effect of copper can be avoided by adding a complexing agent such as EDTA during sample collection as this would inhibit copper (II) catalysis and promote oxidation of iron (II) to iron (III) before analysis. Sulphide can be removed by adding about 0.5 g of zinc acetate and thereafter analyzing the supernatant of the settled sample while thiosulphate can be determined independently using a simple iodometric titration.

It is therefore recommend that the method be employed mainly for the determination of sulphite in wine and sugar, but its application could be extended to relatively clean waters as is always the case with the iodometric

**Table 4.** Comparison of the developed nitroprusside method with iodometric titration method for the determination of sulphite in sugar.

| Type of sugar | Concentration of $SO_3^{2-} mg^{-1} kg^{-1}$ | |
| --- | --- | --- |
| | Nitroprusside method | Iodometric titration |
| Kinyala white sugar | $28.0 \pm 5.4$ | $30.0 \pm 5.5$ |
| | $29.0 \pm 5.4$ | $30.0 \pm 5.5$ |
| | $28.0 \pm 5.4$ | $30.0 \pm 5.5$ |
| | $18.0 \pm 5.4$ | $20.0 \pm 5.5$ |
| | $19.0 \pm 5.4$ | $20.0 \pm 5.5$ |
| Kakira white sugar | $18.0 \pm 5.9$ | $20.0 \pm 5.5$ |
| | $27.0 \pm 5.9$ | $30.0 \pm 5.5$ |
| | $29.0 \pm 5.9$ | $30.0 \pm 5.5$ |
| | $30.0 \pm 5.9$ | $30.0 \pm 5.5$ |
| | $18.0 \pm 5.9$ | $20.0 \pm 5.5$ |
| Kenya Brown Sugar | $30.0 \pm 5.0$ | $30.0 \pm 5.5$ |
| | $21.0 \pm 5.0$ | $20.0 \pm 5.5$ |
| | $19.0 \pm 5.0$ | $30.0 \pm 5.5$ |
| | $28.0 \pm 5.0$ | $20.0 \pm 5.5$ |
| | $29.0 \pm 5.0$ | $30.0 \pm 5.5$ |

*Mean ± SD, at 95% confidence interval.

method for the determination of sulphite in environmental samples (American Public Health Association, 1998).

Fogg et al. (1966) reported that several organic compounds, including thiols, amines, and ketones were also known to form coloured compounds with the NP ion, but usually only in strongly alkaline solution. However, at pH of 6.5 used for the nitroprusside-sulphite reaction, the aforementioned compounds did not interfere. Tin(II), ferrocyanide and arsenates gave white precipitates with the reagents, but these also did not interfere with the sulphite determination.

### Detection and quantification limits

The LoD and LoQ for the NP-$SO_3^{2-}$ reaction method were found to be 0.99 $\mu gSO_3^{2-} mL^{-1}$ (Blank + 3SD) and 2.321 $\mu gSO_3^{2-} mL^{-1}$ (Blank + 10SD), respectively. The obtained detection limits showed that the method could be applied in the detection and quantification of $SO_3^{2-}$ concentrations as low as 1 $\mu g mL^{-1}$.

### Applicability of the developed method for determination of sulphite ion in wine and sugar

The total sulphite content in wines is the sum of the free and bound sulphite. Usually, information about the free sulphite content rather than the total sulphite content is preferred. Araujo et al. (2005) noted that the equilibrium between bound and free sulphite was rather labile and any change in composition of the wine like dilution inevitably shifted the equilibrium making the determination of free sulphite difficult. In this work, relatively high values of sulphite were obtained for red wine and this was attributed to the colour of the wine.

### Conclusion

A manual spectrophotometric method has successfully been developed and validated for the quantification of micro quantities of sulphite ions based on the modified reaction conditions of the nitroprusside-sulphite ion (NP-$SO_3^{2-}$) reaction.

The method in the determination of sulphite ions has also been applied in selected environmental systems such as wine and sugar. The results obtained by the developed method compared well with those obtained using an iodometric titration method for $SO_3^{2-}$ determination.

### ACKNOWLEDGEMENT

The authors acknowledge with gratitude the financial support of the Germany Academic Exchange Service (DAAD).

### REFERENCES

Al-Tamrah SA (1987). Flow injection chemiluminescence determination of sulphite Analyt. P 112.
American Public Health Association (1998). In: Clesceri LS, Greenberg AE, Eaton D (eds.), Standard Methods for the Examination of Water

and Wastewater. pp. 4-173.

Andrade C, Swinehart JH (1972). The Reactions of Pentacyanonitrosylferrate(III) with Bases., The Boedeker Reaction Inorg. Chem. 11(3):648-650.

Araujo AN, Couto CM, Lima JL, Montenegro MC (1998). Determination of $SO_2$ in Wines Using a Flow Injection Analysis System with Potentiometric Detection. J. Agric. Food. 46:168-172.

Araujo CST, de Carvalho JL, Mota DR, Coelho NMM (2005). Determination of sulphite and acetic acid in foods by gas permeation flow injection analysis. Food Chem. 92(4):765-770.

Atanassov GT, Lima RC, Mesquita RBR, Rangel ASS, Tóth IV (2000). Spectrophotometric determination of carbon dioxide and sulphur dioxide in wines by flow injection. Analusis 28:77-82.

Badri B (1988). Spectrophotometric Determination of Sulphite, Sulphate and Dithionate in the Presence of Each Other. The Analyst 113: 351-353.

Claudia RC, Francisco JC (2009). Application of flow injection analysis for determining sulphite in food and beverages: Rev. Food Chem. 112(2):487-493.1

Crompton TR (1996). Determination of Anions, in a Guide for the Analytical Chemist. Springer-Verlag Berlin Heidberg, New York. P 311.

Cunniff P (1995). Official Methods of Analysis of AOAC International, 16 ed. Arlington, Virginia, USA: AOAC International Chapter 47, pp. 27-29.

Edmond LM, Magee EA, Cummings JH (2003). An IEC Method for Sulphite and Sulphate Determination in Wine without Predistillation. LCGC Europe. February:2-6.

Fogg AG, Moser W, Chalmers RA (1966). The Boedeker reaction, Part III. The detection of sulphite ion. Anal. Chim. Acta 36:248-251.

Hasson M, Spohn U (2001). Sensitive and selective flow injection analysis of hydrogen sulfite/sulfur dioxide by fluorescence detection with and without membrane separation by gas diffusion. Anal. Chem. 73:3187-3192.

Humphrey RE, Hinze W (1971). Spectrophotometric determination of cyanide, sulphide, and sulfite with mercuric chloranilate. Anal. Chem. 48:1100.

Isaac A, Livingstone C, Wain AJ, Compton RG, Davis J (2006). Electroanalytical methods for the determination of sulfite in food and beverages. Trends Analyt. Chem. 25(6):589-598.

Koch M, Koppen R, Siegel D, Witt A, Nehls I (2010). BAM Federal Institute for Materials Research and Testing, Richard-Willstaetter-Strasse 11, D-12489 Berlin, Germany. J. Agric. Food Chem. 58 (17):9463–9467.

Koukli II, Sarantoris EG, Calkerinost AC (1988). Continuous flow chemiluminescence determination of sulphite and sulphur dioxide. Analyst 113:603-608.

Leeuwenkamp GR, Vermaat CH, Bult A (1984). Specific cation effect in the reaction of nitroprusside with cystine, acetophenone and sulphide (Legal and Boedeker reaction). Pharmaceutisch Weekblad 6:18-20.

Machado RMD, Toledo MCF, Almeida CAS Vicente E (2008). Analytical determination of sulphites in fruit juices available on the Brazilian market. Braz. J. Food Technol. 11(3):226-233.

Marshall GB, Midgley D (1983). Potentiometric Determination of Sulphite by Use of Mercury(I) Chloride –Mercury(II) Sulphide Electrodes in Flow Injection Analysis and Air-gap Electrodes. Anal. 108:701-711.

Mbabazi J, Yiga S, Ssekaalo H (2011). Evaluation of environmental sulphide by stabilisation of the initial product of the pentacyanonitrosylferrate(II)-sulphide reaction. J. Toxicol. Environ. Health Sci. 3(4):95-100.

McFeeters RF, Barish AO (2003). Sulphite Analysis of Fruits and Vegetables by High-Performance Liquid Chromatography (HPLC) With Ultraviolet Spectrophotometric Detection. Agric. Food Chem. 51(6):1513-1517.

Schubert SA, Wesley J, Fernando Q (1979). Determination of Sulphite and Sulphate in Solids by Time-Resolved Molecular Emission Spectrometry. Anal. Chem 51(8):1297-1301.

Shahine S, Ismael N (1979). Direct Titrimetric Determination of Thiosulphate, Sulphite and Tin(II) With Copper Sulphate Solution. Mikrochica Acta (Wien) 1:371-374.

Smith VJ (1987). Determination of sulphite using a sulphite oxidase enzyme electrode Anal. Chem. 59(18):2256-2259.

Thanh NTK, Decnop-Weever, Kok WT (1994). Determination of sulphite in wine by flow injection analysis with indirect electrochemical detection. Fresenius J. Anal. Chem. 349:469-472.

Xiaoli S, Wei W (1998). Flow injection determination of sulfite in wines and fruit juices by using a bulk acoustic wave impedance sensor coupled to a membrane separation technique. Analyst 123:221-224.

# 13

# Determination of median lethal concentration (LC$_{50}$) of copper sulfate and lead nitrate and effects on behavior in Caspian sea kutum (*Rutilus frisii kutum*)

**Esmail Gharedaashi\*, Mohammad Reza Imanpour and Vahid Taghizadeh**

Department of Fishery, Gorgan University of Agricultural Sciences and Natural Resources, Gorgan, Iran.

The aim of present study was to determine the LC$_{50}$ value of in Caspian sea kutum. The results indicated that median lethal concentration (LC$_{50}$) of Caspian sea kutum (*Rutilus frisii kutum*) for 24, 48, 72 and 96 h of exposure are 2.944, 2.756, 2.562 and 2.310 ppm, respectively and median lethal concentration (LC$_{50}$) of lead to Caspian sea kutum (*R. frisii kutum*) for 24, 48, 72 and 96 h of exposure as 315.841, 298.456, 281.419 and 268.065 ppm, respectively. LC$_{50}$ increased with decrease in mean exposure times for both metals. Physiological responses like rapid opercular movement and frequent gulping of air was observed during the initial stages of exposure after which it became occasional. All these observations can be considered to monitor the quality of aquatic ecosystem and severity of pollution. Hence, concluded that copper is more toxic than lead for Caspian sea kutum (*R. frisii kutum*).

**Key words:** Copper sulfate, lead nitrate, *Rutilus frisii kutum*, Caspian sea kutum, physiological responses, median lethal concentration (LC$_{50}$).

## INTRODUCTION

Heavy metal pollution in water is in large part due to agricultural run-off, industrial waste and mining activities. Mining is by far the biggest contributor to metal pollution. Mine drainage water, effluent from the tailing ponds and drainage water from soil heaps continue to extrude unwanted metals into the aquatic environment (Rani and Sivaraj, 2010). Metal concentrations in aquatic organisms appear to be of several magnitudes higher than concentrations present in the ecosystem (Laws, 2000) and this is attributed to bioaccumulation whereby metal ions are taken up from the environment by the organism and accumulated in various organs and tissues. Metals also become increasingly concentrated at higher trophic levels, possibly due to food-chain magnification (Wyn et al., 2007).

Coastal seawater is easily contaminated by heavy metals due to human activities with heavy metal contamination reported in aquatic organisms (Olojo et al., 2005). The problem has become more serious for aquatic species that live close to the coastline where heavy metals tend to accumulate (Migliarini et al., 2005). The fact that there is increasing use of contaminating chemicals in many industrialized parts of the world makes the development of Eco toxicity measurement techniques an absolute necessity (Brandão et al., 1992). Heavy metal contamination severely interfere with ecological balances of an ecosystem and produces devastating effects on environment quality; anthropogenic inputs like waste disposal directly adds to the burden of environmental degradation (Farombi et al., 2007)

In recent years, toxicological studies have gained a fresh momentum and have emerged as a major field of research owing to the gravity of the situation and increasing diversity of aquatic pollutants. The toxicity of

*Corresponding author. E-mail: gharedaashi.e@gmail.com.

this metal on aquatic organisms is influenced by chemical features of water, such as pH and hardness (Mance, 1987) and its bioaccumulation is directly related to its concentration in seawater (Sadik, 1992).

Assessment of toxicity on particular organism exposed to a particular toxicant will reveal facts regarding the health of given ecosystem and would eventually help us to propose policies to protect the ecosystem. Toxicity tests will reveal the organism's sensitivity to a particular toxicant that would help us to determine the permissible limit of a toxicant in an ecosystem. Heavy metals such as mercury and lead have gained wide interest in the scientific community in recent years due to their potential human health hazards (Shuhaimi-Othman et al., 2010) and Copper is a very important element which could influence the metabolism of the human body and it is also a nutritional element for living beings. But if the intake is too much, it will cause toxicity (Fan et al., 2002).

The toxicity of any pollutant is either acute or chronic. Although the toxicant impairs the metabolic and physiological activities of the organisms, physiological studies alone do not satisfy the complete understanding of pathological conditions of tissues under toxic stress. All toxicants are capable of severally interfering with the biological systems thus producing damage to the structure and function of particular organism and ultimately to its survival (Rani et al., 2011).

These toxicity tests are necessary to predict the safe concentration of the chemicals in the environment (Johnson and Bergman, 1983). The first step is the acute toxicity test on algae, fish, etc; in order to show the potential risks of these chemical materials (OECD, 1993). Acute toxicity test constitute only one of the many tools available to the aquatic toxicologists but they are the basic means of provoking a quick, relatively inexpensive and reproducible estimate of the toxic effects of a test material (Spacie and Hamelink, 1985). The 96 h $LC_{50}$ tests were conducted to measure the susceptibility and survival potential of organisms to particular toxic substances such as heavy metals. Higher $LC_{50}$ values are less toxic because greater concentrations are required to produce 50% mortality in organisms (Hilmy et al., 1985). Majority of the studies concerning the effects of heavy metals on fish have been confined to the acute toxicity test with the death of fish as an end point. Hence in the present study, an attempt has been made to assess the acute toxicity of copper and lead on Caspian sea kutum.

## MATERIALS AND METHODS

### Place, prepared fish and condition to experiment

Metal toxicity tests were conducted in the laboratory conditions. Juvenile Caspian sea kutum selected for this study were obtained from the fish seed hatchery in Gorgan, Iran. Caspian sea kutum

measuring $8 \pm 0.5$ cm in length and weighing $4 \pm 0.5$ g were used for the experiment. They were brought to the laboratory and acclimatized for 14 days and the fish were fed with commercial pelleted food at least once a day during this period. All glassware and aquariums used in this experiment were washed and thoroughly rinsed with deionized water prior to use. Prior to each trial, all aquariums (60 L) capacity, were filled with 50 L of dechlorinated tap water. A total of 27 aquariums, each stocked with 10 fishes were used in our experiments for each metal. Stock solutions of copper sulfate and lead nitrate were prepared by dissolving analytical grade copper sulfate ($CuSO_4.5H_2O$ from Merck) and lead nitrate ($Pb(No_3)_2$ from Merck), respectively in double distilled water. 30 fishes were used per concentration of each heavy metal. Separate groups of 30 fish each served as controls for copper and lead. The physico-chemical characteristics of the water were analyzed as per the procedure of APHA (1998). The mean values for the water qualities tested were as follows, temperature $24 \pm 1\,°C$, pH 7 to 7.5, dissolved oxygen $7.8 \pm 0.2$ mg/L and the experimental medium was aerated in order to keep the amount of oxygen not less than 4 mg/L, hardness $275 \pm 3.58$ mg/L as $CaCo_3$, photoperiodicity 12 L: 12 D, turbidity 2.

### Bioassay $LC_{50}$

Ninety-six hours acute bioassays were performed following in general OECD guidelines for fish acute bioassays (guideline OECD203, 92/69/EC, method C1) (OECD, 1993). For determination of the $LC_{50}$ (lethal concentration) values, following a range finding test, eight Cu (0.5, 0.75, 1, 1.5, 2, 2.5, 3 and 3.5 mg/L) and Pb (100, 200, 220, 240, 260, 280, 300 and 320 mg/L) concentrations were chosen for Caspian sea kutum. For each metal-treated and Control, tree replications were conducted. Metal solutions were prepared by dilution of a stock solution with dechlorinated tap water. A control with dechlorinated tap water only was also used. The number of dead fish was counted every 12 h and removed immediately from the aquaria. The mortality rate was determined at the end of 24, 48, 72 and 96 h. During the toxicity test, the fishes were not fed. Acute toxicity test was conducted in accordance with standard methods (APHA, 1998).

### Statistical analysis

In this study, the acute toxic effect of copper sulfate and lead nitrate on the Caspian sea kutum was determined by the use of Finney's Probit Analysis $LC_{50}$ determination method (Finney, 1971). Confidential limits (Upper and Lower) were calculated and SPSS18 was also used for $LC_{50}$ value of copper sulfate and lead nitrate with the help of probit analysis.

### Behavior observation

Physiological responses like rapid opercular movement and frequent gulping of air was observed during the initial stages of exposure after which it became occasional.

## RESULTS

Acute toxicity of copper and lead showed that mortality is directly proportional to the concentration of the heavy metal copper and lead while the percentage of mortality

**Table 1.** Showing correlation between the copper concentration and the mortality rate on time (24 – 96 h) of Kutum.

| Concentration (mg $L^{-1}$) | N | Mortality rate on time (24 - 96 h) | | | |
|---|---|---|---|---|---|
| | | 24 | 48 | 72 | 96 |
| 0.00 | 30 | 0 | 0 | 0 | 0 |
| 0.50 | 30 | 0 | 0 | 0 | 0 |
| 0.75 | 30 | 0 | 0 | 0 | 0 |
| 1.00 | 30 | 0 | 0 | 0 | 1 |
| 1.50 | 30 | 1 | 1 | 2 | 3 |
| 2.00 | 30 | 4 | 5 | 6 | 8 |
| 2.50 | 30 | 11 | 13 | 15 | 18 |
| 3.00 | 30 | 16 | 19 | 21 | 26 |
| 3.50 | 30 | 21 | 24 | 28 | 30 |

is virtually absent in control (Tables 1 and 3).

## LC$_{50}$ of cooper for kutum

Table 1 shows the relation between the copper concentration and the mortality rate for 24, 48, 72 and 96 h of kutum. Results according to SPSS18 analysis showed that the median lethal concentration (LC$_{50}$) of copper to kutum for 24, 48, 72 and 96 h of exposure are 2.944, 2.756, 2.562 and 2.310 ppm, respectively. A gradual decrease in slope function corresponding to an increase in the exposure period from 24 to 96 h was noticed. Observations on the upper and lower confidence limits revealed a decreasing trend from 24 to 96 h (Table 2).

## LC$_{50}$ of lead for kutum

Susceptibility of kutum to the impact of lead toxicity was found to increase in mortality with an increase in the concentration of lead whereas in the control, mortality was virtually absent (Table 3).

Results according to SPSS18 analysis showed the median lethal concentration (LC$_{50}$) of lead to kutum for 24, 48, 72 and 96 h of exposure as 315.841, 298.456, 281.419 and 268.065 ppm, respectively. There was a gradual decrease in the slope function corresponding to the increase in the exposure period from 24 h to 96 h. Observations on upper and lower confidence limits reveal a decreasing trend from 24 to 96 h. Also evident was that an increase in exposure period influences increase in mortality (Table 4).

## Behavioral changes

The behavior of fish remarkably changed due to the treatment of lead and copper when compared to the control. The various locomotary responses exhibited by fish due to sub lethal concentrations of lead and copper during initial stage of exposure included restlessness, erratic and fast swimming, abrupt change in position and direction, jumping and overall hyperactivity were noticed. The fish showed surfacing tendency throughout the experimental period. Physiological responses like rapid opercular movement and frequent gulping of air was observed during the initial stages of exposure after which it became occasional. Neurological symptoms like jerking movements, frightening and loss of balance were not observed in lead and copper treated kutum. Hence it was concluded that copper is more toxic than lead for kutum.

## DISCUSSION

The present study was initiated to find the susceptibility of the kutum to potentially hazardous heavy metals like copper and lead. Median lethal concentration (LC$_{50}$) of copper to kutum for 24, 48, 72 and 96 h of exposure are 2.944, 2.756, 2.562 and 2.310 ppm, respectively and median lethal concentration (LC$_{50}$) of lead to kutum for 24, 48, 72 and 96 h of exposure were 315.841, 298.456, 281.419 and 268.065 ppm, respectively. Higher percent of mortality occurred with increase in concentration and exposure period of copper and lead.

The median lethal concentration 96 h (LC$_{50}$) value of copper and lead in other aquatic organisms was reported as 300 ppm for lead as in *Tench tinca* by Shah and Altindag (2005), which were higher than present study. The LC$_{50}$ for *R. sumatrana*, for 24, 48, 72 and 96 h for Cu were 54.2, 30.3, 18.9 and 5.6 µg/L and For *P. reticulata*, LC$_{50}$ for 24, 48, 72 and 96 h for Cu were 348.9, 145.4, 61.3 and 37.9 µg/L, respectively (Shuhaimi-Othman et al., 2010), which were lower than present study. The 24 h LC$_{50}$ of Cu was reported as 1.17 mg/L for *P. reticulate* (Park and Heo, 2009), which were lower than present study. Gomes et al. (2009) reported that with juvenil

**Table 2.** Lethal concentration ($LC_{1-99}$) of copper on time (24 - 96 h) for Kutum.

| Point | Concentration (mg L$^{-1}$), (95% confidence limits) | | | | | | | |
|---|---|---|---|---|---|---|---|---|
| | 24h | | 48h | | 72h | | 96h | |
| $LC_1$ | 1.023 | (0.414-1.396) | 1.036 | (0.502-1.375) | 1.038 | (0.572-1.342) | 0.928 | (0.522-1.201) |
| $LC_5$ | 1.586 | (1.158-1.861) | 1.540 | (1.153-1.795) | 1.484 | (1.140-1.717) | 1.333 | (1.028-1.545) |
| $LC_{10}$ | 1.886 | (1.547-2.116) | 1.808 | (1.495-2.025) | 1.722 | (1.438-1.922) | 1.549 | (1.294-1.733) |
| $LC_{15}$ | 2.088 | (1.802-2.295) | 1.989 | (1.720-2.184) | 1.883 | (1.635-2.064) | 1.695 | (1.471-1.862) |
| $LC_{50}$ | 2.944 | (2.753-3.183) | 2.756 | (2.586-2.949) | 2.562 | (2.406-2.727) | 2.310 | (2.165-2.463) |
| $LC_{85}$ | 3.800 | (3.502-4.272) | 3.522 | (3.280-3.884) | 3.240 | (3.037-3.530) | 2.926 | (2.745-3.176) |
| $LC_{90}$ | 4.003 | (3.670-4.539) | 3.703 | (3.433-4.116) | 3.410 | (3.176-3.731) | 3.072 | (2.873-3.355) |
| $LC_{95}$ | 4.303 | (3.915-4.939) | 3.971 | (3.657-4.463) | 3.639 | (3.378-4.032) | 3.288 | (3.058-3.623) |
| $LC_{99}$ | 4.865 | (4.370-5.692) | 4.475 | (4.072-5.120) | 4.085 | (3.751-4.620) | 3.693 | (3.401-4.131) |

**Table 3.** Showing correlation between the lead concentration and the mortality rate on time (24 - 96 h) of Kutum.

| Concentration (mg L$^{-1}$) | N | Mortality rate on time(24 – 96 h) | | | |
|---|---|---|---|---|---|
| | | 24 | 48 | 72 | 96 |
| 0 | 30 | 0 | 0 | 0 | 0 |
| 100 | 30 | 0 | 0 | 0 | 0 |
| 200 | 30 | 0 | 0 | 0 | 0 |
| 220 | 30 | 0 | 0 | 0 | 1 |
| 240 | 30 | 1 | 2 | 3 | 5 |
| 260 | 30 | 3 | 5 | 8 | 13 |
| 280 | 30 | 6 | 9 | 15 | 19 |
| 300 | 30 | 10 | 16 | 21 | 25 |
| 320 | 30 | 16 | 21 | 27 | 30 |

**Table 4.** Lethal concentration ($LC_{1-99}$) of lead on time (24 - 96 h) for Kutum.

| Point | Concentration (mg L$^{-1}$), (95% confidence limits) | | | |
|---|---|---|---|---|
| | 24 h | 48 h | 72 h | 96 h |
| $LC_1$ | 222.078 (185.743-240.943) | 214.293 (185.793-231.077) | 212.018 (190.220-225.967) | 205.120 (185.599-217.770) |
| $LC_5$ | 249.546 (226.223-262.604) | 238.948 (219.198-251.136) | 232.349 (216.356-242.938) | 223.483 (208.958-233.332) |
| $LC_{10}$ | 264.189 (247.095-274859) | 252.092 (236.650-262.186) | 243.187 (230.089-252.185) | 233.330 (221.249-241.790) |
| $LC_{15}$ | 274.068 (260.516-282.789) | 260.960 (248.134-269.932) | 250.499 (239.211-258.568) | 239.973 (229.429-247.610) |
| $LC_{50}$ | 315.841 (304.800-334.010) | 298.456 (290.175-309.199) | 281.419 (274.663-288.673) | 268.065 (261.706-274.525) |
| $LC_{85}$ | 358.615 (338.214-395.102) | 335.953 (322.483-358.199) | 312.338 (303.206-325.688) | 296.156 (288.299-307.123) |
| $LC_{90}$ | 367.494 (345.807-409.860) | 344.821 (329.696-370.216) | 319.651 (309.453-334.945) | 302.800 (294.088-315.333) |
| $LC_{95}$ | 382.137 (356.981-431.814) | 357.964 (340.277-388.136) | 330.489 (318.572-348.806) | 312.647 (302.517-327.653) |
| $LC_{99}$ | 409.604 (377.783-473.152) | 382.620 (359.925-421.952) | 350.489 (335.423-375.062) | 331.118 (318.050-351.041) |

Brazilian indigenous fishes, curimata (*Prochilodus vimboides*) and piaucu (*Leporinus macrocephalus*), 96 h $LC_{50}$ of copper were 0.047 and 0.090 mg/L for curimatã and piauçu, respectively, which were lower than present study. This indicates that different organisms have different sensitivity to heavy metals. The toxicity reported by other studies differs from this study probably due to different species used, age, size of the organism, test methods and water quality such as water hardness, as this can affect toxicity (Hodson et al., 1982; McCahon and Pascoe, 1988). Toxicity of metals may vary depending upon their permeability and detoxification mechanisms (Darmono et al., 1990).

The behavior of fish remarkably changed including restlessness, erratic and fast swimming, abrupt change in position and direction, jumping and overall hyperactivity which were noticed. The fish showed surfacing tendency throughout the experimental period. Physiological responses like rapid opercular movement and frequent gulping of air was observed during the initial stages of exposure after which it became occasional. Neurological symptoms like jerking movements, frightening and loss of balance were not observed in lead and copper treated kutum.

The effect of copper on the gill morphology, directly inducing necrosis, hypertrophy, hyperplasia and high mucus production (Mazon et al., 2002a; Fernandes and Mazon, 2003) and indirectly stimulating proliferation of chloride cells in the secondary lamellae through cortisol (Bonga, 1997), reduces the effectiveness of the respiratory surface, resulting in respiratory impairment.

## Conclusion

In the present study, a comparison of $LC_{50}$ values and behavior change indicated that copper was more toxic than lead to fishes. The results of these studies may provide guidance to selection of acute toxicity to be considered in field biomonitoring efforts designed to detect the bioavailability of lead nitrate and copper sulfate and early warning indicators of this heavy metal toxicity in Caspian Sea kutum.

## ACKNOWLEDGEMENTS

The authors thank the Aquaculture Research Center and Fishery group for the supply of research material. This work was supported by the Gorgan University of Agricultural Sciences and Natural resources. The authors are also grateful to Dr. Imanpour which deeply contributed with valuable comments and suggestions.

## REFERENCES

APHA (1998). Standard Methods for the Examination of Water and Waste Waters. 20th ed. American Public Health Association, Washington DC.

Bonga SEW (1997). The stress response in fish. Physiol. Rev. 77(3):591-625.

Brandão JC, Bohets HHL, Van De Vyver IE, Dierickx PJ (1992). Correlation between the in vitro cytotoxicity to cultured fathead minnow fish cells and fish lethality data for 50 chemicals. Chemosphere 25(4):553-562.

Fan HJ, Zhang W, Lin YX, Lai XK (2002). Speciation analysis for trace copper in Pearl River. J. Anal. Sci. 18(6):496-498.

Farombi EO, Adelowo OA, Ajimoko YR (2007). Biomarkers of oxidative stress and heavy metal levels as indicators of environmental pollution in African cat fish (Clarias gariepinus) from Nigeria Ogun River. Int. J. Environ. Res. Public Health 4(2):158-165.

Darmono D, Denton GRW (1990). The pathology of cadmium and nickel toxicity in the banana shrimp (Penaeus merguiensis dE Man). Asian Fish Sci. 3(3):287-297.

Fernandes MN, Mazon AF (2003). Environmental pollution and fish gill morphology. Fish adaptations. Sci. Publishers, Enfield. pp. 203-231.

Finney D (1971). Probit Analysis Cambridge University Press. Cambridge, UK.

Gomes LC, Chippari-Gomes AR, Oss RN, Fernandes LFL, Magris RDA (2009). Acute toxicity of copper and cadmium for piauçu, Leporinus macrocephalus, and curimatã, Prochilodus vimboides-DOI: 10.4025/actascibiolsci. v31i3. 5069." Acta Sci. Biol. Sci. 31(3):313-315.

Hodson PV, Dixon DG, Spry DJ (1982). Effect of Growth Rate and Size of Fish on Rate of Intoxication by Waterfoorne Lead. Can. J. Fish. Aquat. Sci. 39(9):1243-1251.

Hilmy AM, Shabana MB, Daabees AY (1985). Bioaccumulation of cadmium: toxicity in Mugil cephalus. Comparative Biochemistry and Physiology Part C: Comp. Pharmacol. 81(1):139-144.

Johnson RD, Bergman HL (1983). Use of histopathology in aquatic toxicology-a critic. In: Caims VW, Hadson PV, Nriagy JO (Eds.), Contaminant effects on fisheries. 19-36.

Laws E (2000). Aquatic Pollution - An introductory text. John Wiley and Sons, New York, U.S.A. pp. 309-430.

Mance G (1987). Pollution threat of heavy metals in aquatic environments Pollution monitoring series, Applied Science. Elsevier, Amsterdam/New York. pp. 65-76.

Mazon AF, Cerqueira CCC, Fernandes MN (2002a). Gill cellular changes induced by copper exposure in the South American tropical freshwater fish Prochilodus scrofa. Environ. Res. 88(1):52-63.

McCahon C, Pascoe D (1988). Use of (Gammarus pulex L.) in safety evaluation tests: culture and selection of a sensitive life stage. Ecotoxicol. Environ. Saf. 15(2):245-252.

Migliarini B, Campisi AM, Maradonna F, Truzzi C, Annibaldi A, Scaponi G, carnevali O (2005). Effects of cadmium exposure on testis apoptosis in the marine teleost Gobius niger. Gen. comp. Endocrinol. 142(1-2):241-247.

OECD (1993). (Organization for Economic Co-operation and Development) OECD Guidelines for Testing of Chemicals. OECD, Paris.

Olojo EAA, Olurin KB, Mbaka g, Oluwemimo AD (2005). Histopathology of the gill and liver tissues of the African catfish Clarias gariepinus exposed to lead. Afr. J. Biotechnol. 4:117-122.

Park K, Heo GJ (2009). Acute and Subacute Toxicity of Copper Sulfate Pentahydrate (CuSO_45 · H_2O) in the Guppy (Poecilia reticulata)(Toxicology). J. Vet. Med. Sci. 71(3):333-336.

Rani MJ, John Milton MC, Uthiralingman M, Azhaguraj R (2011). Acute Toxicity of Mercury and Chromium to Clarias batrachus (Linn). Bioresour. Bull. 5:368-372.

Rani AMJ, Sivaraj A (2010). Adverse effects of chromium on amino acid levels in freshwater fish Clarias batrachus (Linn.). Toxicol. Environ. Chem. 92(10):1879-1888.

Sadiq M (1992). Cadmium in toxic metal chemistry in marine environments. In: Sadiq M (ed.), Toxic metal chemistry in marine environments. Marcel Dekker Inc, New York. pp. 106-153.

Shah SL, Altindag A (2005). Effects of heavy metal accumulation on the 96-h $LC_{50}$ values in tench Tinca tinca L., 1758. Turk. J. Vet. Anim. Sci. 29:139-144.

Shuhaimi-Othman M, Nadzifah Y, Ramle NA, Abas (2010). Toxicity of Copper and Cadmium to Freshwater Fishes. World Acad. Sci. Eng. Tech. 65:869-871.

Spacie A, Hamelink J (1985). Bioaccumulation Fundamentals of Aquatic Toxicology: Methods and Applications. Hemisphere Publishing Corporation Washington D.C. pp. 495-525.

Wyn B, Sweetman JN, Leavitt PR (2007). Historical metal concentrations in lacustrine food webs revealed using fossil ephippia from Daphnia. Ecol. Appl. 17(3):754-764.

# 14

# Environmental metals pollutants load of a densely populated and heavily industrialized commercial city of Aba, Nigeria

Tobias I. Ndubuisi Ezejiofor[1]*, A. N. Ezejiofor[2], A. C. Udebuani[1], E. U. Ezeji[1], E. A. Ayalogbu[1], C. O. Azuwuike[1], L. A. Adjero[2], C. E. Ihejirika[3], C. O. Ujowundu[4], L. A. Nwaogu[4] and K. O. Ngwogu[5]

[1]Department of Biotechnology, Federal University of Technology, Owerri, Imo State Nigeria.
[2]Toxicology Unit, Department of Clinical Pharmacy University of Port Harcourt, Rivers State, Nigeria.
[3]Department of Environmental Technology, Federal University of Technology, Owerri, Imo State, Nigeria.
[4]Department of Industrial Biochemistry, Federal University of Technology, Owerri, Imo State, Nigeria.
[5]Department of Chemical Pathology, College of Medicine and Health Sciences, Abia State University Teaching Hospital, Aba, Nigeria.

**Diseases and their associated health defects are most often related to the quality of the total environment which in itself is also related to the quality and quantity of wastes generated in those areas, as partly defined by the nature of activities carried out by the populace. This environment-health relationship dynamics are particularly evident in most tropical environments like Nigeria where various environmental media are laden with sundry pollutants including metals, most of which are often furnished by wastes. This study aims at investigating the environmental metal load of Aba, a major commercial city in South-east Nigeria which is home to many artisanal, small- and medium-scale industrial activities, but presently experiencing waste-related menace. Randomly collected soil samples from different areas of Aba metropolis and a sub-urban community considered less polluted (to serve as control) were analyzed for heavy and non-heavy metals. Results show that while the mean of the estimated heavy metals in the six sites ranged from $0.31 \pm 0$ to $1293.75 \pm 0$ µg/g, for non-heavy metals it ranged between $55.01 \pm 24.88$ and $903.74 \pm 1081.25$. In the control site, the range is between 0 and $1293.75 \pm 0$ for heavy metals while for the non-heavy metals, it is between $72.73 \pm 0$ and $410.50 \pm 0$. The results indicate that the mean concentrations for most of the metals were high with respect to the Nigerian Federal Environmental Protection Agency (FEPA) and World Health Organization (WHO) standards. Findings in this study have serious implications for public health.**

**Key words:** Environmental, heavy metals pollutants load, Aba, Nigeria.

## INTRODUCTION

Heavy metals have been variously defined as those metals with higher atomic number and weight (Norman, 1981); large group of elements with an atomic density of greater than 6 g cm$^{-3}$, which are both biologically and industrially important (Alloway, 1995); any metallic chemical element that has a relatively high density and is toxic or poisonous at low concentration (Holdings, 2004).

They are trace metals that are at least five times denser than water, and as such, they are stable elements (meaning they cannot be metabolized by the body) and bio-accumulative (passed up the food chain to humans). Over 20 different heavy metals are known and include aluminium, antimony, arsenic, cadmium, chromium, cobalt, lead, manganese, mercury, nickel, selenium, tin, vanadium, zinc, platinum, and copper (metallic form versus ionic form) etc. (WHO, 1996b). Heavy metals are natural components of the environment, being present in rocks, soil, plants and animals. They occur in different

*Corresponding author. E-mail: tinezejiofor@gmail.com; favourtine@yahoo.com.

forms: as minerals in rock, sand and soil; bound in organic or inorganic molecules or attached to particles in the air. In today's industrial society, there is no escaping exposure to toxic chemicals and metals. This is so because somehow, our society is dependent upon metallurgy for proper functioning. It is therefore not surprising that human exposure to heavy metals has risen dramatically in the last 50 years as a result of an exponential increase in the use of heavy metals and/or their compounds in industrial and agricultural processes. In the United States for instance, tons of toxic industrial waste are reportedly mixed with liquid agricultural fertilizers and dispersed across America's farmlands (United States Department of Agriculture (USDA), 2000). Mining, manufacturing and the use of synthetic products (for example, pesticides, paints, batteries, industrial waste, and land application of industrial or domestic sludge) can result in heavy metal contamination of urban and agricultural soils. Potentially contaminated soils may occur at old landfill sites (particularly those that accepted industrial wastes), old orchards that used insecticides containing arsenic as an active ingredient, fields that had past applications of waste water or municipal sludge, areas in or around mining waste piles and tailings, industrial areas where chemicals may have been dumped on the ground, or in areas downwind from industrial sites (USDA, 2000). Environmental contamination and exposure to heavy metals is a serious growing problem throughout the world, as both natural sources and anthropogenic processes emit heavy metals into various environmental media.

The rapid industrialization and urbanization of the world have dramatically heightened these emissions, thus increasing the overall environmental load of heavy metals and consequent human exposures to them. Human exposure to aluminium, for instance, is caused by environmental factors such as soil contamination whereby aluminium can be mobilized as a consequence of soil acidification due to the use of certain fertilizers or acid rain; or as a result of aluminium use in some Industrial processes such as metallurgy, food preservation (yokel et al., 2008), water purification (Krewski et al., 2007), and the use of pharmacological and cosmetic products ((Ernst, 2002; Ernst and Coon, 2001; Garvey et al., 2001; Ganrot, 1986).

Arsenic and its compounds is also possessing great human exposure scenario as they are used in pesticides, insecticides, herbicides, and some kinds of alloys. Contaminated food, water, air and cigarette smoking are part of the sources of exposure to arsenic (American Conference of Government Industrial Hygienists (ACGIH), 2003; Namgung and Xia, 2001). As for lead, human exposure also occurs through food, water, air and soil. Food and water lead sources include the use of lead-containing ceramics dishware, metal plumbing, and food cans that contain lead solder (White et al., 2007). People can also be exposed to lead contamination from industrial sources such as lead smelting and manufacturing industries (Goyer, 1996; White et al., 2007). Though the risk of lead contamination from motor vehicle exhaust of leaded gasoline has decreased in the last couple of decades as a consequence of reduction of lead addition to petrol (Stromberg et al., 2008), because lead is a cumulative metal, it still remains a major hazard for human health. Same is the case for many metals because of the problem of non-degradability and accumulation. Occupational activities are part of the environmental factors responsible for environmental distributions and human exposures to metals. Workers that produce or use various metals (for example, arsenic) and their compounds are also at the risk of their exposure. Indeed, many occupations involve daily heavy metal exposures such that over 50 professions entail exposure to mercury alone. It is therefore not quite surprising that despite the efforts of the regulatory agencies, many studies have demonstrated variable concentrations of various heavy metals in various environmental media, thus indicating that the problem of environmental metal pollution is very much around us.

Iwegbue et al. (2006) studied the characteristic levels of heavy metals (Cd, Cr, Cu, Pb, Ni and Zn) of soil profiles of automobile mechanic waste dumps and reported that the concentration of heavy metals decreased with the depth of the profile and lateral distance from the dumpsites; the levels found in this study exceeded background concentrations and limits for agricultural and residential purposes; the distribution pattern of heavy metals in the soil profiles were in the order Pb > Zn > Cu > Cd > Ni > Cr, with the mechanic waste dumps representing a potential sources of heavy metal pollution to environment. Similarly, soils at different dumpsites located within Ikot Ekpene, South Eastern Nigeria and control soil samples taken 10 km away from the dumpsites were all analysed for the concentration of Na, Ca, Pb, Ni, Mn, Mg, Fe, P, N, Cu and Zn (Eddy et al., 2006). The result of the analysis shows that the dumpsites contain significant amount of toxic and essential elements, as significant difference existed between the concentration of these elements in the dumpsites and 10 km away from the dumpsite ($p > 0.05$).

Another study by Inuwa et al. (2007) investigated the concentration of trace metals in soil found around the major industrial areas of the North-western state of Nigeria using absorption spectrophotometry. The results obtained indicated that these metals on dry weight basis in the soil ranged between (0.1 to 0.7 µg/g) Cd, (14.2 to 92.7 µg/g) Cr, (151.5 to 540 µg/g) Pb and (3.5 to 24.7 µg/g) Ni, the results indicating relatively high concentrations of tested metals in industrial areas than those of the control sites. Similarly, the study of Igwilo et al. (2006) showed that the mean metal contents of soil samples from Otuocha agricultural river basin exceeded the WHO guidelines for the parameter in soil. Their result is as follows: Cd (0.07 to 3.345 ppm), Cu (4.38 to 13.54

ppm), Pb (0.59 to 7.34 ppm) and Ni (0.36 to 5.64 ppm), revealing that the obtained values were higher for Pb, Cd, Cu, Ni and other measured parameters in soils from agriculturally active Otuocha river basin relative to the controls just as the study of Adewuyi and Opasina (2011) also showed high concentrations of Fe, Mn, Cu, Zn, Ni, Cd and Pb, and all the parameters were above control, and also exceeded FEPA and WHO guidelines.

However, Aderinola et al. (2009) reported varying concentration of heavy metals in surface water, sediments, fish and periwinkles of Lagos Lagoon. Their report showed that the mean levels of heavy metals in the sediments of Lagos lagoon were generally low and fell within the acceptable limits described by WHO and FEPA. The average concentrations for the heavy metals were (0.083 ± 0.035 mg/kg) As; (1.150 ± 0.090 mg/kg) Cd; (0.867 ± 0.075 mg/kg) Ni; (0.618 ± 0.193 mg/kg) Cr; (0.600 ± 0.272 mg/kg) Cu; (19.393 ± 6.649 mg/kg) Fe; (0.450 ± 0.598 mg/kg) Pb; (2.040 ± 1.049 mg/kg) Mn; (0.730 ± 0.337 mg/kg) Zn, while Iron, Manganese and Nickel were not defined. Equally, a study on Ibeche (Ikorodu) Lagos lagoon area (Ladigbolu et al., 2011) reported higher concentrations of Fe, Cu, Pb, Cd and Zn when compared with the values in unpolluted sediment as well as the standards of the regulatory authorities (WHO, 1982 and FEPA, 1988).

However, apart from the contributions from anthropogenic sources, soils naturally contain trace levels (ppb to ppm) of heavy metals. For example, median concentrations of metals in U.S. soils are 0.2 mg/kg cadmium, 11 mg/kg lead, and 18.2 mg/kg nickel (FEPA, 1999). There are however, considerable variations in these metals concentrations by geographic region and soil type. Thus, threshold limit values (TLV) for metals in soils are determined by some factors including the following: the spatial and temporal changes of geochemical background, the possible effects of deep geological structures and other geochemical factors, natural anomalies exceeding baseline values by order of magnitudes (natural processes may lead to anomalies exceeding baseline values by order of magnitudes) without anthropogenic contamination, and the results of monitoring examinations of metal groups (Sipos and Pokas, 2008).

There are 35 metals that concern us because of occupational or residential exposure; 23 of these are the heavy elements or 'heavy metals': antimony, arsenic, bismuth, cadmium, cerium, chromium, cobalt, copper, gallium, gold, iron, lead, manganese, mercury, nickel, platinum, silver, tellurium, thallium, tin, uranium, vanadium and zinc (Glanze, 1996). Indeed, the body actually has need for approximately 70 friendly trace element heavy metals. Interestingly, small amounts of these elements are common in our environment and diet and are actually necessary for good health but large amounts of any of them may cause acute or chronic toxicity (poisoning). Some of these are trace elements (micronutrients) and perform essential functions for both

plants and animals in which they constitute essential part of the metabolizing and/or detoxifying proteins or enzymes.

Selenium, copper, zinc and iron are examples of this class of metals. Iron for example, prevents anaemia, and zinc is a cofactor in over 100 enzyme reactions. Magnesium and copper are other familiar metals that in minute amounts, are necessary for proper metabolism to occur. They normally occur at low concentrations and are known as trace metals. At high concentrations however, they can be toxic and therefore pose a risk to the health of animals and man. There are 12 others that are very toxic even at levels that are only moderately above background levels (that is, very low concentrations).

These are the toxic heavy metals and include arsenic cadmium, lead, mercury, Nickel, etc. (WHO, 1996a; Carpenter, 2001). They act as poisonous interference to the enzyme systems and metabolism of the body. No matter how many good health supplements or procedures one takes, heavy metal overload will be a detriment to the natural healing functions of the body.

The most common problem-causing cationic metals (metallic elements whose forms in soil are positively charged cations for example, $Pb^{2+}$) are mercury, cadmium, lead, nickel, copper, zinc, chromium and manganese, while the most common anionic compounds (elements whose forms in soil are combined with oxygen and are negatively charged for example, $MoO_4^{2-}$) are arsenic, molybdenum, selenium, and boron (USDA, 2000).

Toxic heavy metals have no function in the body and can be highly toxic. The metals are taken into the body through inhalation, ingestion and skin absorption. If heavy metals enter and accumulate in body tissue faster than the body's detoxification pathways can dispose of them, a gradual build-up of these toxins will occur. High-concentration exposure is not necessary to produce a state of toxicity in the body tissues and over time, toxic concentration levels may be reached. Heavy metals are dangerous not only because of their inherent nature but also because of their bioaccumulative tendency and problem of biomagnifications with increasing trophic levels, and therefore can cause permanent damage to health. While the inorganic form of the metal may not be easily taken up, the organic (alkylated) forms are readily taken up by body tissues and can be retained for a considerable length of time (Berlin and Ulberg, 1963; Garrett et al., 1992).

Consequently, organometals regardless of which base it is, are in general more toxic to humans than the inorganic form, primarily because they are lipid soluble and therefore penetrate the body more easily, especially the brain (Carpenter, 2001). Methyl mercurials for instance, can easily pass the placental barrier and the blood brain barrier, while inorganic mercury passes these barriers less easily. When the rate of introduction of toxic metals into the body system is greater than that of removal, the metals accumulate and this could play a major role in heavy metal toxicity. Accumulated heavy metals at different levels of body burdens exhibit corresponding levels of toxicity that may be more acute on juvenile stages of organisms than on adults (WHO, 1996a, b; Carpenter, 2001). The negative impacts of heavy metals result from their toxicity to biological processes including those catalyzed by micro organisms (Kelly and Tate, 1998). The degree to which a system, tissue, organ or cell is affected by a heavy metal toxin depends on the toxin itself and the individual's degree of exposure to the toxin (Extreme Health, 2004). For any metal, the toxicity is a function of its molecular configuration and physico-chemical properties such as solubility, particle size, and other determinants of the extent to which a material can intrude into biochemical processes (Norman, 1981).

Excess heavy metal accumulation in soils is toxic to humans and other animals. Exposure to heavy metals is normally chronic (exposure over a longer period of time) due to food chain transfer. Acute (immediate) poisoning from heavy metals is through ingestion or dermal contact. Chronic problems associated with long-term heavy metal exposures are: lead, mental lapse; cadmium, affects kidney, liver, and gastrointestinal tract; arsenic, skin poisoning, affects kidneys and central nervous system. Heavy metal toxicity can result in damaged or reduced mental and central nervous function, lower energy levels, and damage to blood composition, lungs, kidneys, liver, and other vital organs. Long-term exposure may result in slowly progressing physical, muscular, and neurological degenerative processes that mimic Alzheimer's disease, Parkinson's disease, muscular dystrophy, and multiple sclerosis. Allergies are not uncommon and repeated long-term contact with some metals or their compounds may even cause cancer (International Occupational Safety and Health Information Centre, 1999).

Most human load of toxic metals is acquired from the ambient concentrations of these metals through inhalation of dust and fumes, ingestion of food and drink and/or absorption through skin in extreme cases (Occupational Safety and Health Administration (OSHA), 1991; Agency for Toxic Substance and Disease Registry (ATSDR), 2003). However the most classical toxicities associated with these metals have come through massive pollution of the environment through industrial activities. Apart from the raw materials or process chemicals, production of wastes is an integral part of industrial activities that can increase the metal load of the ambient environment through pollution of the environment with metal-bearing wastes in the form of solids, liquids, gases and air-borne particulate matter; and quite often, facilities needed for their proper disposal are not adequate. This underscores the need for investigation into the toxic metal load of our immediate environment, given the massive degradation suffered by most of our cities from various forms of wastes.

Aba, a commercial city east of the Niger is one of those cities of Nigeria currently under very heavy load of waste related degradation, with very high possibilities of heavy metal pollution. Again, this city plays host to large army of artisans (technicians of various sorts) engaged in all forms of industrial ventures, particularly numerous small-scale artisanal activities involving use, conversion and reuse of metal-containing compounds with consequent release of metal-bearing wastes. They include masons, electricians, paint makers and painters, petroleum pump attendants and the street-side freelance retailers of petroleum products, automobile repairers of various categories (including mechanics, panel beaters, etc.), welders, tinkers and other metal smelting artisans, etc. To mention this few. Consequently, all forms of metal-containing wastes and metal junks form part of the massive heaps of wastes that litter every nook and crannies of the town. Indeed the city plays host to uncountable small- and medium-scale industrial ventures that generate equally uncountable varieties of wastes which are eventually released to human surroundings.

Thus, there is indeed very high possibility of soil, water and air pollution from several environmental pollutants that most probably include some metal forms with equally very high possibilities of human contaminations through the environmental media. The extent of metals pollutant load of this city is presently unknown, thus making this study a very highly warranted one.

## MATERIALS AND METHODS

### Samples and preparation

Specifically, the study examined the heavy metal load of soils from randomly selected areas of Aba metropolis. Soil samples from a particular area of Aba (a housing estate) considered less polluted because of its low industrial activities as at the time of this study, and a sub-urban area within the outskirt of Aba were also included to serve as controls. In all, soil samples were collected from six different locations in Aba metropolis and its environs as follows:

Site 1: Ndiegoro Area (beyond the popular Ngwa road);
Site 2: Aba Main River (popularly called Aba Waterside);
Site 3: Ogbor Hill Area (area beyond the Aba river);
Site 4: Aba Main Area (Aba Central Township area);
Site 5: World Bank Housing Estate, Aba;
Site 6: Umuamacham village, a sub-urban area within the outskirt of Aba.

Accordingly, soil samples from these sites constitute samples 1 to 6 analyzed in the study. At each sampling site, about $9 \times 9$ cm$^2$ hole was dug on the ground to the depth of about 15 to 20 cm from the soil surface, and a generous amount of soil sample was collected. The samples were collected in sterile plastic containers previously rinsed in a solution of 95% ethyl alcohol, and then distilled water. At the base laboratory, one gram of each of the samples was carefully weighed out, with caution not to contaminate the samples with extraneous metal sources.

### Metal analysis

Pre-metals analyses treatments of soil samples involve acid digestion of the soil samples. The weighed out gram of each of the samples was dried, sieved and transferred into a 100 mL beaker.

**Table 1.** Concentrations of the heavy metals at the various soil samples (µg/g) compared with FEPA and WHO standards.

| Metal | Soil sample | | | | | Mean ± S.D | Control soil sample | WHO (1984)/FEPA (1991) standards in sediments (mg/l) | WHO (1984)/FEPA (1991) standards in water (mg/l) |
|---|---|---|---|---|---|---|---|---|---|
| | 1 | 2 | 3 | 4 | 5 | | | | |
| Cd | 4.17 | N.d | 6.25 | N.d | N.d | 5.21±1.47 (n=2) | N.d | 0.003 | 0.005 (WHO) 0.01 (FEPA) |
| Co | 2.78 | 2.78 | 2.78 | 6.94 | 2.78 | 3.61±1.66 (n=5) | N.d | NA | NA |
| Cr | 25.0 | N.d | N.d | 25.0 | N.d | 25.0±0 (n=2) | N.d | 2.000 | 0.05 |
| Zn | 13.4 | 4.6 | 26.6 | 1.4 | 8.6 | 10.92±8.77 (n=5) | 0.12 | 3.000 | 1.5 |
| Cu | 80.0 | 19.5 | 18.0 | 66.5 | 11.0 | 39.0±28.43 (n=5) | 17.5±0 | 1.000 | 0.05 |
| Pb | 131.25 | 35.0 | 32.5 | 150.0 | 6.25 | 71.00±64.89 (n=5) | N.d | 0.010 | 0.05 |
| Ni | 25.0 | 10.0 | 40.0 | 7.50 | 17.50 | 20.00±13.11 (n=5) | N.d | 0.02 | 0.1-0.2 |
| V | 0.31 | 0.31 | 0.31 | 0.31 | N.d | 0.31±0 (n=4) | N.d | NA | NA |
| Mn | 391.94 | 33.07 | 106.45 | 133.06 | 66.13 | 146.13±127.55 (n=5) | 28.42±0 | 0.050 | NA |
| Fe | 778.03 | 741.67 | 731.82 | 815.90 | 853.03 | 784.09±45.47 (n=5) | 1293.75±0 | 0.300 | 0.3-1.00 |

N.d - Not detected :NA=Not Available

**Table 2.** Concentrations of the non-heavy metals in the various soil samples compared with those of the control (µg/g).

| Metal | Soil sample | | | | | Mean ± S.D | Control soil sample |
|---|---|---|---|---|---|---|---|
| | 1 | 2 | 3 | 4 | 5 | | |
| Mg | 108.63 | 52.63 | 104.25 | 96.63 | 81.50 | 88.73±20.27 (n=5) | 186.00±0 |
| Ca | 1337.5 | 12.50 | 2843.75 | 125.00 | 200.00 | 821.54±1081.25 (n=6) | 410.50±0 |
| K | 81.82 | 34.09 | 88.66 | 38.64 | 31.82 | 58.00±24.88 (n=6) | 72.73±0 |
| Na | 287.50 | 125.00 | 381.25 | 75.00 | 100.00 | 194.78±119.64 (n=5) | 200.00±0 |

Concentrated nitric acid ($HNO_3$) (10 mL) and concentrated perchloric acid ($HClO_4$) (5 mL) (that is, acid volume ratio of 2:1) were added and the mixture was placed on a hot plate under fume cupboard and heated to near dryness, until its colour changed to white.

This residue was allowed to cool and then dissolved with 20% nitric acid (5 mL). After filtering the mixture with filter paper, the filtrate was made up to 20 $cm^3$ volumes with distilled water. The various concentration standards for the various elements as well as the blank were equally prepared. Each of these preparations was then analyzed for heavy metals using elemental atomic absorption spectrophotometric method (Dawson, 1978) (AAS-UNICAM 919 model was used) while analyses of monovalent metals like sodium (Na) and potassium (K)

was performed using low temperature flame photometric method (Ramsay et al., 1953) (Jenway Flame photometer, PFP.7 model was used).

## RESULTS

Results of the metals analyzed in the study are shown in Table 1 (heavy metals) and Table 2 (non-heavy metals). Results show that the means of the estimated heavy metals in the six sites ranged from 0.31 ± 0 µg/g for vanadium to 1293.75 ± 0 µg/g for iron. The non-heavy metals

ranged between 55.01 ± 24.88 µg/g (potassium) and 903.74 ± 1081 µg/g (calcium). While between 0.31 ± 0 µg/g (vanadium) and 784.09 ± 45.47 µg/g (iron) remained the mean concentration range for heavy metals in the test site, that of the non-heavy metals was between 55.01 ± 24.88 µg/g (potassium) and 903.74 ± 1081.25 µg/g (calcium). In the control site, the range was between zero (for the non-detected heavy metals-cadmium, cobalt, chromium, lead, nickel and vanadium) and 1293.75 ± 0 µg/g for iron; while for the non-heavy metals, the range was between 72.73 ± 0 µg/g (potassium) and 410.50 ± 0 µg/g

**Table 3.** Concentration ranges and means of Heavy Metals in the various Soil Samples compared with the Mean Concentrations in the Control (µg/g), and with FEPA and WHO standards.

| Metal | Concentration range for soil samples 1 to 6 (µg/kg) | Mean ±S.D | Average concentration in Hungarian soil[a,b,c] | Hungarian baseline value (Threshold limit)[d] | Average concentration in Nigerian soil and FEPA threshold limit value |
|---|---|---|---|---|---|
| Cd | 0-6.5 | 5.21±1.47 (n=2) | 0.01-2 | 1 | NA |
| Co | 2.78-6.94 | 3.61±1.66 (n=5) | 50-200 | 30 | NA |
| Cr | 0-25.0 | 25.0±0 (n=2) | 10-15 | 75 | NA |
| Zn | 0.12-26.6 | 10.92±8.77 (n=5) | 50-100 | 200 | NA |
| Cu | 11.0-80.0 | 39.0±28.43 (n=5) | 10-15 | 30 | NA |
| Pb | 0-150.0 | 71.00±64.89 (n=5) | 15-30 | 100 | NA |
| Ni | 0-40.0 | 20.00±13.11 (n=5) | 15-30 | 40 | NA |
| V | 0-0.31 | 0.31±0 (n=4) | NA | NA | NA |
| Mn | 28.42-391.94 | 146.13±127.55 (n=5) | NA | NA | NA |
| Fe | 731.82-1293.75 | 784.09±45.47(N=5) | NA | NA | NA |

a: Aubert and Pinta (1978), b: Kabata-Pendias (1984), c: Andriano (1986), d: Hungarian Government Regulation number10/2000 (NA = not available).

**Table 4.** Mean concentrations of heavy metals of the various soil samples (µg/kg) compared with both Hungarian and Nigerian threshold limit.

| Metal | Concentration range for soil samples 1-6 (µg/kg) | Mean ±SD | Average concentration in Hungarian soil[a, b, c] | Hungarian baseline value (Threshold limit)[d] | Average concentration in Nigerian soil and FEPA threshold limit value |
|---|---|---|---|---|---|
| Cd | 0-6.5 | 5.21±1.47 (n=2) | 0.01-2 | 1 | NA |
| Co | 2.78-6.94 | 3.61±1.66 (n=5) | 50-200 | 30 | NA |
| Cr | 0-25.0 | 25.0±0 (n=2) | 10-15 | 75 | NA |
| Zn | 0.12-26.6 | 10.92±8.77 (n=5) | 50-100 | 200 | NA |
| Cu | 11.0-80.0 | 39.0±28.43 (n=5) | 10-15 | 30 | NA |
| Pb | 0-150.0 | 71.00±64.89 (n=5) | 15-30 | 100 | NA |
| Ni | 0-40.0 | 20.00±13.11 (n=5) | 15-30 | 40 | NA |
| V | 0-0.31 | 0.31±0 (n=4) | NA | NA | NA |
| Mn | 28.42-391.94 | 146.13±127.55 (n=5) | NA | NA | NA |
| Fe | 731.82-1293.75 | 784.09±45.47 (n=5) | NA | NA | NA |

[a]Aubert and Pinta (1978), [b]Kabata-Pendias(1984), [c]Andriano(1986), [d]Hungarian Government Regulation number 10/2000 (2000), NA= not available.

In Table 3, the concentration ranges and means of the evaluated heavy metals for both the test and control soil samples are presented with the standards (permissible limits) prescribed by the regulatory agencies, FEPA and WHO. In Table 4, the mean concentrations recorded in our study for Aba, a typical tropical environment, are presented for comparison with the average concentrations (calcium).

documented for Hungary, a typical European country.

## DISCUSSION

A study of the results generally revealed that the mean concentrations for most of the metals including such toxic ones as lead, nickel, chromium, cadmium, etc. were all found to be quite high relative to those of the control sites where (except iron, manganese and copper) most of them were either not detected at all or the concentration of the detected ones were relatively low (Table 1). However, the same could not be said for the non-heavy metals since the control site recorded higher mean concentrations for most of them (except calcium) than was the case for the mean concentrations of equivalent metals in other sample sites (Table 2). The mean metal contents of all the soil samples that is, sites 1 to 6, were quite high with respect to Fe, Ca, Na and Mg in that order (Tables 1 and 2).

Among the toxic metals, Pb (71.00 ± 64.89 µg/g), Cu (35.42 ± 29.76 µg/g), Cr (25.0 ± 0 µg/g) and Ni (20.00 ± 13.11 µg/g) recorded worrisome mean concentrations. Even Zn, Cd and Co at mean concentrations of 9.12 ± 9.5, 5.21 ± 1.47 and 3.61 ± 1.86 µg/g, respectively cannot simply be glossed over, since the least of these concentrations has surpassed the limits allowable in soil by regulatory agencies including the Nigerian Federal Environmental Protection Agency (FEPA, 1991) and World Health Organization (WHO, 1984) (Table 3). However, it was observed that regulatory data were not wholesomely available for all the metals for the different environmental media, particularly soil, thus making data comparisons for safe limits a fairly difficult task.

An alternative approach was to compare our data with average concentrations of these metals in Nigerian soil (that is, baseline value for Nigerian soil) and this too was not available. Consequently, in addition to comparing our data with the availables from the regulatory bodies, we also compared them with those of a typical European country (Hungary) with fairly complete soil metal data (Table 4). As can be easily gleaned from the table, the mean concentrations for Cd, Pb, Cr and Cu for Aba metropolis also surpassed the average concentration range for Hungarian soils, while the concentration range for Zn (50 to 100) and Co (50 to 200) in Hungarian soil surpassed by far the level of these metals in Aba soils. For other metals however, the values in our study are nearly close to the available concentration ranges, as not all of these were available also for the Hungarian soil.

A close look at the results of the individual sites (Tables 1 and 2) revealed that apart from Fe, Ca, Na, Mg and K that were generally high, soil samples from the control community (site 6) had relatively lower values for most of the heavy metals, although as high as 17.5 µg/g of copper was obtained therein, just as 17.5 µg/g of nickel was also recorded for its immediate neighbour, site 5- the

World bank housing estate. In the former, most of these metals were not detected at all, whereas at site 5, Cu (11.0 µg/g), Zn (8.6 µg/g), Pb (6.25 µg/g) and Co (2.78 µg/g) were also detected. As at the time of this study, Site 5 remained a newly mapped out residential estate, planned and being developed as a world bank assisted housing project, and housing construction and related activities including block moulding businesses were the most dominant activities here; while site 6 was a sub-urban community few kilometres away from the heart of Aba metropolis, but rapidly being annexed by the expanding metropolis since residents of the metropolis are now resorting to this and other nearby communities in quest for solution to their housing/accommodation problems now becoming tighter in the main areas of the metropolis. But for the encroaching urbanization, this was just a quiet community, farming being the major activity of note in this area.

In areas of sites 5 and 6, anthropogenic activities such as high human density, high industrial and vehicular activities/associated emissions, high turnover of wastes, etc., were still occurring at a very low level, though at a relatively faster rate in the former than the later. This might explain the metal distribution patterns found in these areas whereby comparatively lower values of toxic heavy metals like Cd, Co, Cr, Pb, Ni and V were found relative to what is obtained in other areas in the heart of the metropolis.

However, the world bank housing estate (site 5) recorded higher values for these toxic metals relative to the situation in the sub-urban community (site 6) where except for Cu detected in fairly reasonable amount (17.5 µg/g), and Zn detected in quite insignificant quantity (0.12 µg/g), most of the toxic heavy metals were not detected at all (Table 1). When this is taken along with the unfavourable situation in the other areas of the city with regard to metal pollution, it is clearly evident that indeed anthropogenic variables may actually be responsible for the high metal load of Aba metropolis generally, as noted in other studies (National Institute of Occupational Safety and Health (NIOSH), 1991; Egbuna, 1992; WHO, 1996b; Gazso, 2002; Agency for Toxic Substance and Disease Registry (ATSDR), 2003; Iwegbue et al., 2006; Eddy et al., 2006; Igwilo et al., 2006; Inuwa, 2007; Ladigbolu et al., 2011; Adewuyi and Opasina, 2011).

Despite the seemingly lower values recorded for these other metals, the higher values recorded for iron in the control site, and the reckonable presence of such heavy metals as Cu, Zn, Pb and Co at site 5 (Table 1) showed that neither the so-called control site nor the world bank housing estate (considered to have very close characteristic to the control site in terms of human activities) was totally free of the hazards of metal pollution. The reason for the high iron contents in these two sites is not presently known. However, for most of the soil samples, the levels of Fe, Ca, Na, Mg, and Mn (Tables 1 and 2) were also quite high and could be attributable to the fact that this group of metals constitute over 99% of the total element content of the earth's crust (Mitchell, 1964). Again, this could be as a result of climatic factors as observed by Aubert and Pinta (1977) that soils in the tropical climate zones (which Nigeria is among) are mostly ferrallitic, and the levels of Mn in such soils ranges from 20 to 5000 ppm.

Though the central area of the metropolis (site 4) recorded formidable presence of most metals, this was in no way comparable to the situation observed within Ndiegoro area of the city (site 1), for although various metals' presence were observed for most sites in the heart of this commercial city, Ndiegoro area in particular showed remarkable presence of most metals at varying concentrations, including those of Pb, Cu, Zn, Ni, Cr, etc. (Tables 1 and 2). The findings in this area expectedly rivalled what was obtained in a Romanian study (Lacatusu et al., 2001) in which most of the heavy metals were found to be many times in excess of their maximum allowable limits (MAL), and in some instances also surpassed the Hungarian data (Table 4).

As can be easily gleaned from the table, the mean concentrations for Cd, Pb, Cr and Cu for Aba metropolis surpassed the average concentration range for Hungarian soils. For other metals however, the values in our study are nearly close to the available concentration ranges. However, the findings in Ndiegoro area are not surprising at all, since this is a very squalid urban slum beyond the popular Ngwa road. In this area of Aba noted for its in-glorious status as the "Aba erosion disaster area" of the 1980s, there are multiple erosion sites with stagnant pools following erosion menace in most of the area. In spite of this, this area is very thickly populated, bringing about a very high population density. In terms of activity, there are legions and varieties of it, for apart from trade on many commercial items including textiles products of various descriptions, all forms of artisanal works such as tinkering and other metal works are found in this area. The popular "New market" noted for the sale of mainly food items occupies almost the length of the equally popular Ngwa road through which many gateways in terms of streets opened into this urban slum. In the central part of the city (site 4), though trading activities also thrive in most of the roads, electronics and other household items are the major items of trade here; in this area is located the Aba shopping centre where apart from the sale of textiles and electronics, repairs and engineering conversions of various electrical and electronic wares take place, and accordingly, various categories of artisans are found in this centre. Auto-repair related activities and their workshops are also predominant just as the resultant metal junks/wastes from mechanic and panel beating workshops.

This area of Aba is also predominantly characterized by public and private office works. It is therefore not quite surprising that the findings here mirrored the situation in one of the home studies (Iwegbue et al., 2006) in which

the characteristic levels of heavy metals of soil profiles of automobile mechanic wastes were found to exceed background concentrations and limits for agricultural and residential purposes.

Apparently, the electronic and mechanic waste junks represented potential sources of heavy metal pollution in this area's environment. This is in spite of the fact that what could be considered as organized municipal waste collection and evacuation program is found in this area where also there are city network of roads. As in the World Bank housing estate, sanitation is fair here relative to the situation in other parts of the metropolis. A striking observation made in this study was the fact that most of the metals (apart from iron and cadmium) were relatively low in soil sample 3 (from Aba Main River). The findings here though similar to that of Aderinola et al. (2009) in which the mean levels of heavy metals in the sediments of Lagos lagoon were generally low and fell within the acceptable limits described by WHO and FEPA, they were however quite contrary to expectations, given the fact that the river receives discharges/washouts from sundry activities being carried out along its banks including those of a soap/detergent industry, a major brewery producing assorted drinks, and the city animal slaughterhouse/meat processing centre. However, the reason for these lowered metals concentrations may be due to dilution effect, since this is a free-flowing stream. Though the area immediately bordering this river, the Ogbor Hill area (site 3) showed reduced metal loads relative to Ndiegoro area (site 1), toxic metals such as Nickel, lead, Zinc, Copper and Cadmium, in addition to other metals like iron, sodium, Magnesium and Manganese showed noticeable presence in this area (Tables 1 and 2). Also in this same area, there was a profuse presence of calcium, thus indicating the need for geological studies of Ogbor hill area for mineral deposits, particularly, those of calcium compounds. Apart from trading and artisanal activities common to almost every street in Aba, this area houses the slaughter and the nearby meat market; meat processing and other related ventures including tanning, burning/conversion of cow bones and horns are carried out just at the basement of the bridge across the river into this area of the town. Glass and bottle-producing industries are housed in this area also. Who knows, these might be contributory to the high calcium content of this area.

The peculiar terrain that characterized both Ndiegoro and Ogbor hill areas, their very thick population density with the associated high waste turnover, coupled with the near total absence of any form of waste management program make these area the most dirty areas of the metropolis. These factors might explain the pattern of results obtained for these areas. Following these findings, it is most likely that the bore holes sunk in most parts of Aba, particularly Ndiegoro and Ogbor Hill areas would be yielding hard water. This might explain the observation by this lead/corresponding author that water from bore holes from Ndiegoro area of Aba (site 1) usually has nauseating taste, and presents with somewhat milky/turbid sediments on standing, some moments after collection. The need for qualitative studies of water supply sources of Aba and its environs is thus highly warranted, particularly as it concerns Ndiegoro and Ogbor hill areas (sites 1 and 3) not only as a quality control measure, but also to establish a good basis for policy options that will properly address the public health implications of these findings, thus stemming the tide of the potential health problems associated with these.

Soil metal contents as high as revealed in this study for some areas have serious public health implication for the residents of this metropolis, given the catalogue of health effects associated with exposure to metals generally, and toxic heavy metals in particular. Several substances including metals, enter the body through gastrointestinal tract, and after absorption are transported by the hepatic portal vein to the liver; thus, the liver is the first organ perfused by chemicals that are absorbed in the gut (Adedara et al., 2011). Metals act by attacking nervous tissues as protoplasmic poisons, or as histotoxins by destroying liver, kidneys and/or other body tissues. In general, heavy metals are systemic toxins with specific neurotoxic, nephrotoxic, fetotoxic and teratogenic effects. Heavy metals can directly influence behavior by impairing mental and neurological functions, influencing neurotransmitter production and utilization, and altering numerous metabolic body processes. Systems in which toxic metal elements can induce impairment and dysfunction include the blood, cardiovascular, eliminative pathways (colon, liver, kidneys, skin), endocrine (hormonal), energy production pathways, enzymatic, gastrointestinal, immune, nervous (central and peripheral), reproductive, and urinary systems. These toxic impacts of Metals are affected through the formation of free radicals (highly reactive oxygen species), a disturbance of the pro- and antio-xidant balance of the body, both of which can cause cell damage by destruction of proteins, degradation of nucleic acids, or lipid peroxidation (Gutteridge, 1995; Adedara et al., 2011). Toxic heavy metals have been implicated in many health defects arising due to their toxic effects involving several body tissues, organs and systems (WHO, 1996b). The health effects range from contact dermatitis to other damages involving these body parts and functions. Such include neurological and behavioral disorders, haematological defects, liver and kidney damage, cardiovascular defects including increased blood pressure/hypertension, skeletal tissue damage, reproductive defects, genetic defects and induction of many types of cancers often effected through membrane lipid peroxidation, induction of oxidative stress and damage to cellular components of the affected organs, and possibly, the activation of oxidant-sensitive trans-cription factor, among others (WHO, 1983, 1996a, b; Schwartz, 1995; Ankra et al., 1996; Reeves and Vanderpoo, 1997; Ezeonu and Ezejiofor, 1999, 2002; Hu, 2001; Noonan et al., 2002; Zeitz et al., 2002; ATSDR, 2003; Dioka et al., 2004; Aimo and Oteiza, 2006; Alimba et a.l, 2006; Adedara, 2011). The deleterious effect of lead for instance, can involve both reactive oxygen species (ROS) and reactive nitrogen species. Oxidative stress has been associated with Pb exposure in humans and in experimental animal models.

In humans occupationally exposed to lead, biomarkers of oxidative stress such as malondialdehyde, reduced glutathione (GSH) status, glutathione peroxidase (GSPx) and catalase are known to have exceeded the mean value concentration of the control population (Costa et al., 1997; Garcon et al., 2004; Devi et al., 2007). Against the backdrop of these possible health effects, a subsequent study is highly indicated to investigate the Aba residents for these and other markers of metal-facilitated toxicity and indeed insipient health defects that may for now be silent but ravaging away lives secretly among the populace. Equally indicated is a study of both flora and fauna (indeed, the biodiversity) within this study area, since effects of metal toxicity stops not only at the door steps of humankinds, but also usually cut across all life forms. As observed by Nriagu (1988), environmental metal poisoning is a silent epidemic, and based on the findings of this study, the residents of Aba may already be at great risk of this epidemic.

## Conclusion

This is the first study that documented the metal load of Aba metropolis (also known as, Enyimba city), a comercial nerve centre South-east of Nigeria. The study revealed that the mean concentrations for most of the metals including such toxic ones as Pb, Ni, Cr, Cd, etc. were quite high with respect both to those of the control sites and the standards prescribed by Nigerian Federal Environmental Protection Agency (FEPA, 1991) and World Health Organization (WHO, 1984), and these findings have serious implications for public health. The high metal load revealed in this study showed that this city presently under siege by wastes, particularly refuse, is under very heavy load of metal pollution and therefore under serious threat of metal epidemic. A direct relationship existing between anthropogenic activities, waste load, and metal pollutant load of a place as shown by this study meant that a greater number of Nigerian urban cities may already be under this threat of metal pollution, given the mountainous heaps of refuse and other wastes that are common sights in most of these cities. Since environmental metal poisoning is a silent epidemic, the findings of this study suggest that every resident of Aba and perhaps most other Nigerian urban cities may already be at great risk of this epidemic, which possibly is already ravaging away lives silently.

## ACKNOWLEDGEMENTS

Miss Uloma Eke (R.I.P) of the Department of Industrial

Microbiology, Federal University of Technology, Owerri and a former student of this lead/corresponding author, is hereby remembered for her field assistance, while the staff of Central laboratories, University of Uyo, Nigeria is also appreciated for their laboratory support.

## REFERENCES

Adedara IA, Teberen R, Ebokaiwe AP, Ehwerhemuepha T, Farombi EO (2011). Induction of oxidative stress in liver and kidney of rats exposed to Nigerian bonny light crude oil. Environmental Toxicol. DOI 10, 1002/tox.20660 (In press).

Aderinola OJ, Clarke E.O, Olarinmoye O.M, Kusemiju V, Anatekhai M.A (2009). Heavy Metals in Surface Water, Sediments, Fish and Periwinkles of Lagos Lagoon. American-Eurasian J. Agric. Environ. Sci. 5 (5): 609-617.

Adewuyi GO, Opasina MA (2011). Physicochemical and Heavy Metals Assessments of Leachates from Aperin Abandoned Dumpsite in Ibadan City, Nigeria. E-Journal of Chemistry. http://www.e-journals.net, 7(4), 1278-1283.

Agency for Toxic Substance and Disease Registry (ATSDR) (2003).Case studies in Environmental Medicine, Lead toxicity (http//www.atsdr.cdc.gov/hec/csem/Lead exposure pathways.html).

Aimo L, Oteiza PI (2006). Zinc deficiency increases the susceptibility of human neuroblastoma cells to lead-induced activator protein-1 activation. Toxicol. Sci. 91:184-191.

Alimba CG, Bakare AA, Latunji CA (2006). Municipal landfill leachate induces chromosome aberrations in rat bone marrow cells. Afr. J. Biotechnol. 52053-2057.

Alloway BJ (1995). Heavy Metals in Soils 2nd edition, Blackie academic and Professional, Glasgow, p. 500.

American Conference of Government Industrial Hygienists(ACGIH) (2003). Documentation of the arsenic, elemental and inorganic compounds except arsine TLV, in threshold limit values for chemical substances and Physical Agents and Biological Exposure Indices. Cincinnati, Ohio, ACGIH Worldwide.

Ankrah N-A, Kamiya Y, Appiah-Opong R, Akyeampong YA, Addae MM (1996). Lead levels and related biochemical findings occurring in Ghanaian subjects occupationally exposed to lead. East Afr. Med. J. 375 – 379.

Aubert H, Pinta M (1977). Trace Elements in soils, Elsevier Scientific Publishing Company, Amsterdam, p.395.

Berlin M, Uberg S (1963). Accumulation and retention of mercury in mouse 111: An auto radiographic compensation of methyl mercury dicyanidiamide with organic mercury Arch. Environ Health 6: 610-616.

Carpenter DO (2001). Effects of Metals on the Nervous System of Humans and Metals. Int. J. Occup. Med. Environ. Health 14(3): 209-218.

Costa CA, Trivelato GC, Pinto AM, Bechara EJ (1997). Correlation between plasma 5-aminolevulinic acid concentrations and indicators of oxidative stress in lead exposed workers. Clin Chem 43:1196-1202.

Dawson JB (1978). Analytical Atomic Spectroscopy. In: Williams DL, Nunn RF and Marks V (Eds), Scientific Foundations of Clinical Biochemistry Vol.1. William Heinemann Medical books Ltd; Pp95-120

Devi SS, Biswas AR, Biswas RA, Vinayagamoorthy N, Krishnamurthi K, Shinde VM, Hengstler JG, Hermes M, Chakrabati T (2007). Heavy metal status and oxidative stress in diesel engine tuning workers of central Indian population. J Occup Med 49: 1228-1234.

Dioka CE, Orisakwe OE, Adeniyi FAA, Meludu SC (2004). Liver and Renal Function Tests in Artisans occupationally exposed to lead in mechanic Village in Nnewi, Nigeria. Int. J. Environ. Res. Pub, Hlth. 1: 21 – 25.

Eddy NO, Odoemelem SA, Mbaba A (2006). Elemental Composition of Soil in Some Dumpsites. Electron. J. Environ. Agr. Food Chem. 5 (3): 1349-1365.

Egbuna DO (1992). Understanding Refining and Petrochemicals Operations. Obingoz African Holdings Publishers Ltd Lagos. vol.1, pp. 25–26.

Ernst E (2002). Heavy Metals in traditional Indian remedies. Eur. J. Clin. Pharmacol. 57:891-896.

Ernst E, Coon JT (2001). Heavy metals in traditional Chinese medicines: A systematic review. Clin. Pharmacol. Therapeut. 70 (6): 497-504.

Extreme Health (2004). Toxic Metal- Sources and specific effects HYPERLINK "mailto:Info@%20extreem"Info@ extreme healthusa.com.

Ezejiofor TIN, Ezeonu FC (2002). Biochemical indicators of occupational health hazards in Enugu coal miners, Nigeria. Inter. J. Environ. Hlth Hum. Dev. 3 (2): 31 – 35.

Ezeonu FC, Ezejiofor TIN (1999). Biochemical indicators of occupational health hazards in Nkalagu cement industry workers, Nigeria. Sci. Total Environ. 228: 275 – 278.

Federal Environmental Protection Agency (FEPA) Act (1991). Guidelines and standards for Industrial effluent, gaseous emissions and hazardous waste management in Nigeria. National Environmental Protection Regulations, Federal Republic of Nigeria. Supplement to Official Gazette Extraordinary - Part B. 78 (42): B15 – 31.

Federal Environmental Protection Agency (FEPA) Guidelines and Standards for environmental pollution control in Nigeria. Decree 58 of 1988; p. 238.

Ganrot PO (1986). Metabolism and possible health effect of aluminium. Environ. Health, Perspect, 65: 363-441.

Garcon G, Leleu B, Zerimech F, Marez T, Haguenoer JM, Furon D, Shiral P (2004). Biologic markers of oxidative stress and nephrotoxicity as studied in biomonitoring of adverse effect of occupational exposure to lead and cadmium. J. Occup. Environ. Med. 46:1180-1186.

Garrett NE, Garrett RJB, Archdeacon JW (1992). Placental transmission of mercury to foetal rat. Toxicol. Appl. Pharmacol. 22: 649-654.

Garvey GJ, Hahn G, Lee RV, Harbison RD (2001). Heavy Metal of Hazards Asians traditional remedies Inter. J. Environ. Hlth Res. 11:63-71.

Gazso LG (2002). The key microbial process in the removal of toxic metals and radionuclides from Environment. Minireview. National Center for Public Health, Hungary 7:3-4.

Goyer RA (1996). Results of lead research: Prenatal exposure and neurological consequences. Environ. Health Perspec,t 104:1050-1054.

Gutteridge JMC (1995). Lipid peroxidation and antioxidants as biomarkers of tissue damage. Clin. Chem. 41:1819-1828.

Holding BV (2004). Heavy Metals (http: www. Lenntech. Com/heavy metals.htm).

Hu H (2001). Heavy Metal Poisoning. In: Fauci AS, Braunwald E, Isselbacher KJ, Kasper DL, HU Howard. Poorly controlled Hypertension in a painter with chronic Lead Toxicity. Environ.l Health Perspective, 109(1): 95- 99.

Igwilo IU, Afonne OJ, Maduabuchi UJ, Orisakwe OE (2006). Toxicological study of the Anam River in Otuocha, Anambra State, Nigeria. Arch Environ. Occup. Health, 61(5):205-208.

International Occupational Safety and Health Information Centre(1999). Metals. In Basics of Chemical Safety, Chapter 7. Geneva: International Labour Organization.

Inuwa M, Abdulrahman FW, Birnin-Yauri UA, Ibrahim SA (2007). Analytical Assessment of Some Trace Metals in Soils around the Major Industrial Areas of North-western Nigeria. Trends, Appl. Sci. Res. 2:515-521.

Iwegbue CMA, Isirima NO, Igwe C, Williams ES (2006). Characteristic levels of heavy metals in soil profiles of automobile mechanic waste dumps in Nigeria. Environmentalist, 123-128, DOI: 10.1007/s10669-006-7482-0.

Kelly JJ, Tate RL (1998). Effect of Heavy Metal Contamination and Remediation on Soil Microbial Communities in the Vicinity of a Zinc Smelter. J.Environ. Qual. 27:609-617.

Krewski D, Yokel RA, Nieboer E, Borchelt D, Cohen J, Harry J, Kacew S, Lindsay J, Mahfouz AM, Rondeau V (2007). Human health risk assessment for aluminium, aluminium oxide and aluminium hydroxide. J. Toxicol. Environ. Health B Crit. Rev. 10(Suppl1) 1-269

Ladigbolu IA, Balogun KJ, Shelle RO (2011). Hydrochemistry and levels of some heavy metals in samples of Ibeshe, Lagos Lagoon Complex,

Nigeria. J. Am. Sci. 7(1):625-632.

Mitchell RL (1964). Chemistry of the soil 2nd edition. John Willey and sons, New York: USA. Pp.268-320.

Namgung U, Xia Z (2001). Arsenic induces apoptosis in rat cerebellar neurons via activation of JNK3 and P38 MAP kinases. Toxicol. Appl. Pharmacol. 174:130.

National Institute of Occupational Safety and Health (NIOSH) (1991) Criteria for a recommended standard- occupational exposure to ethylene glycol Monoethyl ether and their acetates. Washington, DC: NIOSH (Publication number 91–119).

Noonan CW, Sarosua SM, Campagna D, Kathman SJ, Lybarger JA, Mueller PW (2002). Effects of exposure to low levels of environmental cadmium on renal biomarkers. Environ. Health Perspectives, 110(2): 151 -155.

Norman MT (1981) Environment and Health. Ann Arbor Inc. Michigan Pp.367-406.

Nriagu JO (1988). A Silent epidemic of environmental metal Poisoning. Environ. Pollut. 50: 139-161.

Occupational Safety and Health Administration (OSHA) (1991). Substance data sheet for Occupational exposure to Lead (W.W.W. OSHA. Gov).

Ramsay JA, Brown RHJ, Falloons SWHW (1953). Simultaneous determination of sodium and potassium in small volumes of fluid by flame photometry. J. Exptl. Biol. 30:1.

Reeves PG, Vanderpoo RA (1997). Cadmium burden of men and woman who report regular consumption of confectionery sunflower kernels containing a natural abundance of cadmium. Environ. Health Perspect. 105(10): 1098 -1104.

Schwartz J (1995). Lead, blood pressure and cardiovascular disease in men. Arch Environ. Health. 50(1): 31-37.

Sopos P, Poka K (2008) Threshold Limit Values for Heavy metals in Soils is the function of spatial and temporal variation of geochemicalScience, H-1112, Budapest, Hungary, p. 1-8.

Stromberg U, Lundh T, Skerfving S (2008). Yearly measurements of blood lead in Swedish children since 1978: the declining trend continues in the petro-lead-free period 1995-2007. Environ. Res. 107:332-335.

United States Department of Agriculture (USDA) (2000). Heavy Metal Soil Contamination. Soil Quality Institute, Natural Resources Conservation Service, Urban Technical Note No.3; pp1-7.

White LD, Cory-Slechta DA, Gilbert ME, Tiffany-Castiglioni E, Zawia Nh, Virgolini M, Rossi-George A, Lasley SM, Qian YC, Basha MR (2007). New and evolving concepts in the neurotoxicology of lead. Toxicol. Appl. Pharmacol. 225:1-27.

WHO (1982). Rapid Assessment of sources of Air, Water and Land pollution. Offset Publication; 62: 7.

WHO (IPCS) (1983) Environmental Health Criteria 27: Guidelines on Studies in Environmental Epidemiology. WHO, Geneva. Pp. 351

World Health Organization (WHO) (1996a). Trace elements in human nutrition and health. WHO, Geneva. pp. 343.

World Health Organization (WHO) (IPCS) (1996b). Environmental Health Criteria 171: Diesel Fuel and Exhaust Emissions. WHO, Geneva. 389pp.

Yokel RA, Hicks CL, Florence RL (2008). Aluminium bioavailability from basic sodium aluminium phosphate, an approved food additive emulsifying agent, incorporated in cheese. Food Chem. Toxicol. 46:2261-2266.

Zeitz PA, Orr MF, Kaye WE (2002). Public Health consequences of Mercury Spills: Hazardous Substances Emergency Events Surveillance System, 1993-1998. Environ. Health Perspect. 110(2): 129 -13.

# 15

# Water and sanitation situation in Nima and Teshie, Greater Accra Region of Ghana

Doris A. Fiasorgbor

Faculty of Development Studies, Presbyterian University College, Ghana, Akuapem Campus, Ghana.
E-mail: daffias@gmail.com.

**This study examined the water and sanitation (WATSAN) situation in Nima and Teshie, Greater Accra Region of Ghana. A number of research instruments and methods of primary and secondary data collection were employed. These were focus group discussions (FGD), field observation, and interviews. Data collected were edited, coded, and analyzed with the aid of Statistical Package for Social Scientists (SPSS) to generate tables. The residents of Nima and Teshie communities reported that they fetched water from a number of sources. Among these sources are Ghana Water Company Limited (GWCL) pipe connections, rainwater harvest and hand dug wells for Nima. Contrary to Nima, Teshie had additional sources from tanker services, streams and the sea. The price of various containers of water varied depending on the season of the year, the source of water and storage system. Also, most adult women in Nima do not patronise public toilet facilities. All the FGDs conducted in Nima indicated that residents pay a fee to dispose of their garbage into the public refuse containers but residents do not pay to do so in the Teshie community. It is recommended that the taps should be opened frequently during the day time to ease the acute water supply to the urban poor. The urban poor (especially women and children) should be informed on the proper disposal of solid waste at designated places by providing more refuse containers at vantage points, and appropriate sanitation facilities that would not exclude any group of the society should be designed.**

**Key words:** Urban poor, water, sanitation, waste management, health.

## INTRODUCTION

Providing water for the poor is one of the Millennium Development Goals (MDGs). The aim is to reduce the number of people living in poverty and the proportion of people without access to water and sanitation. More than one tenth of the world's population still relied on un-improved drinking water sources in 2010 (UNICEF/WHO, 2012). The provision of water and sanitation services in deprived urban settlements is a challenge faced by many developing countries (Boadi, 2004).The lack of these services threatens public health and the integrity of the environment in both peri-urban areas and formal urban areas (McGranahan, 2007; Mulenga et al., 2004).

In Ghana today, at least 50% of the population resides in urban areas of which only 18% have access to improved sanitation and 90% to improved drinking water sources (Water Aid, 2008). Although accessibility to improved drinking water sources look encouraging, only 40% have access to piped water, which in most cases is supplied intermittently. The remaining 60% depends on other improved sources such as standpipes, protected dug wells, protected springs and rainwater harvesting. According to the Government of Ghana, inadequate water supply and sanitation services contributed to over 70% of diseases in Ghana, costing the country significant financial resources for health care and productivity (United Nations, 2005; UNDP, 2006). The aim of this research is to examine the ways in which the urban poor deal with in-adequate access to water and sanitation services in Nima and Teshie, Greater Accra Region. The specific objectives are:

(1) To identify the water and sanitation issues of the urban poor in Accra and; (2) To investigate the effects of water and sanitation challenges on the urban poor.

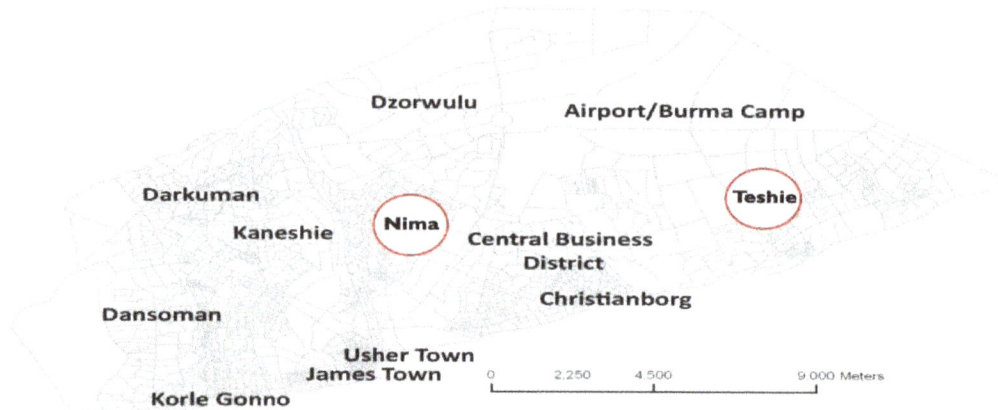

**Figure 1.** A map showing the two study communities.

## Study area

Nima and Teshie (Figure 1) were chosen for the study because they were considered poor communities in Accra by Water Aid Ghana. This paper is written out of a large study on 'Poverty and Water'. Nima is densely populated with 82,329 people in 2008 (UNESCO-IHE/SWITCH, 2010). The population is made up of various ethnic groups but predominantly Muslims from the northern parts of Ghana and neighbouring West African countries. Water supply in Nima is served by Accra North District of Ghana Water Company Limited (GWCL) while the collection of solid waste is contracted to ABC waste management company.

Teshie is located 14 km to the east direction of Central Accra, bordered with Nungua (East) and Gulf of Guinea (South). It is categorized as urban as well, with high density of indigenous people. Population in Teshiewas is about 171,875 (LEKMA, 2012). The water supply service in this area is provided by Accra East District of GWCL. Ledzokuku Krowor Municipal Assembly (LEKMA) provided waste management services where the collection of solid waste is contracted to Daben and Zoomlion waste management companies.

## METHODOLOGY

This study is part of a large study on water and poverty of two selected poor communities in Accra (Nima and Teshie). The analysis is based on data generated from various methods, namely secondary sources, FGDs, interviews (key informants), and field observations. The use of multiple methods ensured triangulation of data by allowing for the cross-checking of information, with the basic aim of validating answers and conclusions reached in the study. Poor urban communities of Accra (Nima and Teshie) were selected on the basis of several criteria, including their migrant or indigenous status, location, age of settlement, and demographic composition. Nima is largely a migrant community, while Teshie (along the coast) can be described as an indigenous community.

Twelve and five FGDs were conducted in Nima and Teshie communities, respectively. This involved men, women, the youth (between the ages of 17 and 28 years, and single), school children (between the ages of 12 and 16), and members of a Community Based Organization (CBO) in the Teshie community; key informant interviews (mainly the health and sub-metropolitan council officials responsible for the Nima and Teshie communities). With the help of a local guide, a transect walk/observation was conducted in each community, which provided an opportunity to observe directly the community water and sanitation infrastructure and services. Data collected were edited, coded, and analyzed with the aid of Statistical Package for Social Scientists (SPSS) to generate tables.

The result of this study has been published as a technical report by Water Aid Ghana. Table 1 shows the number of participants from Nima and Teshie while Table 2 shows the demographic profile of participants. Even though the population of Teshie is bigger than the population of Nima; majority of the residents are poor, whereas in Teshie, it is basically those living along the coast that were considered to be poor. Besides this, Teshie has a mixed-income population made up of fishermen, fishmongers, traders, drivers and office workers. Also, there are some parts of Teshie (Teshie Estates) that house some rich residents of Accra.

## FINDINGS

The residents of Nima and Teshie communities reported that they fetched water from a number of sources. Among these sources are GWCL pipe connections, rainwater harvest and hand dug wells for Nima. Contrary to Nima, Teshie had additional sources from tanker services and streams. Recently, "sachet water" (500 ml of water packed in a plastic sachet which is sold either cold or hot) has become one of the most preferred sources of drinking water. None of the two communities had a borehole even though the Nima community indicated it as one of the preferred sources of water. Table 3 shows the sources of water in the study areas.

Access to water was found to be difficult in that the taps flowed briefly at midnight when the people were asleep. Taps were also closed for as long as between 2 to 6 months. Whenever the taps were in good conditions, they were opened once in a week. For instance, the poor (especially women and children) spent a minimum of 5

**Table 1.** Study participants.

| Study area | Respondent | No. of respondents | No. of meetings held | Method |
|---|---|---|---|---|
| Nima | Nima East | 72 | 6 | FGD |
| | Nima West | 65 | 6 | Interview |
| | Water vendors | 15 | 16 | Interview |
| | Toilet and bath attendants | 14 | 8 | Interview |
| | Sub-Metropolitan Council | 10 | 1 | Interview |
| | Biostatistician Mamobi Poly- clinic | 1 | 1 | Interview |
| | Building Inspector | 1 | 1 | Interview |
| | Cleansing Officer | 1 | 1 | Interview |
| | GWCL | 2 | 1 | Interview |
| | Refuse dump attendant | 3 | 3 | Interview |
| | ABC Waste truck driver | 4 | 1 | Interview |
| | Assembly members | 2 | 2 | Interview |
| Teshie | Teshie Concerned Citizens Association | 59 | 1 | FGD |
| | Water vendors | 21 | 19 | Interview |
| | Toilet and bath attendants | 25 | 26 | Interview |
| | Water vendors | 21 | 19 | Interview |
| | Sub-Metropolitan Council | 3 | 1 | Interview |
| | BiostatisticianLa General Hospital | 1 | 1 | Interview |
| | Environmental Health Officers | 3 | 1 | Interview |
| | Daben Cleansing staff | 8 | 1 | Interview |
| | Traditional rulers/opinion leaders | 28 | 2 | FGD |
| | Sub-metro council | 5 | 2 | Interview |
| | Assembly members | 4 | 4 | Interview |

**Table 2.** Demographic profile of participants.

| Community | Men | Percentage (%) | Women | Percentage (%) | Total | Percentage (%) |
|---|---|---|---|---|---|---|
| Nima | 91 | 48 | 99 | 52 | 190 | 100 |
| Teshie | 86 | 48 | 92 | 52 | 178 | 100 |
| Total | 177 | 48 | 191 | 52 | 368 | 100 |

**Table 3.** Source Reliance

| Source | Percentage reliance on source usage |
|---|---|
| **Nima** | |
| GWCL | 64 |
| Rainwater | 30 |
| Sachet Water | 6 |
| **Teshie** | |
| GWCL | 40 |
| Tanker Services | 18 |
| Rainwater | 9 |
| Hand dug well | 12 |
| Streams | 6 |
| Sachet Water | 7 |
| Seawater | 8 |

minimum and a maximum 4 h in looking for water by travelling over long distances ranging between a few metres to about 10 km. They used water supplied by GWCL sources, "sachet water" and rainwater harvest. The price of various containers of water varied depending on the season of the year, the source of water and storage system. Table 4 shows the varying cost of water for Nima and Teshie, respectively.

The trotter (a container in which trotter/pig feet is stored and exported to the country), also known and commonly called "pig feet container" by respondents is not applicable to the Nima community. Also, about 48% of respondents in Nima said the toilets were far from their homes, thus it is inconvenient for them to always go to the public toilets. In this case, it did not apply to only women but all members of the community. The people of Nima and Teshie who used public toilets and facilities outside their homes reported that they would prefer having

**Table 4.** Price of water per container type.

| Container type (litres) | Rainy season | | Dry season | |
|---|---|---|---|---|
| | Direct from tap | Reservoir | Direct from tap | Reservoir |
| | USD (¢) | USD (¢) | USD (¢) | USD (¢) |
| **Nima** | | | | |
| Bucket (18) | 15-20 | 20-25 | 25-30 | 25-35 |
| Gallon (36) | 25-30 | 25-30 | 30-40 | 40-50 |
| Bowls (27) | 20-25 | 20-40 | 25-30 | 40-50 |
| | | | | |
| **Teshie** | | | | |
| Bucket (18) | 10-20 | 15-25 | 15-30 | 15-35 |
| Gallon (36) | 20-30 | 25-30 | 30-40 | 40-50 |
| Bowls (27) | 15-25 | 20-40 | 25-30 | 40-50 |
| Trotter containers (72) | 40-50 | 60-70 | 65-75 | $1.00 |

USD = United States Dollars, ¢ = cents

## Levels of access to water and sanitation services

All the participants from both communities have indicated that service charges were not affordable. In the light of unaffordability of WATSAN services, some of the residents looked for alternatives knowing the implications though. All the FGDs conducted in Nima indicated that residents pay a fee to dispose their garbage into the public refuse containers but residents do not pay to do so in the Teshie community. People who could not afford the fees threw refuse into public drains. All the participants indicated that the quantity of water used for both domestic and commercial purposes in Nima and Teshie was insufficient. All respondents at FGDs conducted in Nima complained that "sachet water" producers use water-pumping machines to divert water to their homes/storage facilities, thus depriving other community members from getting water. The respondents indicated that even though they have reported the situation to the appropriate authorities (GWCL), nothing seemed to have been done about it. According to the residents of Teshie, insufficient water supply was attributed to undue delay in the completion of the rehabilitation of the under-ground reservoir and overhead tank at the "Cold Store Area" (a suburb of the Teshie community), development of large companies and residential estates, insufficient production of water at the water source and low pressure.

It is ironic to note that in the face of all the problems enumerated by the communities, the findings revealed that there were over 50 and 40 "sachet water" producers in Nima and Teshie, respectively. A sachet of water cost 5¢ and 30 pieces packed in a bag is sold at 50¢ by distributors and retailed between 75¢ and $1.50 (US dollars) depending on the brand and location. Access to WATSAN services is considered a major problem in the two communities. The factors that account for this are presented and these factors are ordered according to their importance as expressed by the participants from the FGDs. Most mentioned problems of access to WATSAN Services in Teshie are:

1. Irregular water supply;
2. Increase industrial use of water by Coca Cola, Printexe etc.;
3. Damaged old pipelines (faulty valves and leakages);
4. Limited number of public toilets and refuse dumps;
5. Unplanned and poor access to Nima, resulting in the inability for tanker services to reach many parts of the community;
6. Springing up numerous sachet water production units;
7. Pay-as-you-use system.

## Quantity and quality of water used by the poor

The daily quantity of water used is not sufficient (especially during the dry season) as reported above. The responses revealed that even though the various households would have preferred to use more water than what is served, they have learned over the years to manage water and also adopt measures to access water in difficult times. An average of 5 persons in a family use between 2 and 5 buckets (about 18 litres) of water a day for all their domestic water needs in Nima. In Teshie, a family of 4 persons used an average of 8 buckets of water daily. However, 50% of the respondents from Nima reported that water from GWCL is the most polluted due to the rupture of some pipelines and resultant seepage of foreign materials into the water. Even though the study participants said they promptly reported burst in pipelines to their Assembly men (they represent local government electoral areas within a district) in the local government system of Ghana, they are elected by universal adult suffrage who also report to the GWCL; the company normally did not act on time. The GWCL officials however maintained when problems were reported, they responded promptly if they had the materials to repair it and delayed if the materials were not unavailable.

Also, the research found out that the GWCL did not conduct any end-user quality check. The officials of the Company said they only check the quality of water at the treatment plant, and only supply water to consumers if the water was safe to use. A participant from one of the Nima FGDs has this to say about the quality of water supplied by GWCL:

*"The pipe water is not good. I will never drink water from the taps. It is full of faeces. One day we were fetching from the taps, all of a sudden we realised that the water was dirty and smelly. Thinking it was dirty because the pipes have not been opened for a while. But in no time, we had the information that a burst pipeline was filled with human excreta from a neighbour's over flowing toilet. We had already collected human excreta into our containers. No, I will never drink that thing again".*

The research found that those living along the coast in Teshie wash with seawater and rinse themselves with sachet water or a cup of fresh water. Washing of bowls and other utensils was done with seawater.

## DISCUSSION

The poor in the two urban slums in this survey had poor access to water by not having their settlements well connected to the GWCL. A high proportion of about 46% of consumers in urban areas in the country do not have house connections, and thus obtain water through different supply options (yard connections, neighbours, private/public standpipes, trucks and domestic vendors). The residents of Teshie used water from tanker, stream and sea; these water sources were peculiar to Teshie because they were not available in Nima. For instance, the Teshie community is a coastal community and this thus explains why they used sea water. The use of the trotter container is not applicable to the Nima community. This may be due to the fact that a majority of the community residents are Moslems who are generally averse to pork and anything associated with it. The gallons and the trotter containers in Teshie were the most used containers for carrying water over long distances because these receptacles were covered and thereby minimized the incidence of pollution during transportation. Water from unconventional sources like streams and drainage gutters is used in the absence of water from safe sources.

Patronage of "sachet" or "pure" water becomes unusually high. The quantity for the various uses is also reduced to ridiculous levels like bathing with water from 'booters' (a plastic or metal kettle which the Moslems fetch water in to perform ablution before prayers). It can contain between 1.5 and 3 litres of water.

The quantity used by the households was not fixed, it changes as there occurred prolonged periods of no supply from the GWCL. Dry and rainy seasons also influenced the use of water. During the raining season, the quantity used was higher than what has been given above. The source of the rain water was polluted. The respondents indicated that the quality of water was very important to them as polluted water has health implications; thus they are very careful when it comes to quality of water, especially for drinking purposes. Some community members considered water from the GWCL source and "sachet water" as the safest for drinking

The people of Nima and Teshie do not have adequate access to sanitation facilities including an adequate waste collection system. Most of the bathrooms in the various houses do not have gutters connecting them to a nearby public drainage system. Thus when they wash, water spills all over the place. It is a common sight in both communities to see *Spirogyra* over a big area behind almost all houses. This was especially common in Teshie. This situation can have health implications for children in the areas, as they play around bare-footed in this fungal growth environment. Also, high cost, lack of privacy, inconvenience and long distance limited direct access to toilets for many households in the two study communities. The user fees charged for the use of the toilets were deemed unaffordable by the residents. However, they have no choice but to use them, but for the more vulnerable ones like the aged, children and the unemployed youth, it only makes them expose the communities and the general public to health risks by throwing their waste into public drains, opening their bowels along the beaches and throwing them into the refuse dumps. The practice of throwing faeces into refuse containers was not common in Teshie. However, 61% of the respondents living near the beach said they defecate at the beach when they are seriously pressed.

Many countries are off track in meeting the MDG sanitation target, including much of sub-Saharan Africa and several of the most populous countries in Asia (UNICEF/WHO, 2012). This study also found that about 95% of adult women in Nima do not go to public toilets due to lack of privacy at the toilets. They indicated that they feel uncomfortable in going to queue with men at the facilities, thus they prefer to defecate into carrier bags and later dump them in the public refuse containers. This practice was re-emphasised by one of the refuse dump attendants in Nima. The practice of adult women not patronising public toilets in Nima as indicated in the findings exposed the community and the general public to health problems. However in Teshie, the story was a bit different as only 12% of women expressed the same feeling but the others did not really mind queuing with their male counterparts at the facilities, though they agreed that it is not comfortable doing so. Furthermore, all the toilet facilities visited, except the metro facility in Teshie, revealed that soap and towels are not provided for hand washing. The attendants at the toilets explained that they had previously provided soap and towels at the toilets but residents were fond of stealing them hence the decision to stop providing them. This happens inspite of the fact that both the attendants and the users know the implications of not washing their hands with soap and water after using the toilet.

The residents of these two poor communities in the city of Accra indicated that they will prefer having toilets in their homes to the public facilities. This is in line with what Arku (2010) found in his study in rural communities in the Volta Region of Ghana, that community members who used toilet facilities outside their homes reported that they wished to have the facilities located in their homes and at vantage points across their communities.

## CONCLUSION AND RECOMMENDATIONS

The quality of water used by the residents of the two study areas under survey posed great deal of health concern to them. The situation was more pronounced in Nima Township where pipe borne water was usually polluted through the rupture of pipes. As a survival instinct, the people used various means to fill in the gaps created by insufficient water supply by GWCL, which resulted in perennial problems in the two areas. Those who lived along the coast in Teshie had to contend with bathing with seawater and rinsing themselves with sachet water or a cup of fresh water. Washing of bowls and other utensils was done with seawater. Indiscriminate defecation is commonplace in the urban poor townships. The beach of Teshie for instance was littered with human excreta and solid waste in much the same way as the gutters and open spaces in Nima.

The data supports the suggestion that women and girls often bear the brunt of problems associated with water and sanitation in poor urban settlement in developing countries (WEDO, 2003). Women and girls are burdened with fetching and carrying water over long distances, leaving them little time for education or to make a living. In city slums where sanitation facilities are poor or non-existent, going to the toilet at night or in the early morning puts women at risk of rape and sexual harassment.

It is recommended that the taps should be opened frequently during the day time to help ease the acute water supply to the urban poor. There should also be the cooperation between GWCL and secondary service providers to safeguard the quality of service rendered to the residents of poor urban dwelling places. The urban poor (especially women and children) should be informed on the proper disposal of solid waste at designated places by providing more refuse containers at vantage points, also daily collection of refuse by private refuse collectors operating in poor urban settlements should be ensured and appropriate sanitation facilities that would not exclude any group of the society should be designed.

## ACKNOWLEDGEMENTS

The author is grateful to Water Aid Ghana and the Presbyterian University College, Ghana for supporting this study. Thanks to Mr. Raymond Goka and Mr. Elikem Mensah who were part of the team of researchers who collected and organized the data.

### REFERENCES

Arku FS (2010). May be more intricate than you think: Making rural toilet facilities possible using the demand responsive approach. Journal of African Studies and Development Vol. 2(7), pp. 184-190.

Boadi KO (2004). Environment and Health in the Accra Metropolitan Area, Ghana. Jyväskylä Stud. Biolog. Environ.Sci. 145.

change.Government of Ghana (2005). Ghana Poverty Reduction Strategy (GPRS 2006-2009)

Ledzokuku Krowor Municipal Assembly (2012). Water and Sanitation http://lekma.lakesidegh.com/health/582/waste-management

McGranahan G (2007). Urban environments, wealth and health: shifting burdens and possible responses in low and middle-income nations. Working Paper, International Institute for Environmentand Development, London. pp. 4-8.

Mulenga M, Manase G, Fawcett B (2004). Building links for improved sanitation in poor urban settlements: recommendations from research in Southern Africa, Institute of Irrigation and Development Studies, University of Southampton, Southampton, England. pp.15-28.

SNV (Netherlands Development Organisation) (2005). Ghana Water and Sanitation Situation Analysis Report. Prepared by MAPLE Consult.

UNDP (2006). Human Development Report 2006; Beyond scarcity. Power, poverty and the global water and sanitation crisis. UNDP, New York.

UNESCO-IHE/SWITCCH (2010). Mapping Pro-Poor Water Supply Services in Accra City, Ghana.

UNICEF and WHO (2004). Meeting the MDG Drinking Water and Sanitation Target: A Mid-Term Assessment of Progress. UNICEF/WHO, Geneva, Switzerland.

UNICEF AND WHO (2012). Progress on Drinking Water and Sanitation. 2012 Update. 59 p. UNICEF AND WHO, USA. Accessed from: http://www.unicef.org/media/files/JMPreport2012.pdf

United Nations (2005). The Millennium Development Goals Report 2005. United Nations Department of Public Information, New York.

New York Water Aid (2008). Urban Sector Assessment Report. http://unstats.un.org/unsd/mi/pdf/mdg%20book.pdf

Water Aid (2009). Water, sanitation and hygiene for development: Advocacy for Women's Environment and Development Organization (2003), Gender, Water and Poverty: Key Issues, Government Commitments and Actions for Sustainable Development. Accessed from http://www.wedo.org/wp-content/uploads/untapped_eng.pdf

# 16

# Faecal and heavy metal contamination of some freshwaters and their vicinities in Ijebu-north, Southwestern Nigeria

O. M. Agbolade[1]*, O. O. Adesanya[2], T. O. Olayiwola[3] and G. C. Agu[2]

[1]Department of Plant Science and Applied Zoology, Parasitology and Medical entomology laboratory, Olabisi Onabanjo University, P. M. B. 2002, Ago-Iwoye, Ogun State, Nigeria.
[2]Department of Microbiology, Olabisi Onabanjo University, P. M. B. 2002, Ago-Iwoye, Ogun State, Nigeria.
[3]Department of Chemical Sciences, Olabisi Onabanjo University, P. M. B. 2002, Ago-Iwoye, Ogun State, Nigeria.

**Freshwater contamination poses several serious risks to human health. This study was designed to determine the level of faecal and heavy metal contamination in and around some freshwaters in Ijebu North, southwestern Nigeria. Soil samples collected from the vicinities of the freshwater bodies were examined parasitological using test tube floatation method. Soil and water samples were analyzed for Cu, Pb, Cd and Zn. Total viable count and faecal coliform count (FCC) were determined in Omi and Areru streams. *Ascaris lumbricoides* was most frequent around the water bodies. At Konigba pond, Cu, Cd and Zn had mean concentrations 2.10 ± 0.55, 0.50 ± 0.23 and 4.98 ± 2.25 mg/kg, respectively, while at Ajeri pond, Cu, Pb, Cd and Zn had mean concentrations 6.19 ± 1.56, 2.51 ± 1.99, 0.41 ± 0.15 and 58.07 ± 39.29 mg/kg, respectively. In Omi and Areru TCC ranged 5.2 - 15.4 and 4.8 - 12.2 cfu / ml × $10^4$, respectively. The ranges of Cu, Pb, Cd and Zn in Omi were 1.8 - 5.9, 0.12 - 1.18, 0.09 - 0.74, and 11.0 - 23.44 mg/l, respectively, while they were 2.1 - 5.6, 0.09 - 1.36, 0.05 - 0.79 and 11.24 - 17.34 mg/l, respectively in Areru. The study showed the need to provide regular potable water and educate the inhabitants of the study area.**

**Key words:** Freshwater, parasitic helminthes, heavy metals, coliform bacteria, Nigeria.

## INTRODUCTION

On a global scale, contamination of aquatic environments has been regarded as a significant problem with several actual and potential risks to human health. Contamination of water bodies could be through point-source pollution which includes industrial effluents, municipal sewage treatment plants, resource extraction (mining) and combined sewage-storm-water overflows. It also occurs through non-point-source pollution including agricultural runoff (pesticides, fertilizers and pathogens), storm-water and urban runoff, and atmospheric deposition of persistent organic pollutants such as mercury (Nriagu, 1979; Ross, 1994; WHO, 1995; Ritter et al., 2002). Numerous studies have been done on contamination of water bodies in many parts of the world, particularly with a view to reversing the trend and/or the resultant effects on human health (Hill et al., 2005, 2006; Meyer et al., 2005; Mishra et al., 2008; Rai, 2008). However, to the best of knowledge, no information exists in literature on water bodies' contamination from any part of Ijebu North, southwestern Nigeria despite the fact that dumping of refuses from households and mechanic workshops and indiscriminate defaecation near water bodies is common in the area.

---

*Corresponding author. E-mail: agbolmos@yahoo.com.

**Abbreviations: TVC,** Total viable count; **FCC,** faecal coliform count; **MPN,** most probable number; **EMB,** eosin methylene blue.

The immediate and future usefulness of water bodies to man cannot be overemphasized. It is therefore imperative to regularly assess the quality vis-à-vis the safety of water bodies which are frequented by humans. In view of the foregoing, this study was designed to determine the level of faecal and heavy metal contamination in some important freshwater bodies in Ijebu North, southwestern Nigeria. In addition, soil samples from the immediate vicinities of the water bodies were examined for parasitic helminth eggs and heavy metals. It is common knowledge that presence of parasitic helminth eggs is indicative of faecal contamination and that at least partly due to surface run-off, such faeces and heavy metals in the vicinities of water bodies may eventually get into the affected water bodies. It is hoped that the findings of this study will stimulate formulation of suitable strategies towards sustainable control/prevention of contamination of water bodies in the study area.

## MATERIALS AND METHODS

### Study area and sites

The study area consisted of Ago-Iwoye and Oru in Ijebu-North Local Government area of Ogun State, southwestern Nigeria. The area lies within latitudes 6° 55′ and 6° 58′ N, longitudes 3° 50′ and 3° 54′ E. Ago-Iwoye is the main seat of the Olabisi Onabanjo University while Oru is a rapidly developing town located about 3.8 km from Ago-Iwoye. The populations of the two towns have been described earlier (Agbolade and Odaibo, 1996; Agbolade et al., 2008a). The freshwater bodies visited in Ago-Iwoye are Omi stream, Ajeri pond, and Konigba pond, while only Areru stream was visited in Oru. The water bodies are important sources of water for laundry and other domestic purposes particularly during prolonged water shortage periods. In addition, Omi stream is important for bathing, swimming and edible water snail collection while Areru stream is also a source of water for manual oil-palm fruit processing, bathing and irrigation on nearby farms. Areru has a refuse dump site at one side of its bank.

### Soil sample collection

Soil samples were collected from each of two accessible sides of each freshwater body along two transects which were 50 m apart. Samples for parasitological examination were collected monthly between August 2007 and February 2008. During each working visit to each of the water bodies, two surface samples were collected along each transect, the first was 3 m (labeled 1A, 1C, 2A and 2C) while the second was 9 m (labeled 1B, 1D, 2B and 2D) away from the bank of each water body. For heavy metal analysis, soil samples were collected at the surface and at 30 cm depth on each transect once for each of Konigba and Ajeri ponds.

### Parasitological examination of soil samples

The test tube floatation method was used. Five grammes of each soil sample were mixed thoroughly with distilled water. The suspension was strained through a net mesh to remove coarse particles. The filtrate was centrifuged for three minutes and the supernatant was decanted. The resultant sediment was further broken-up by shaking and tapping the tube. The sediment was

mixed with Zinc Sulphate ($ZnSO_4$) solution (specific gravity of 1.2). A test tube was filled with the mixture, which was then allowed to stand for few minutes with a cover slip on top to collect any floating eggs. The cover slip was then removed and examined under the microscope.

### Heavy metal analysis of soil sample

Two grammes of each soil sample were digested using 2 M $HNO_3$ and then tested for Cu, Pb, Cd and Zn using atomic absorption spectrophotometer (APHA, 1998).

### Bacteriological examination and heavy metal analysis of water samples

Water samples were collected according to the routine method at four different points (25 m apart) along the course of each of Omi and Areru streams once per month in November 2007, January and February 2008. Total viable count (TVC) of each sample was determined using the routine serial dilution method. Faecal coliform count (FCC) of each water sample was determined using the most probable number (MPN) technique (APHA, 1998). This involved the presumptive test using MacConkey broth with Durham tube, confirmatory test using Eosin Methylene Blue (EMB) agar. The tubes and plates were incubated at 37°C for 24 - 48 h. Gas and turbidity in the tubes as well as metallic slum or pink dark centre colonies on EMB agar indicated positive. All isolates that produced gas at 37°C, stained gram negative and were non-spore forming and rod shaped were regarded as faecal coliform. Potato dextrose agar was used to isolate the fungi after incubation for four to seven hours at 28°C. Each of the water samples was analyzed for Cu, Pb, Cd and Zn using atomic absorption spectrophotometer (APHA, 1998).

### Statistical analysis

Mean values and standard deviations were calculated to summarize replicate values of heavy metal concentrations and microbial counts. The chi-square ($\chi^2$) test was used to compare frequencies of occurrence of helminth eggs (Frank and Althoen, 1994).

## RESULTS

The helminthes recorded in the immediate vicinities of the water bodies visited in this study are shown in Table 1. At Omi stream, Ascaris lumbricoides had statistically highest frequency of occurrence (100%, 24/24), while Schistosoma haematobium and Trichuris trichiura had similar frequency (8.3%, 2/24 each) ($\chi^2$ = 144.24, p < 0.001). 82.5% (198/240) of the A. lumbricoides eggs counted were fertile. The total number of eggs recorded for each of S. haematobium and T. trichiura was four. At Ajeri pond, A. lumbricoides had statistically highest frequency of occurrence (93.8%, 15/16), while each of S. haematobium and Schistosoma mansoni had a frequency of 6.3% (1/16) ($\chi^2$ = 143.92, p < 0.001). 1.1% (4/379) of the A. lumbricoides eggs counted was fertile. Only one egg was recorded for each of S. haematobium and S. mansoni. At Konigba pond, A. lumbricoides had statistically highest frequency of occurrence (79.2%, 19/24),

**Table 1.** Helminthes in the vicinities of some water bodies in Ijebu North, southwestern Nigeria.

| Water body | Month/Year | No of eggs per 5 g of soil[*] | | | | | | | |
|---|---|---|---|---|---|---|---|---|---|
| | | 1A | 1B | 1C | 1D | 2A | 2B | 2C | 2D |
| Omi | Aug 2007 | As (1) | As (2) | na | na | As (7) | As(6) | na | Na |
| | Sep 2007 | As (5) | As (6) | na | na | As (10) | As(7) | na | Na |
| | Oct 2007 | As (8) | As(10) | na | na | As(16) | As(12) | na | Na |
| | Nov 2007 | As (10) | As (9) Tt(12) | na | na | As(16) | As(9) | na | Na |
| | Dec 2007 | As (4) | As (6) | na | na | As(6) Sh(12) | As(4) | na | Na |
| | Jan 2008 | As(33) | As(7) | na | na | As(24) Tt (2) | As(23) Sh(2) | na | Na |
| Ajeri | Dec 2007 | As(15) | 0 | na | na | As(4) | As(50) | na | Na |
| | Jan 2008 | As(1) | As(3) | na | na | As(30) | As(11) Sm(1) | na | Na |
| | Feb 2008 | As(41) | As(10) | na | na | As(13) | As(90) | na | Na |
| | Mar 2008 | As(30) | As(36) Sh(1) | na | na | As(20) | As(25) | na | Na |
| Koni-gba | Dec 2007 | As(2) | As(25) Sh(2) | As(3) | As(20) | na | na | As(1) | 0 |
| | Jan 2008 | 0 | As(6) Sh(1) | As(1) | 0 | na | na | As(3) St(1) | As(3) |
| | Feb 2008 | As(20) | As(44) | As(39) | As(50) Tt(1) | na | na | As(16) | As(50) Sh(1) |
| | Mar 2008 | As(24) Sh(1) | As(43) | 0 | 0 | na | na | As(5) | As(30) |
| Areru | Aug 2007 | As(6) Tt (2) | As(14) | As(2) Tt (5) | na | na | na | As(21) Sh(3) | St(1) |
| | Sep 2007 | As(8) | As(11) Ta(1) | As(11) Hk(2) | As(6) Sh(5) | na | na | As(9) | As(3) |
| | Oct 2007 | As(2) | As(16) | As(12) | na | na | na | As(20) | As(1) |
| | Nov 2007 | As(11) | As(1) | As(20) | As(25) | na | na | As(27) | As(7) St(2) |
| | Dec 2007 | As(23) | 0 | Tt (27) | As(7) Hk(1) | na | na | As(13) | Sh(1) |
| | Jan 2008 | As(64) | As(94) | As(65) St(5) | As(47) | na | na | As(78) | As (301) |
| | Feb 2008 | As(5) | As(1) Sh(1) | As(5) Sh(1) | As(27) Sh(1) | na | na | na | As(33) Sh(1) |

[*] As = A .lumbricoides, Tt = T. trichiura, Sh = S. haematobium, Sm = S.mansoni, St = S. stercoralis, Ta = Taenia, Hk = Hookworm, na = Not assessible.

followed by *S. haematobium* (16.7%, 4/24), *Strongyloides stercoralis* (4.2%, 1/24) and *T. trichiura* (4.2%, 1/24) ($x^2$ = 148.31, p < 0.001). 11.2% (43/385) of the *A. lumbricoides* eggs counted were fertile. The total number of eggs for *S. haematobium* was five, while each of *T. trichiura* and *S. stercoralis* had one.

At Areru stream, *A lumbricoides* hadstatistically highest frequency of occurrence (92.3%, 36/39), followed by *S. haematobium* (17.9%, 7/39), *T. trichiura* (7.7%, 3/39), *S. stercoralis* (7.7%, 3/39), hookworm (5.1%, 2/39) and *Taenia* (2.6%, 1/39) ($x^2$ = 271.41, p < 0.001). The total eggs for *A. lumbricoides, T. trichiura, S. haematobium,* hookworm, *S. stercoralis* and *Taenia* were 996, 34, 13, 3, 8, and 1, respectively.

**Table 2.** Microbial counts from Omi and Areru streams, Ijebu North, southwestern Nigeria

| Stream | Month/Year | Total viable count ($\times 10^4$ cfu/ml)[*] | Total coliform count ($\times 10^4$ cfu/ ml)[*] |
|--------|-----------|-------------------------------|--------------------------------|
| Omi | Nov. 2007 | 24.7 ± 3.0 | 10.2 ± 5.2 |
| | Jan. 2008 | 18.2 ± 1.6 | 10.3 ± 2.5 |
| | Feb. 2008 | 16.0 ± 2.6 | 7.6 ± 1.9 |
| Areru | Nov. 2007 | 22.8 ± 4.5 | 6.5 ± 1.5 |
| | Jan. 2008 | 13.8 ± 1.9 | 8.5 ± 2.9 |
| | Feb. 2008 | 10.7 ± 1.5 | 8.7 ± 1.8 |

[*] Values are Mean ± S.D of four replicates.

**Table 3.** Concentrations of heavy metals in Omi and Areru streams, Ijebu North, southwestern Nigeria

| Stream | Month/Year | Concentration (mg / l)[*] | | | |
|--------|-----------|------|------|------|------|
| | | Cu | Pb | Cd | Zn |
| Omi | Nov. 2007 | 2.60 ± 0.70 | 0.25 ± 0.10 | 0.18 ± 0.09 | 18.20 ± 4.27 |
| | Jan. 2008 | 5.15 ± 0.66 | 0.94 ± 0.18 | 0.58 ± 0.16 | 15.66 ± 3.18 |
| | Feb. 2008 | 4.90 ± 0.53 | 0.64 ± 0.36 | 0.60 ± 0.09 | 12.59 ± 1.39 |
| Areru | Nov. 2007 | 2.65 ± 0.47 | 0.19 ± 0.09 | 0.10 ± 0.04 | 15.91 ± 1.28 |
| | Jan. 2008 | 4.50 ± 0.93 | 1.10 ± 0.23 | 0.53 ± 0.20 | 15.27 ± 2.02 |
| | Feb. 2008 | 3.48 ± 0.56 | 0.85 ± 0.13 | 0.35 ± 0.06 | 13.72 ± 2.09 |

Values are Mean ± S.D of four replicates.

At Konigba pond, Cu, Cd and Zn were detected in 33.3, 58.3 and 100% of the soil samples with concentrations within the ranges of 1.55 - 2.80, 0.30 - 0.95 and 0.45 - 8.45 mg/kg with means of 2.10 ± 0.55, 0.50 ± 0.23 and 4.98 ± 2.25 mg/kg, respectively. Pb was not detected at this pond. At Ajeri pond, Cu, Pb, Cd and Zn were detected in 87.5, 50.0, 87.5 and 100% of the soil samples with concentrations within the ranges of 4.60 - 9.40, 1.45 - 5.50, 0.15 - 0.55 and 2.55 - 114.50 mg/kg with means of 6.19 ± 1.56, 2.51 ±  1.99, 0.41 ± 0.15 and 58.07 ± 39.29 mg/kg, respectively.

The results of the bacteriological analysis of water samples from Omi and Areru streams are shown in Table 2. The ranges of TVC in Omi and Areru were 13.6 - 28.2 and 8.6 - 27.8 $\times 10^4$ cfu / ml, respectively. The ranges of TCC in Omi and Areru were 5.2 - 15.4 and 4.8 - 12.2 $\times 10^4$ cfu / ml, respectively. The microbes recorded were *Escherichia coli, Staphylococcus aureus, Bacillus cereus, Bacillus subtilis, Proteus vulgaricus, Proteus morganii, Pseudomonas aeruginosa, Streptococcus faecium, Klebsiella aerogenes, Aspergillus fumigatus, Aspergillus parasiticus, Aspergillus tamarii, Fusarium oxysporium,* and *Penicillium oxalicum.* 57.1% (8) of the 14 different microbial species recorded occurred consistently along the water course of both streams throughout the study. These include *E. coli, S. aureus, B. cereus, B. subtilis, P. vulgaricus, P. aeruginosa* and *S. faecium. A. fumigatus* was the only consistent fungi isolated. 14.3% (2/14) of the microbial species isolated were associated with Areru only and these were *P. morganii* and *K. aerogenes.* The

ranges of Cu, Pb, Cd and Zn in water samples from Omi were1.8 - 5.9, 0.12 - 1.18, 0.09 - 0.74, and 11.0 - 23.44 mg/l, respectively, while they were 2.1 - 5.6, 0.09 - 1.36, 0.05 - 0.79 and 11.24 - 17.34 mg/l, respectively in Areru. Table 3 shows the mean concentrations of the heavy metals in the streams.

## DISCUSSION

The occurrence of eggs of parasitic helminthes in the immediate vicinities of water bodies in this study is indicative of high level of contamination of their surroundings. The practice of indiscriminate defaecation is common around virtually all the water bodies studied. The presence of a refuse dump site at one side of Areru is a source of concern. A previous study showed heavy faecal and parasitic helminthes contamination of dump sites in the study area (Agbolade et al., 2009). Observations also showed that dumping of refuse and/or human faeces into water drainages is a common practice in the study area. The public health significance of this is that nearby water bodies are contaminated with human faeces, particularly in rainy season. Much of the faecal contamination in the water bodies might have been from the immediate vicinities of the water bodies. The observation that substantial percentage of *A. lumbricoides* eggs recorded from the vicinities of three of the water bodies that was fertile is important. Many of such fertile eggs may be blown by wind or flushed by surface runoff into the water bodies. Transmission of *A. lumbricoides* through the water bodies cannot be ruled out. This is because they enjoy regular human contact and, therefore, direct and/or indirect ingestion of the contaminated water seems almost inevitable. The study area has long been known as being endemic for human schistosomiasis, especially the urinary type and the water snail vectors are present (Agbolade et al., 1996; 2004; Okunuga and Agbolade, 1998). Due to frequent contamination of the water bodies, regular supply of the vectors with miracidia from human schistosome eggs seems almost guaranteed.

The microbial analysis of the streams included in this study shows that their TVC exceeded the recommended limit of $1.0 \times 10^2$ cfu / ml (WHO, 1971). The presence of faecal coliforms is an index of bacteriological quality of water. Their presence in the studied streams further lent support to the inference that faecal deposits and waste dumps find their way into them (Hill et al., 2006). It was noticed that the lowest faecal count was observed at sampling point closest to the road. It may be that humans who defaecate directly into the water bodies do so far away from the open, while faecal deposits are carried to other points of the water bodies by runoff (Crowther et al., 2001). Nevertheless, smaller mammals (particularly rodents) might have been coming around the streams to drink water in the course of which they defaecate into them (Belton et al., 1999; Banwo, 2006).

Generally, there is a gradual reduction in the mean TVC in both Omi and Areru streams from November through February. In Omi stream, the mean TCC increased slightly in January, but decreased sharply in February. The period of this study (November - February) coincided with the end of the rainy season. The practice of faeces disposal into drainages seems more prevalent during the rainy season. The belief among the autochthonous population is that the faeces would be washed away by rain water. This probably accounts for the gradual decrease in the mean TVC in both streams and mean TCC in Omi stream. However, in Areru, the mean TCC increased appreciably from in November and reached the peak ($8.70 \times 10^4$ cfu/ml) in February. It is possible that the human defaecation directly into the stream was not reduced as the rains subsided, or animals were responsible for the continued faecal contamination during this period. The gradual increase in mean TCC in Areru might have been due partly to the emergence of *P. morganii* and, more importantly, *K. aerogenes*. These bacterial species seemed to have replaced some fungal species including *A. tamarii*, *A. parasiticus*, *P. oxalicum* and *F. oxysporium*. Possibly *P. morganii* and/or *K. aerogenes* are highly repulsive to the growth of those fungal species. This requires further studies.

The results of this study show that, *E. coli*, *S. aureus*, *B. cereus*, *B. subtilis*, *P. vulgaricus*, *P. aeruginosa* and *S. faecium* are the predominant bacterial species in the streams. The presence of *E. coli* (a coliform bacteria species) in the streams depicts their being unsafe for human use,

particularly for drinking and cooking. The presence of heavy metals in the vicinities and surface water of water bodies in this study is another source of concern. It is known that the occurrence of many heavy metals, such as lead, are often a consequence of anthropogenic and, sometimes, natural processes (WHO, 1995). Refuse dumping directly and/or indirectly into the water bodies are part of the anthropogenic processes which might have contributed to the presence of the metals in and/or around the water bodies. Increased environmental pollution from exhausts of automobiles in the towns is also anthropogenic and this might have contributed to the quantity of Pb available in the water bodies and their surroundings. Gradual release of heavy metals from contaminated surroundings into water bodies cannot be ruled.

Zn and Cu are some of the heavy metals that have been reported to be of useful importance to man and plants (Brady and Weil, 1999; Wardlaw, 2003). However, there concentrations in surface water are often a major factor. For instance, in this study Zn exceeded the maximum tolerable concentration (WHO, 1971; Duruibe et al., 2007). On the other hand, Pb and Cd are toxic even at extremely low levels (WHO, 1971; Wardlaw, 2003) and yet both occurred at unbearably high concentrations in this study.

Unfortunately, apart from the possibility of direct ingestion of heavy metals by humans, studies have shown that some fishes and freshwater snails, including edible ones such as *Lanistes libycus*, innately accumulate heavy metals in their bodies (Adewunmi et al., 1996; Agbolade et al., 2008b; Ekpo et al., 2008). It has earlier been noted that specimens of *L. libycus* from the study area sometimes have *Fusarium sp* attached to their shells (Agbolade et al., 2008b). The combined effect of all these is that humans who frequent the water bodies do so at the detriment of their health.

The findings of this study show that the studied water bodies are unsafe in their present polluted conditions. There is urgent need to ensure adequate and regular provision of safe potable water and refuse management facilities in the study area. Provision and adequate maintenance of household and public toilets by landlords and the government, respectively, in the area is desirable. Adequate education of the inhabitants of the study area on the significance of personal and environmental hygiene particularly with regard to the dangers of indiscriminate refuse and faecal disposal is also urgently needed.

## ACKNOWLEDGEMENTS

The authors appreciate the Technologists of the Institute of Agricultural Research and Training, and Rotas Soilab Limited, Ibadan, Nigeria, for their useful suggestions and technical assistance on heavy metal analysis of water and soil samples.

## REFERENCES

Adewunmi CO, Becker W, Kuchnast O, Oluwole F, Dorfler G (1996). Accumulation of copper, lead and cadmium in freshwater snails in southwestern Nigeria. Sci. Total Environ., 193: 69-73.

Agbolade OM, Odaibo AB (1996). *Schistosoma haematobium* infection among pupils and snail intermediate hosts in Ago-Iwoye, Ogun State. Nigerian J. Parasitol., 17: 17-21.

Agbolade OM, Akinboye DO, Fajebe OT, Abolade OM, Adebambo AA (2004). Human urinary schistosomiasis transmission foci and period in an endemic town of Ijebu North, Southwest Nigeria. Trop. Biomed. 21 (Suppl), 15-22.

Agbolade OM, Akintola OB, Agu NC, Raufu T, Johnson O (2008a). Protection practices against mosquito among students of a tertiary institution in southwest Nigeria. Wld. Appl. Sci. J., 5 (1): 25-28.

Agbolade OM, Olayiwola TO, Onomibre EG, Momodu LA, Adegboyegun-King OO (2008b). Trado-medicinal and nutritional values and biosafety of *Lanistes libycus* in Ijebu-North, southwest Nigeria. Wld. Appl. Sci. J., 3(6): 921-925.

Agbolade OM, Oni TT, Fagunwa OE, Lawal KM, Adesemowo A (2009). Faecal contamination of dump sites in some communities in Ijebu-North, south-western Nigeria. Nigerian J. Parasitol., 30 (2): 57-60.

APHA (1998). Standard methods for the examination of water and wastewater. 20th ed. American Public Health Association, Washington DC.

Banwo K (2006). Nutrient load and pollution study of some selected stations along Ogunpa river in Ibadan, Nigeria. M.Sc. Dissertation. University of Ibadan, Ibadan, Nigeria.

Belton D, Ryan T, Irwin G, Cameron C, Dugan-Zich D (1999). National stock drinking water telephone survey. June 1988. MAF Quality Management (Ministry of Agriculture and Forestry), Ruakura Research Centre, Hamilton.

Brady NC, Weil RR (1999). The Nature and Properties of Soils. 12th ed. Prentice-Hall Inc., New Jersey.

Crowther J, Kay D, Wyer MD (2001). Relationship between microbial water quality and environmental conditions in coastal recreational waters: the Fylde coast, UK. Water Res., 35(17): 4029-4038.

Duruibe JO, Ogwuegbu MOC, Egwurugwu JN (2007). Heavy metal pollution and human biotoxic effects. Int. J. Phys. Sci., 2(5): 112-118.

Ekpo KE, Asia IO, Amayo KO, Jegede DA (2008). Determination of lead, cadmium and mercury in surrounding water and organs of some species of fish from Ikpoba river in Benin city, Nigeria. Int. J. Phys. Sci., 3 (11): 289-292.

Frank H, Althoen SC (1994). Statistics: Concepts and Applications. Cambridge University Press, Cambridge.

Hill DD, Owens WE, Tchounwou PB (2005). Comparative assessment of the physico-chemical and bacteriological qualities of selected streams in Louisiana. Int. J. Environ. Res. Public Hlth., 2(1): 94-100.

Hill DD, Owens WE, Tchounwou PB (2006). The impact of rainfall on fecal coliform bacteria in Bayou Dorcheat (North Louisiana). Int. J. Environ. Res. Public Hlth., 3(1): 114-117.

Meyer KJ, Appletoft CM, Schwemm AK, Uzoigwe JC, Brown EJ (2005). Determining the source of fecal contamination in recreational waters. J. Environ. Hlth., 68(1): 25-30.

Mishra VK, Upadhyaya AR, Pandey SK, Tripathi BD (2008). Heavy metal pollution induced due to coal mining effluent on surrounding aquatic ecosystem and its management through naturally occurring aquatic macrophytes. Bioresource Technol., 99(5): 930-936.

Nriagu JO (1979). Global inventory of natural and anthropogenic emissions of trace metals to the atmosphere. Nature, 279: 409-411.

Okunuga AO, Agbolade OM (1998). Urinary schistosomiasis among school children in Oru, Ogun State. Nigerian J. Sci., 32: 71-74.

Rai PK (2008). Heavy metal pollution in aquatic ecosystems and its phytoremediation using wetland plants: an ecosustainable approach. Int. J. Phytoremediation, 10(2): 131-158.

Ritter L, Solomon K, Sibley P, Hall K, Keen P, Mattu G, Linton B (2002). Sources, pathways, and relative risks of contaminants in surface water and groundwater: a perspective prepared for the Walkerton inquiry. J. Toxicol. Environmental Hlth., 65(1): 1- 142.

Ross SM (1994). Toxic metals in soil-plant systems. Wiley, Chichester.

Wardlaw GM (2003). Contemporary Nutrition: Issues and Insights. 5th ed. McGraw Hill, New York, pp. 321-325, 555-558.

WHO (1971). International Standard for Drinking Water. 3rd ed. Geneva.

WHO (1995). Environmental Health Criteria 165. International Programme on Chemical Safety. Geneva.

# 17

# Effects of gas flaring on rainwater quality in Bayelsa State, Eastern Niger-Delta region, Nigeria

**E. E. Ezenwaji[1], A. C. Okoye[2] and V. I. Otti[3]**

[1]Department of Geography and Meteorology, Nnamdi Azikiwe University, Awka, Nigeria.
[2]Department of Environmental Management, Nnamdi Azikiwe University, Awka, Nigeria.
[3]Department of Civil Engineering, Federal Polytechnic, Oko, Nigeria.

The aim of the paper was to study the spatial effects of gas flaring on rainwater quality in Bayelsa State, Eastern Niger-Delta, Nigeria. The physicochemical variables were isolated from the analysis of rainwater samples from eight locations in the State, while mean monthly rainfall data for 2011 of the selected areas was collected from the Port Harcourt International Airport which has a distance of 15 km to the farthest location and 6 km to the nearest location. The analysis of physicochemical elements was variously done using the relevant methods. The result shows that all the eight physicochemical elements have values above the World health Organisation (WHO) 2004 maximum allowable concentration level, but in varying degrees with $NO_3^-$ achieving the highest. Pearson's product moment correlation was employed to establish relationship between the eight variables and the quantity of gas flaring in the area, then principal component analysis (PCA) was utilized to collapse the eight variables into significant and orthogonal components. After these two analyses, the principal component regression (PCR) analysis was used to calculate the relative contributions of the physicochemical variables with $NO_3$ contributing the highest of 38.44% to poor quality of rain water in the area, while temperature contributed the least of 0.19%. After the PCR calculation of the entire State, it was further performed in individual locations. Result showed that all local government areas in South-West zone of the State have high rate of nitric acid accumulation in rainwater sample with those in the central zone having average rate of accumulation, while those in north-western zone exhibiting a low rate. Conclusions were drawn from the result while the major recommendation was on the need to develop a home grown gas flaring policy option that will address the problem.

**Key words**: Eastern, Delta, analysis, elements, rainfall.

## INTRODUCTION

Gas flaring is an unavoidable part of the petroleum process confirming that there is hardly any oil producing nation which do not flare some percentage of her gas. However, statistics from various countries show that none of the affected countries flares as much quantity of gas as Nigeria. Atevure (2004) gave statistics of the percentage of gas flaring in some known world producers of oil. According to him, Libya for instance flares about 21% of its natural gas, while Saudi Arabia, Canada and Algeria flare 20, 8 and 5% respectively, conversely Nigeria flares over 90% of its gases.

Since after the Nigerian Civil War (1967 to 1970), considering this level of flaring there had been continuous concern by Nigerians over the ways the gases are flared in the Niger-Delta region (Enehero, 1973; Aggrey 1983; Obadina, 2000; Oghifo, 2001). Olukoya (2008) noted that Nigeria is the world's biggest flarer of Associated Gas (AG) with more than 1,000 gas flaring points that release over 23 billion/m$^3$ of gas per annum. Surprisingly, notwithstanding this magnitude of flaring, over 80% are not recovered (Evo, 2002). Since the discovery of oil in Nigeria, particularly in the Niger-Delta region in 1956 at Oloibiri, Enete and Ijioma (2011) noted that gas has constantly been flared in Nigeria, polluting the Niger Delta environment. This is indeed serious, because of the large deposit of gas estimated at 120 trillion cubic feet which made Nigeria the ninth largest concentration in the world (Atevure, 2004). The deleterious effect of gas flaring on the environment is wide and varied. For example Abube (1988) noted that gas flaring destroys the aquatic environment, Okezie (1989) investigated its effect at Izombe field on the growth potentials, productivity and yield of selected farm crops and established that heat radiation results in micro bacteria decline of the affected areas which gives rise to poor farm yields.

Recent studies have investigated the impact of gas flaring on micro-climate and vegetation in which their findings indicate that over 10 hectares of vegetal land were destroyed in the area in 1998 (Odilison, 1999; Efe, 2003), on soil, air and water quality; the results show that virtually all rivers in the area are polluted (Ekanem, 2001), This is same with plant growth and vegetation in general (Dengimo, 2008; Abara, 2009). Some authors have closely associated gas flaring to the increasing poverty in the affected communities. For example Uge (2009) established a correlation between the quantity of gas flared and poverty level in Otakeme area of Bayelsa State and concluded that a strong positive correlation of 0.60 existed between the two variables. It is, however, no longer surprising that the long political agitation in the Niger-Delta region is largely attributable to the environmental degradation of oil exploitation which gas flaring is a major component. Gas flaring results in acid rain within the flared micro environment. This has been extensively discussed by Adesanya (1984), Odjugo (2002), Efe (2002), and Ogunkoya and Efe (2003), and its effect on rainwater resources have been extensively investigated in the Western Niger-Delta region (Efe, 2002; Efe et al., 2005; Rim-Rukeh et al., 2005; Lawan, 2010), but the degree and geographical extent of gas flaring on rainwater in the eastern parts of the Niger-Delta region has remained relatively scanty. This paper, therefore, seeks to examine the effects of continuous gas flaring on rainwater in Bayelsa State, Eastern Niger Delta which is one of the largest gas flaring areas in the country. The result of the paper will enable relevant government authorities to appreciate the magnitude of the flaring in the area and ensure that many laws already in place to address the problem are properly enforced.

## MATERIALS AND METHODS

### Area of study

Bayelsa State is situated in eastern Niger Delta region located between latitudes 04°15`N and 05°23`N and between longitudes

longitudes 05°22`E and 06°45`E. It is one of the major oil producing States in Nigeria contributing over 40% of the daily production in the country. It has 17 on-shore oil flow stations which flare an average of 800,00 m$^3$ of gas per day for each of the stations and a total of 13,700,000 m$^3$ of gas per day for the entire state (Bayelsa State Government, 2010). It is one of the six States that make up the south-south geopolitical region of Nigeria and has boundaries with Rivers State in the east, Delta State in the west and Gulf of Guinea in the south. It has a total land area of 9,059 km$^2$ with a population density of 188 persons/km$^2$ (Bayelsa State Government, 2008). The state is drained by so many rivers which are at their advanced stage and such rivers include Orashi which forms the State's eastern border with Rivers State, Nun which forms the western border with Delta State, Apoi and Kugba and numerous creeks which drain the state's hinterland. Vegetation is mainly the mangrove and salt water swamps, but a major part had largely been destroyed by oil exploration. Annual rainfall amounts range from 2,500 mm in the northern parts of the State to 4,000 mm in its southern areas (Olowoyo, 2011), while mean temperature is generally 28.0°C. The State has eight local government areas (LGAs), namely; Brass, Ekeremor, Kolokuma/Opukuma, Nembe, Ogbia, Sagbama, Southern Ijaw and Yenogoa and has developed in many sectors since its creation in October 1, 1996 (Bayelsa State Government, 2008).

### Data collection

Water samples for this study were collected from rainwater at the beginning of the rainy season in all the eight LGAs of the state between March and May, 2012 as this was the time when gaseous impurities in the air reached their highest concentration having not been optimally removed by the rainwater on account of the dry season that lasted for some months. The first major rainfall was recorded in most of these areas between February 24 and 28, 2012, but our water samples were collected from the middle of March, 2012 to allow impurities to be removed by the early rains. The method used in collecting the water samples followed that employed by Okoye et al. (2011) in which free fall rainwater was collected with a rainfall collector mounted on a support 1.5 m above the ground to avoid rain splashing effect. The funnel of the collector was filled with fibre to screen out inserts and trap debris. The rainwater samples were stored for not more than 24 h at a temperature of about 4°C in a refrigerator prior to analysis. Mean monthly rainfall amounts of the area were collected from the nearby Port Harcourt International Airport.

### Data analysis

#### Physicochemical analysis

The analysis was done to determine the level of atmospheric pollutants, mainly temperature (T°C), total dissolved solids (TDS), conductivity, nitric acid (NO$_3^-$), sulphuric acid (SO$_4^{2-}$), carbonic acid (CO$_3^-$), lead (pb), and pH level. Some unstable parameters such as temperature, pH and conductivity were determined on the spot, using thermometer and pH metre (Jenway model No 3520) and conductive metre (Jenway model No. 470), respectively. The concentration of Pb and iron (Fe) in the water samples were determined with atomic absorption spectrophotometer (AAS). The wave lengths used for measuring were 217.0 nm for Pb and 204 nm for Fe. The investigation of chemical parameters (CO$_3$, NO$_3^-$, SO$_4^{2-}$) were done using a range of techniques including AAS methods (APHA, 1992), while electrical conductivity was measured with a conductivity meter which gave the readings directly in microsiemens per centimetre (μ.Scm$^{-1}$) at 25°C.The conductivity metre (K constant = 0.1) was standardised from time to time using stand solutions.

## Statistical analysis

The major statistical techniques employed were Pearson's product moment correlation technique to establish relationships between the eight physicochemical parameters associated with the quantity of gas flaring in the area. Principal component analysis (PCA) was utilized to collapse the eight variables into significant and orthogonal components while principal component regression (PCR) was employed to develop the regression model that would have stable coefficients. It is necessary to state that the use of PCA and PCR arose as a result of the severe autocorrelation noticed in our data. All the statistical analyses were performed with the aid of Statistical package for social sciences (SPSS) version 20 running on Windows PC.

## RESULTS AND DISCUSSION

The result of the physicochemical and statistical analysis is presented in Tables 1 to 3. The inconsistence units in which various parameters were recorded were standardized by the model for the PCA and PCR analysis.

The relationship between the eight physicochemical parameters which were the result of gas flaring and rain water in Bayelsa State was established with the use of Pearson's product moment correlation technique. The result of the statistics is presented in Table 3.

Table 3 reveals a high association between some variables as well as serial autocorrelation as many factors show strong and significant positive correlation with each order. For example $X_1$ (temp) is strongly and positively correlated with $X_2$ (Pb) (0.69), same for $X_1$ (Temp) and $X_8$ (pH) (0.72), $X_3$ (Condu) and $X_2$ (Pb) (0.60) and $X_4$ (TDS) (0.66). Furthermore, $X_4$ (TDS) is very highly correlated with $X_3$ (Cond.), then $X_5$ (NO$_3$) with $X_8$ (pH) (0.74). Same for $X_6$ (CO$_3$) and $X_8$ pH (0.81), $X_7$ (SO$_4$) and $X_8$ (pH) (0.88). With these very serious autocorrelations that characterized our data, there was no other alternative than to subject our correlation result to PCA, so as to transform them into defined orthogonal components that can be employed further for PCR. When PCA was transformed, the primacy of two components manifested (Table 4). Varimax rotation however maximizes the covariance of loadings on each component in order to achieve as many high and as many low loadings as possible while maintaining the orthogonality (that is, the uncorrelation) of the original component.

The two aforementioned components provided more information on the variables as well as exhibit high stability of the variables and were therefore used to perform the Multiple Regression Analysis (MRA). Component I stood out clearly in this regard and was therefore selected. The use of the values of the variables achieved by the PCA for the performance of PCR was to ensure that the variables are as orthogonal as possible so as to make their coefficients are strong, thereby ensuring that unnecessary interference or 'noise' arising from severe

autocorrelation is highly minimized or completely removed. This calculation which was done for the entire State was extended to each of the eight LGAs for us to ascertain their spatial disposition. The result for the entire state is as follows:

$$Y = 286.4 - 1.88X_1 + 047X_2 - 0.76X_3 + 1.02X_4 + 5.64X_5 + 3.02X_6 + 4.1X_7 - 2.60X_8 \quad (1)$$

From Equation 1, it was revealed that the combined contribution of the independent variables to the overall variation was 91.6%. The significance of each parameter in the model was formulated and the result presented in Table 5.

Analysis of variance (ANOVA) was further performed to determine the adequacy or otherwise of the formulated model (Table 6) usually, the p-value less than 0.05 indicates the adequacy of the model while the contrary means that the model is inadequate. When the model is inadequate, it goes to explain that it cannot sufficiently explain the variations in the dependent variable. In this analysis, our model offered a good explanation of the dependent variable. This is because in Table 6, the p-value is less than 0.05 which implies that the model can adequately explain variation in the dependent relationship between the amount of monthly rainfall and the eight physicochemical variables.

To determine the relative importance of the physicochemical variables, successive values of the multiple correlation coefficients were calculated by introducing successive independent variables at each computation that is, Ry. $X_1$, Ry. $X_1$, $X_2$, Ry. $X_1$, $X_2$, $X_3$, etc. The difference between the squared multiple correlations ($R^2$) are regarded as the contribution of each variable.

Table 7 shows the relative contribution of the variables to poor rainwater quality in Bayelsa State.

The earlier calculation was repeated for each of the eight locations used in the study and the result is presented in Table 8. Bivariate Linear Regression (BLR) analysis performed between the mean amount of monthly rainfall and nitric acid which is the most important variable for each of the eight sample locations (towns) shows a startling revelation

## DISCUSSION

Table 2 shows that the level of concentration of all the analysed physicochemical parameters were above the WHO maximum allowable level, but the worse affected are the nitric acid, carbonic acid and sulphuric acids which show marked concentration above the maximum standard required. This result supported the long view held by the villagers in Niger Delta that gas flaring was damaging their health, reducing their crop production and destroying their homes. Edem (2011) noted that in 2010,

**Table 1.** Physicochemical parameters employed in the analysis.

| Label | Unit of measurement | Description | Max. allowable concentration (WHO, 2004) |
|---|---|---|---|
| $X_1$ (Temp) | °C | Average temperature of water for the period of data collection | 32.5 |
| $X_2$ (pb) | mg/L | Presence of lead in water | 0.10 |
| $X_3$ (condu) | μs/cm | Level of conductivity | 1000 |
| $X_4$ (TDS) | mg/L | Level of solid content in water | 500 |
| $X_5$ ($NO_3^{2-}$) | mg/L | Level of nitric acid in water | 30 |
| $X_6$ ($CO_3$) | mg/L | Content of carbonic acid in water | 600 |
| $X_7$ ($SO_4^{2-}$) | mg/L | Presence of sulphuric acid in water | 500 |
| $X_8$ (pH) at 25 °C | units | Acidity or alkalinity level in water | 6.5 – 9.2 |
| $X_9$ (Amount) | mm | Quantity of water collected | - |

$X_8$ is the total quantity of water collected during the field work which was used as the dependent variable in the regression analysis, while physicochemical parameters where employed as the dependent variable.

**Table 2.** Laboratory result of rainwater in Bayelsa State.

| Sample location | Variable | | | | | | | | |
|---|---|---|---|---|---|---|---|---|---|
| | Amount (mm) | Temp (°C) $X_1$ | Pb (mg/L) $X_2$ | Condu (μs/cm) $X_3$ | TDS (Mg/L) $X_4$ | $NO_3$ (Mg/L) $X_5$ | $CO_3$ (Mg/L) $X_6$ | $SO_4$ (Mg/L) $X_7$ | pH $X_8$ |
| Bessambiri | 325 | 30 | 0.49 | 61 | 501 | 46 | 640 | 586 | 5.3 |
| Ekeremor | 188 | 29 | 0.31 | 54 | 570 | 44 | 720 | 610 | 5.4 |
| Kaiama | 186 | 30 | 0.28 | 45 | 514 | 41 | 739 | 629 | 5.3 |
| Ogbia | 215 | 31 | 0.24 | 76 | 686 | 49 | 802 | 631 | 5.0 |
| Oporoma | 339 | 30 | 0.38 | 50 | 549 | 40 | 700 | 602 | 5.1 |
| Sagbama | 210 | 30 | 0.20 | 65 | 563 | 39 | 680 | 626 | 5.4 |
| Twon | 350 | 29 | 0.30 | 74 | 504 | 41 | 710 | 600 | 5.3 |
| Yenogoa | 198 | 29 | 0.30 | 51 | 558 | 40 | 704 | 644 | 5.3 |

**Table 3.** Correlation matrix of physicochemical variables.

| Variable | $X_1$ | $X_2$ | $X_3$ | $X_4$ | $X_5$ | $X_6$ | $X_7$ | $X_8$ | $X_9$ |
|---|---|---|---|---|---|---|---|---|---|
| $X_1$ (Temp) | 1.00 | | | | | | | | |
| $X_2$ (Pb) | 0.69* | 1.00 | | | | | | | |
| $X_3$ (Condu) | 0.02 | 0.60* | 1.00 | | | | | | |
| $X_4$ (TDS) | 0.34 | 0.66* | 0.83* | 1.00 | | | | | |
| $X_5$ ($NO_3$) | 0.28 | 0.20 | 0.61* | 0.44 | 1.00 | | | | |
| $X_6$ ($CO_3$) | 0.17 | 0.36 | 0.04 | 0.64* | 0.38 | 1.00 | | | |
| $X_7$ ($SO_4$) | 0.23 | 0.11 | 0.15 | 0.50 | 0.21 | 0.44 | 1.00 | | |
| $X_8$ (pH) | 0.72* | 0.24 | 0.08 | 0.10 | 0.74* | 0.81* | 0.88* | 1.00 | |
| $X_9$ (Amount) | 0.40 | 0.63* | 0.81* | 0.55 | 0.61* | 0.66* | 0.39 | 0.60* | 1.00 |

*coefficients that are significant at 0.05 level

about 2,000 persons were treated of various respiratory diseases in Ekeremor LGA, while over 20 hectares of farm land were destroyed by acid rain of that year. As already known, $SO_2$, $NO_2$, and $CO_2$ are released during gas flaring and their presence in the atmosphere of Niger Delta were studied by Olabaniyi and Efe (2007) which made them to conclude that this resulted in high concentration values of $NO_3^-$ and $SO_4^{2-}$ noticed in rainwater of the area. Ogunkoya and Efe (2003) findings show that acid rain caused about 30% of all respiratory diseases reported in the region and further noted that rainwater in Warri area contain over 75% of $SO_4^{2-}$ and $NO_3^-$. Again,

**Table 4.** Varimax rotated component matrix of the variables.

| Variable | Component | | Communality |
| | I | II | |
|---|---|---|---|
| $X_1$ (Temp) | -0.06500 | 0.11321 | 0.32011 |
| $X_2$ (pb) | 0.01462 | 0.95682* | 0.84302 |
| $X_3$ (condu) | -0.18332 | -0.43018 | 0.02246 |
| $X_4$ (TDS) | 0.14628 | 0.05697 | 0.038864 |
| $X_5$ ($NO_3^-$) | 0.93200* | -0.05128 | 0.92306 |
| $X_6$ ($CO_3$) | 0.88614* | 0.02350 | 0.95710 |
| $X_7$ ($SO_4^{2-}$) | 0.94661* | 0.00649 | 0.90690 |
| $X_8$ (pH) | 0.96018* | 0.04263 | 0.89550 |
| $X_9$ (Amount) | 0.3290 | 0.64600* | 0.91033 |
| Eigen value | 5.64 | 2.88 | - |
| % of Variance explained | 72.8 | 25.6 | - |
| Cumulative % explained | 72.8 | 98.4 | - |

*Highly loaded variables

**Table 5.** Significance of physicochemical parameters.

| Predictor | Coefficient | Se | T | P |
|---|---|---|---|---|
| Constant | 181.4010 | 6.3930 | 12.77 | 0.020 |
| $X_1$ (Temp) | -0.88 | 0.1714 | 2,32 | 0.44 |
| $X_2$ (pb) | 0.47 | 0.0286 | -0.76 | 0.042 |
| $X_3$ (Condu) | -0.76 | 0.1516 | -0.93 | 0.039 |
| $X_4$ (TDS) | 1.02 | 0.4188 | 4.41 | 0.022 |
| $X_5$ ($NO_3^-$) | 5.40 | 0.6076 | 2.43 | 0.002 |
| $X_6$ ($CO_3$) | 7.02 | 9.4057 | 2.89 | 0.013 |
| $X_7$ ($SO_4^{2-}$) | 13.1 | 11.1390 | 4.70 | 0.016 |
| $X_8$ (pH) | -2.60 | 6.3262 | 2.20 | 0.48 |

regarding pH values, we know that the result from all the LGAs is above the required maximum of 6.5. Figures as high as 5.0 and 5.1 were recorded in Ogbia and Oporoma areas, both of them with very high gas flaring areas. This is however in line with the result of acid rain of 5.4 obtained in Eket in the neighbouring Akwa Ibom State. The high cases of leukaemia reported in this area are largely attributed to both the inhaling of these gases by villagers and consumption of acid infested rain water. Other parameters $X_1$ (Temp), $X_2$ (Pb), $X_3$ (Condu), and $X_4$ (TDS), although with somewhat high values, but are closely related to the WHO approved maximum desired level. The finding is in agreement with those of Igili (2006) and Abua (2010) who reported high incidence of leukaemia in Bassambiri and Ogbia areas of Bayelsa State.

In Table 3, the correlation of the physicochemical parameters with each other show that $X_5$ ($NO_3$), $X_6$ ($CO_3$) and $X_7$($S_4$) have high correlation coefficients and positively correlated with $X_8$ (pH). Also, it was found that $X_1$ (Temp) correlated with $X_8$ (pH). These were however expected,

because the heavy presence of these gases in the atmosphere combines with the rainwater to form various acids which heavily reduce the potential hydrogen in water, thereby polluting it with acid content. Our visit to the area shows that even surface river bodies contain high level of acids and in many localities the inhabitants do not have any other source of water to depend on, except on water from those acid polluted streams. The people of the area, however, know that something was wrong with the water they consume from these source, because of objectionable odour observed from them. Our investigation in Oloibiri in Ogbia area from some hospital sources show that about 26% of all the people suffer from either bronchitis or chronic coughing. These are respiratory diseases associated with acid rain (Igoniwari, 2012).

In Table 4 on PCA, we were able to establish that all the acidic substances together with pH loaded highly. This is an indication that the substances that are acidified are closely associated with pH. It has been indicated earlier that this association is clearly the reason why the rainwater in the area is of poor quality even to the degree of being harmful. Again the high level of acidity in rain has corroded corrugated iron sheets in almost all houses in the state. Furthermore, the strength of these highly loaded variables have been corroborated by the high value of communality associated with them.

Table 5 shows again that the variables $X_5$ ($NO_3^-$), $X_6$($CO_3$) and $X_7$ ($SO_4^{2-}$) are all strongly significant in the study area as could be revealed from their low p-value which are far less than the threshold of 0.05. The overall strength of the model is seen in Table 6 which indicates a p-value of 0.021 showing that the regression model produced is very adequate in the explanation of acid rain in the area. The low residual error value of I is an indication that the unexplained variables are weak, attesting to the high level of 91.6% attributed to the combined effects of

**Table 6.** Result of analysis of variance.

| Source | Df | Sum of squares (SS) | Mean square (MS) | F | P |
|--------|-----|---------------------|------------------|-------|-------|
| Regression | 8 | 21295.0 | 3042.1 | 17.36 | 0.021 |
| Residual Error | 1 | 9.8 | 170.8 | | |
| Total | 9 | 21304.9 | | | |

**Table 7.** Relative contributions of the physicochemical variables.

| Variable | Multiple R | $R^2$ | $R^2$ Change |
|----------|-----------|-------|--------------|
| $X_5$ ($NO_3$) | 0.620 | 0.3844 | 38.44 |
| $X_6$ ($CO_3$) | 0.790 | 0.6241 | 23.97 |
| $X_7$ ($SO_4$) | 0.845 | 0.7242 | 10.01 |
| $X_8$ (pH) | 0.901 | 0.8118 | 8.70 |
| $X_2$ (pb) | 0.936 | 0.8761 | 6.73 |
| $X_4$ (TDS) | 0.950 | 0.9025 | 12.64 |
| $X_3$ (Condu) | 0.960 | 0.9216 | 0.91 |
| $X_1$ (Temp) | 0.961 | 0.9235 | 0.19 |
| - | - | - | 91.6% |

**Table 8.** Spatial structure of the variables contributing to rainwater quality in Bayelsa State.

| Sample site | Most important variable | Contribution (%) |
|-------------|------------------------|------------------|
| Bessambiri | $X_5$ (heavy presence of nitric acid in rain water) | 42.6 |
| Ekeremor | $X_7$ (heavy presence of sulphuric acid in rain water) | 16.8 |
| Kaiamah | $X_6$ (heavy presence of carbonate acid in rain water) | 21.5 |
| Ogbia | $X_5$ (heavy presence of nitric acid in rain water) | 28.2 |
| Oporoma | $X_5$ (heavy presence of nitric acid in rain water) | 38.2 |
| Sagbama | $X_8$ (low potential of hydrogen (pH) in rain water) | 13.8 |
| Twon | $X_5$ (heavy presence of nitric acid in rain water) | 48.9 |
| Yenagoa | $X_8$ (low potential of hydrogen (pH) in rain water) | 18.2 |

the explanatory variables leaving only 8.4% to be explained by variables not included in the analysis.

Furthermore, the importance of nitric acid in the rainwater was brought out clearly when considering high percentage contribution of 38.44% out of 91.60% attributed to all explanatory variables (Table 7). This result is in line with that obtained by earlier researchers that the nitric acid was found as the chief contaminant of rain water in Ibesikpo in Uyo region of the Niger Delta. In the area, whenever it rains, people feel a severe type of irritation and even burnt when water enters their eyes and concluded that this is a clear indication of heavy acid content in rain water.

The spatial disposition of the study is shown in Tables 8 and 9. Table 8 indicates that nitric acid is the most important rainwater contaminant out of four variables which were isolated by the model in the state. The towns where it is dominant are Bassambiri (Nembe LGA (42.6%)), Ogbia (Ogbia LGA (28.2%)), Oporoma (Southern Ijaw LGA (38.2%)), and Twon (Brass LGA (48.9%)). A close study of the location of these towns where nitric acid is dominant show that all of them are at the south western end of the state. The percentage contribution of nitric acid in all of them is an indication that the area was heavily affected by gas flaring and nitric oxide which is regarded as the greatest contributor of acidity of rainwater is formed during lightning storms by the reaction of two common atmospheric gases, namely nitrogen and oxygen gases. In air, nitric oxide is oxidized to nitrogen dioxide ($NO_2$) which in turn reacts with water to give nitric acid ($HNO_3$). This acid dissociates in water to yield hydrogen ions and nitrate ions ($NO_3^-$) in a

**Table 9.** Intercepts slopes and regression equation of the relationship between monthly rainfall amounts and nitric acid.

| Sample site | a (intercept) | b (slop) | Regression line equation |
|---|---|---|---|
| Bessambiri | 288.40 | 7.86 | Y = 288.4 + 7.8x |
| Ekeremor | 194.10 | 2.49 | Y = 194.10 + 2.49x |
| Kaiamah | 189.92 | 1.96 | Y = 189.92 + 1.36x |
| Ogbia | 223.45 | 9.02 | Y = 223.45 + 7.02x |
| Oporoma | 249.16 | 8.66 | Y = 249.16 + 8.66x |
| Sagbama | 184.88 | 1.89 | Y = 184.88 + 1.29x |
| Twon | 320.32 | 6.34 | Y = 320.32 + 6.34x |
| Yenagoa | 190.76 | 2.32 | Y = 190.76 + 2.32x |

reaction. In many streams and rivers in this area, surface water had become so acidic that fish can no longer live in them. Aggrey (2006) noted that gas flaring in the area has almost destroyed all the fishes in the water. One of the substances that causes this is nitric acid. The water samples collected from domestic sources of water supply showed that a lot of toxic ions especially $Al^{3+}$ were present in them. Efe (2011) stated that it should be noted the target water quality range (TWQR) for pH for domestic water use is 6.5. In the Niger Delta area, water supply sources available to the residents have values far lower than TWQR value and this made it to be of low quality. Also, it was observed that a lot of water receptacles in these four areas are daily corroded on account of their reaction with acid contents of the rain. The low hydrogen potential (pH) was observed as one of the important factors responsible for low rainwater quality in Sagbama (Sagbama LGA (13.8%)) and Yenagoa (Yenagoa LGA (18.2%)). In these two LGA, it was observed during field work that the submersible pumps and the borehole casing are corroded shortly after they were installed. In fact, in Yenagoa, the capital city, it was observed in Opolo Housing Estate that water supply installation do not last more than one year, because of corrosion. In another area in Yenogoa, it was observed that water supply pipe network together with metal water reservoirs were easily destroyed by the acid rain heightening the problem of water supply in the area. In areas such as Adagbabiri, Agbere, and Ebedebiri in Sagbama LGA, a lot of agricultural lands were found almost charred, because of the high acid content of the rain as revealed in the low pH of the area causing damage to plant leaves which slows or even stop the process of photosynthesis in plants. It was equally observed in Mile 2 Sagbama that the soil of the area looked very friable indicating that the low pH in the area has changed the once alluvial and fertile soil into acid soil which has affected its fertility and by extension of agricultural productivity.

The statistical model has isolated sulphuric acid in rainwater in Ekeremor as the most important variable with 16.8% contribution to the poor quality of acid rain in the area. The effect of this on the water samples in some communities in the area indicated an unacceptable water odour. The various water sources are still being used by the people for their domestic needs, a situation that has endangered the health of many in the area.

Finally, carbonic acid has been identified in Kaiama in Kolokuma/Opokuma LGA as the most important contaminant in rain water with 21.5%. Some communities in the area such as Opokuma and Igbedi were heavily affected by carbonic acid. Samples of water investigated from the area showed that most water bodies in the area were now bereft of fish population, a problem that has dealt with the main occupation of the people. In fact almost every economic activity of the people impaired as crops were damaged while drinking water sources were heavily polluted.

Furthermore, the bivariate regression analysis between monthly rainfall amounts and nitric acid, the most important contaminant isolated in rain water in Bayelsa State indicated that all the towns used in the study exhibited upward trends in their nitric acid contribution to rainwater pollution in the area. In spatial terms, it could be seen that Oporoma with the coefficient of 8.66 mg/L, Ogbia with 7.02 mg/L and Twon with 6.34 mg/L all in Southwest of the state possessed the steepest slopes in the analysis, showing that Oporoma has the greatest rate of nitric acid concentration by one unit increase in rainwater (8.66 mg/L), followed by Bassambiri (7.8 mg/L), Ogbia (7.02 mg/L) and Twon (6.34 mg/L) (Table 8). Then on the second category are Ekeremor (2.49 mg/L) and Yenagoa (2.32 mg/L), and the third group are Kaiamah (1.36 mg/L) and Sagbama (1.2 mg/L). The categorization of LGAs in the state on the level of nitric acid concentration is as follows:

(1) South Eastern Zone (highly concentrated): Southern Ijaw, Ogbia, Nembe and Brass.
(2) Central Zone (averagely concentrated): Ekeremor and Yenagoa
(3) North Western Zone: Kolokuma/Opokuma and Sagbama.

This disposition of nitric acid concentration areas in Bayelsa State clearly gives a good picture of the geographical areas of greatest affliction where government should intervene as a matter of urgency.

## Conclusion

The destruction of the environment and human health are two of the many effects of acid rain on any area. The exploitation of crude oil and its associated gas flaring which had resulted in the acid rain in the Niger Delta region of Nigeria have had a fairly long history. The agitation by the people for a safe environment had provoked many researchers in the area into investigating the extent of environmental degradation as a result of gas flaring. Bayelsa State with 17 onshore oil flow stations flares an average of 800,000 $m^3$ of gas per day for each of these stations. The effect of this activity will undoubtedly be much on the environment. This study has indeed revealed that almost all physiochemical parameters identified are above the WHO recommended minimum level. The study was able to establish that nitric acid alone contributed 38.16% out of the combined contribution of 91.6% attributed to other eight variables. Further analysis of the spatial concentration of nitric acid on water samples revealed that areas in the south eastern parts of the state had the highest concentration, which can be explained by the fact that many of these flow stations are located there. Other areas where water samples show high concentration of nitric acid were categorized into zones. This zonation pattern should be used as a guide to establish the extent of mitigation measures that could be adopted for various zones of the state.

## RECOMMENDATIONS

Based on the foregoing, the following recommendations were suggested. The formulation of a practical home-grown policy option that will comprehensively tackle the gas flaring problem in the Niger Delta should be the central measure to be adopted. It is indeed not enough to provide funds in the derivation account to the state government which have not been satisfactorily used for the purpose. However, based on the result of our investigation, the policy will address the following areas:

(1) Controlling the flared natural gas through the enforcement of laws aimed at limiting the quantity of gas to be flared by oil companies.
(2) Upstream petroleum producers that flare gas should be heavily taxed in accordance with the measure of gas flared.
(3) Tax holidays may be offered to companies that show manifest compliance to the law as an incentive.
(4) Houses and other structures to be coated with anti-acidic chemicals to reduce the impact of acid on them.
(5) Government should always treat the acidified water

bodies before they are consumed by the people. This can be done with the derivation funds monthly released to them as well as use the funds to help farmers in the areas of fishing and crop production.
(6) Continuous studies will be encouraged in the area to ascertain the temporal harmful effects of acid rain as well as its spatial disposition on the environment.
(7) The Niger Delta Development Commission (NDDC) should not only be involved in the provision of physical infrastructure but should be additionally meant to shoulder the responsibility of monitoring gas flaring and ensuring that oil companies comply with the provisions of the existing law in that regard.
(8) Serious regional development plan that will make use of the products of the aforementioned actions should be started as a way of improving the life of the people. Such plan will incorporate health, agricultural, commercial industrial water supply and biodiversity dimensions.

## REFERENCES

Abara  D (2009). "Chemical Analysis of Soil in parts of the Niger Delta". J. Environ. Chem. 2(1):28-34.

Abua T (2010). On the relationship between acid rain and the number of cases of leukaemia disease in some Niger Delta communities. Comm. Health 8:21-29.

Abube M (1988). "Gas flaring and Environmental Deterioration in parts of the western Nigeria Delta, Nigeria". J. Environ. Stud. 9(3):102-111.

Aggrey MM (1983). Gas Flaring and Future life in the Niger Delta. Paper delivered at the National Conference on Petroleum Production in Nigeria, Lagos. 7th – 8th, October.

APHA (1992). Standard Method for the Examination of Water and Waste Water, 18th Edition. American Public Health Association, Washington DC.

Atevure BSV (2004). Processes of Oil Production and Environmental Degradation: An Overview. J. Environ. Anal. 2(1):76-85.

Bayelsa State Government (2008). Bayelsa State, Glory of All lands.www.bayelsa.govt.ng

Edem BI (2011). Effect of gas flaring in Ekeremor area of Bayelsa State, Nigeria. J. Environ. Ecol. 3:38-51.

Efe SI (2002). Urban Warming in Nigeria Cities: The case of Warri Metropolis.' Afr. J. Environ. Stud. 3:160-168.

Efe SI (2003). "Effect of Gas Flaring on Temperature and Adjacent Vegetation in Nigeria Delta Environment". Int. J. Environ. 1(1):91-101.

Efe SI, Ogban FE, Horsfall M, Akpomonor (2005). Seasonal Variation of Physicochemical Characteristics in water resources Quality in Western Niger Delta Region, Nigeria.

Enehoro, A. (1973). Oil Production in Nigeria: Need for early reflection. Government Printer, Lagos.

Enete IC, Ijioma MA (2011). "Effects of Gas Flaring on Soil Nutrients in Ekpan, Ogunu and Ekurede Itsekiri Communities, Delta State, Nigeria". Trop. Built Environ. J. 1(2):163-170.

Evo J (2002). The Initial Mistakes in Nigeria Oil Exploration: Need for a Rethink. National Light Newspaper, 1051:6.

Obadina V (2000). "Gas flaring in Nigeria: Matters Arising." Oil Exploration J. 2:181-193.

Odilison K (1999). Issues in the purification of Rainwater from Oil Exploration area. Ecol. Environ. 5(3):204 -215.

Ogunkoya OO, Efe EJ (2003). Rainfall Quality and Sources of Rainwater acidity in Warri area of the Niger Delta, Nigeria. J. Min. Geol. 39(2)125-130.

Okezie M (1989) "Water Quality of some surface Rivers in the Nigeria's Niger Delta." Environ. Health 2(2):84-91.

Okoye AC, Oluyemi EA, Oladikpo AA, Ezeonu FC (2011). "Physicochemical Analysis and Trace Metal Levels of Rainwater for

Environmental Pollution Monitoring in Ile-Ife, South-western Nigeria".J. Environ. Appl. Serv. 6(3):326-331.

Olabaniyi SB, Efe SI (2007). "Comparative Assessment of Rainwater and Groundwater Quality in an oil producing area: Environmental and Health Implications" J. Environ. Health Res. 6(2).

Olowoyo DN (2011). Physicochemical Characteristics of rainwater Quality of Warri axis of Delta State in Western Niger Delta region of Nigeria. J. Environ. Chem. Ecotoxicol. 3(2):320-322.

Olukoya M (2008). "Gas Flaring and the preservation of Biodiversity in the Niger Delta" in Aghabu (ed.) Environmental Issues in Niger Delta Region of Nigeria. Johnson Publishers, Lagos.

Rim-Rukah M, Ikiafa D, Okokoyo J (2005). Monitoring Air Pollution due to Gas Flaring using rainwater.

Uge T (2009). "Acid Rain Menace in Niger Delta Region of Nigeria". Ecol. Biodivers.3:71-80.

# 18

# Effect of some pesticides on growth, nitrogen fixation and *nif* genes in *Azotobacter chroococcum* and *Azotobacter vinelandii* isolated from soil

**Aras Mohammed Khudhur[1,2] and Kasim Abass Askar[2]**

[1]Soil and Water Department, College of Agriculture, University of Salahaddin, Erbil, Iraq.
[2]School of Biomedical and Biological Sciences, University of Plymouth, Plymouth, PL4 8AA, United Kingdom.

This study was designed to evaluate the effects of three pesticides (Imazetapir, Dimethoate and Bayleton 50) at the recommended concentration (in the field), on the growth of pure cultures of *Azotobacter chroococcum* and *Azotobacter vinelandii*, on the amount of fixed nitrogen and *nif* genes. Herbicide Imazetapir had no negative effect on nitrogen fixing bacteria, while Dimethoate and Bayleton 50 exhibited inhibitory effect on growth. Same effects were obtained on fixed nitrogen obtained when treated with studied pesticides. *nif*H1, *nif*H2, *nif*H3, *nif*U and *nif*V from *A. chroococcum*, and *nif*H, *nif*K, *nif*D, and *nif*M gene in *A. Vinelandii* were lost when pots were cultivated with wheat and treated with both Dimethoate and Bayleton 50, therefore, be deemed highly susceptible to them. While herbicide did not affect the *nif* genes, the bands on gel electrophoresis appeared as normal sample.

**Key words:** *Azotobacter chroococcum, Azotobacter vinelandii,* fungicide (Bayleton 50), herbicide (Imazetapir), insecticide (Dimethoate).

## INTRODUCTION

Pesticides are used to control specific fungi, herbs, insects and other pests in crops (Johnsen et al., 2001). Pesticide application is still the most effective and accepted means for the protection of plants from pest (Bolognesi, 2003), but the extensive use of pesticide over the past four decades has resulted in tribulations caused by interaction with natural biological system (Ayansina and Oso, 2006).

In the last few decades, numerous soil microorganisms have been found to have a positive effect on plant development. Besides the well-known symbiotic nodular bacteria, free nitrogen fixers in the rhizosphere (*Azotobacter*) can also stimulate plant growth or reduce the damage caused by soil-borne plant pathogens (Kloepper et al., 1989) and has been used as a potential nitrogenous fertilizer to increase crop yield (Steinberga et al., 1996; Mrkovaaki et al., 2001). Some pesticides used

in agriculture can be harmful to *Azotobacter,* not only to inhibit the nitrogen fixation process in *Azotobacter,* but also to reduce the bacterium's respiration rate and hence preclude its positive effects (San-Tos and Flores, 1995; Mrkovaaki et al., 2001).

The objective of the present study was to determine the effects of one herbicide (Imazetapir), one insecticide (Dimethoate) and one fungicide (Bayleton 50) on the growth, nitrogen fixation and *nif* gene of pure cultures of *Azotobacter chroococcum* and *Azotobacter vinelandii* isolated from Erbil City soil, Iraq.

### MATERIALS AND METHODS

#### Pesticides

The pesticides were commonly used for wheat crop as recommended by Agriculture Research Centre in Erbil City, Iraq to the farmers where Imazetapir serves as herbicide (100 g/L), Dimethoate (Rogar) as insecticide (3 g/L) and Bayleton 50 as fungicide (2 g/L).

## Wheat variety

Wheat variety *Triticum aestivum var. Aras* was used in the experiment, kindly provided by Agriculture Research Centre in Erbil City.

## Bacteria used

The bacterial isolates used in the study were *A. chroococcum* and *A. vinelandii* isolated from the soil of Erbil City, Iraq, on the basis of cultural, morphological and biochemical characteristics as described by Forbes et al. (2002).

## Plant-Microbial interactions

Pot experiment was carried out in green house; each pot was filled with 8 kg Ainkawa soil of 60% moisture content. Wheat seeds were planted at a rate of 10 seed/pot. After sowing, the pesticide was applied by spraying on the soil surface at known quantity/unit area of the pots. After 5, 15, 30 and 70 days, soil samples were taken under plant rhizosphere zone. The total number of *Azotobacter* was determined by MPN using nitrogen free Jensen's broth medium. The *Azotobacter* was isolated from the wheat rhizosphere of pots and identified depending on cultural, morphological and bioche-mical characteristics as described (Forbes et al., 2002), the *Azotobacter* were identified as *A. vinelandii* and *A. chroococcum*.

## Total bacterial count

Soil samples were taken from each pot in duration of 5, 15, 30 and 70 days. The size of *Azotobacter* population of wheat rhizosphere was determined with MPN (Johnsen et al., 2001) using Jensen's broth medium.

## Effect of pesticide on bacterial nitrogenase activity

The effect of pesticide on $N_2$-fixation of two *Azotobacter* species (*A. vinelandii* and *A. chroococcum*) that were used as inoculum in biofertilizer were examined by studying their nitrogenase activity in their respective growth inoculum with a 100 ml of individual bacterial culture containing $35 \times 10^5$ cells/ml of Jensen's broth medium. The flasks were prepared by adding stock solution of pesticides at the rate of 2 g/L for Bayleton 50 (fungicide), 100 g/L for Imazetapir (herbicide) and 3 g/L for Dimethoate (insecticide), with 3 replications for each treatment. The flasks were incubated at 28°C for 10 days. Then, the total nitrogen was calculated for each flask using micro-Kjeldhal method as described in Rowell (1996).

## DNA extraction

Genomic DNA was extracted and purified from *A. vinelandii* and *A. chroococcum* cells isolated from rhizosphere of wheat plants in pot experiment using the QIA amp DNA Mini Kit (Jayashreet et al., 2007). Polymerase chain reaction (PCR) amplification of nitrogenase genes nifH1, nifH2, nifH3, nifU, nifV and FV genes for *A. chroococcum*, and nifK, nifD, nifM, nifH and FV genes for *A. vinelandii* was performed. Gene's sequences were obtained from NCBI site, and were designed in OPERON diagnostic Ltd, Germany. The primer length and melting temperature were design-ed with coordination between forward and reverse primers. The melting and annealing temperature were calculated following Womble (2000). Primers amplification was completed using the protocol and reagents followed by Rajeswari and Kasthuri (2009). The programmed temperature sequence was 96°C followed by 55°C for 1 min, and 72°C for 1 min, the temperature sequence was run for 30 cycles, the final product extension was conducted at 72°C for 6 min followed by 4°C temperature hold. The primers acquired from Operon Biotechnologies, Germany were used. The forwards and reverse primers are shown in Table 1.

## Gel electrophoresis

DNA amplification was checked by electrophoresis of each PCR product in a 1.5% (w/v) agarose gel, in Tris-borate-EDTA (TBE) buffer for 1 h at 3.2 V/cm. Gels were stained in ethidium bromide for 15 min and thereafter washed for 5 min. DNA fragments were visualised at 312 nm with a UV-trans illuminator Image Master VDS (Amersham Biosciences) (Helmut et al., 2004).

## RESULTS

The study has shown that the herbicide Imazetapir involved had no negative effect on the growth of nitrogen fixing bacteria in the soil as compared to the control, $3.1 \times 10^6$ cell/g of the soil from $34 \times 10^6$ cells/g survived. The insecticide (Dimethoate) and fungicide (Bayleton 50), however, did have inhibitory effect on the growth of nitrogen fixing bacteria (Table 2), $0.001 \times 10^6$ and $0.2 \times 10$ cells/g soil survived after 70 days after application with Dimethoate and Bayleton 50, respectively, as compared with the control $4.2 \times 10^6$ cells/g. While $3.1 \times 10^6$ cells/g survived when Imazetapir was applied.

The effect of pesticides on nitrogen fixation by nitrogen fixing bacteria in the nitrogen free Jensen's broth medium has been studied (Table 3), the results indicated that nitrogenase activity decreased in the presence of Dimethoate and Bayleton 50 and the amount of fixed nitrogen were 0.09 and 0.009 mg/ml when *A. chroococcum* was present, and 0.008 and 0.005 mg/ml when *A. vinelandii* was inoculated, while for control treatment 1.80 and 1.19 mg/ml for *A. chroococcum* and *A. Vinelandii*, respectively. Moreover, Imazetapir was found to have less inhibitory effect on nitrogenase when two nitrogen fixing bacteria were used as inoculum on the broth culture. The amount of fixed nitrogen was 1.09 and 0.82 mg/ml for *A. chroococcum* and *A. Vinelandii*, respectively (Table 3). In general, the obtained value of fixed nitrogen after 70 days of incubation was less than the control treatment, with negative significant differences of 0.55 and 0.54 for *A. chroococcum* and *A. Vinelandii*, respectively (Table 3).

To investigate the effect of the tested pesticides on nitrogenase activity through amplification of nifH1, nifH2, nifH3, nifU, nifV and FV genes for *A. chroococcum* nifH, nifD, nifK, nifM and FV genes for *A. vinelandii* by PCR technique, the PCR products are as shown in Figures 1 and 2. Band generation of PCR amplified nif genes fragments were used to evaluate the effect of tested

**Table 1.** The forwards and the reverse primers used.

| Primer | Sequence (5'3') | Nucleotide | Reference |
|--------|-----------------|------------|-----------|
| | *Azotobacter vinelandii* | | |
| *nif*H-F- | (taccgatacgcagttacgcggt) | 22 | Setubal et al. (2009) |
| *nif*H-R | (tcagacttcttcggcggtttttg) | 22 | |
| *nif*D-F- | (taccgatacgcagttacgcggt) | 22 | Setubal et al. (2009) |
| *nif*D-R | (tcagacttcttcggcggtttttg) | 22 | |
| *nif*K-F- | (tactcggtcgttcagctattttag) | 24 | Setubal et al. (2009) |
| *nif*K-R- | (ttagcgtaccaggtcgtggttgta) | 24 | |
| *nif*M-F- | (tcctattttgcctgtttgggac) | 23 | Setubal et al. (2009) |
| *nif*M-R- | (teagaggtcggccgacagcgcgg) | 23 | |
| V-F- | (tacagtagcggaaggttagggt) | 22 | Setubal et al. (2009) |
| FV-R- | (tcagccgccgaccttgatgccg) | 22 | |
| | *Azotobacter chroococcum* | | |
| nifH1-F- | (cagacacgaagaagccgggc) | 20 | Setubal et al. (2009) |
| nifH1-R- | (gaccagcagcttgttgttga) | 20 | |
| nifH2-F- | (cgccggcgcagtgtttgcgg) | 20 | Setubal et al. (2009) |
| nifH2-R- | (cactcgttgcagctgtcggc) | 20 | |
| nifH3-F- | (cgatgactgaagactgaacgag) | 22 | Setubal et al. (2009) |
| nifH3-R- | (aaggtgcggtcaggagagaa) | 20 | |
| nifU-F- | (atgtgggattattcggaaaaa) | 21 | Setubal et al. (2009) |
| nifU-R- | (tcagcctccatctgccgtggg) | 22 | |
| nifV-F- | (gatggctagggtgatcatcgacga) | 24 | Setubal et al. (2009) |
| nifV-R- | (gccattcctcctgccgccagttcg) | 24 | |
| FV-F- | (tacagtagcggaaggttagggt) | 22 | Setubal et al. (2009) |
| FV-R- | (tcagccgccgaccttgatgccg) | 22 | |

**Table 2.** Cell/g of nitrogen fixing bacteria applied with pesticides during 70 days incubation.

| Days from infection | Control | Dimethoate | Bayleton 50 | Imazetapir |
|---------------------|---------|------------|-------------|------------|
| 5 | $36 \times 10^6$ | $5.6 \times 10^6$ | $0.1 \times 10^6$ | $34 \times 10^6$ |
| 15 | $23.5 \times 10^6$ | $2.4 \times 10^6$ | $0.06 \times 10^5$ | $21 \times 10^6$ |
| 30 | $14 \times 10^6$ | $0.05 \times 10^6$ | $0.04 \times 10^3$ | $12.5 \times 10^6$ |
| 70 | $4.2 \times 10^6$ | $0.001 \times 10^6$ | $0.2 \times 10$ | $3.1 \times 10^6$ |

pesticides on *A. chroococcum* and *A. vinelandii* isolated from soil cultivated with wheat plants and applied with pesticides. The number of bands present or absent is used to estimate the influence of pesticides on the bases of the number of shared amplification products. Figure 1, lane 5 shows the amplified nif genes in *A. chroococcum*, lanes 1, 2 and 3 are pesticide treatment, lane 1 applied with Dimethoate, lane 2 treated with Bayleton 50 and lane 3 treated with Imazetapir are as shown in Figure 1. Both insecticide and fungicide treated pots affected

**Table 3.** Total nitrogen of Jensen's broth cultures inoculated with *Azotobacter* and treated with pesticide after 15 days incubation.

| Bacterial inoculum | Total nitrogen (mg/ml) | | | |
|---|---|---|---|---|
| | Control | Dimethoate | Bayleton 50 | Lmazetapir |
| *A. chroococcum* | 1.80 | 0.09 | 0.009 | 1.09 |
| *A. venilandii* | 1.19 | 0.008 | 0.005 | 0.82 |

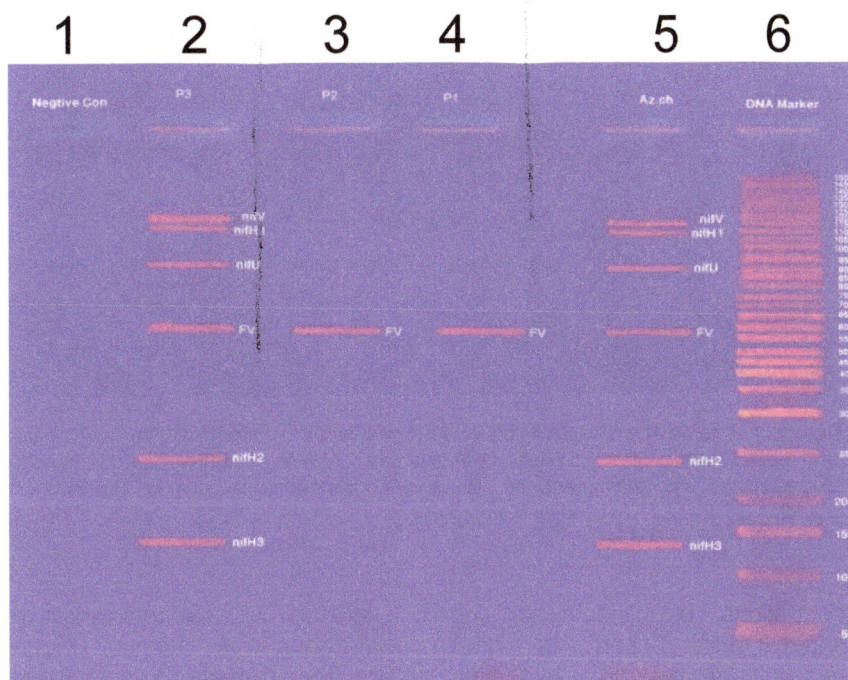

**Figure 1.** The PCR product of *Azotobacter chroococcum* isolated from the soil treated with pesticide as follows: Lane1, Negative control; Lane 2, Soil treated with Imazetapir; Lane 3, Soil treated with Bayleton 50; Lane 4, Soil treated with Dimethoate; Lane 5, Untreated *A. Chroococcum*; Lane 6, DNA marker.

negatively the nif genes, and *nif*H1, *nif*H2, *nif*H3, *nif*U and *nif*V were lost from the gel, while herbicide used in lane 3 did not have affected on the nif genes, and the bands appeared as normal sample of *A. chroococcum* in lane 4. Effect of tested pesticides on *nif* genes in *A. vinelandii* (Figure 2) was the same as *A. chroococcum* except *nif*M was not affected when Dimethoate was applied. FV gene was not affected when tested with pesticide applied in both isolated *Azotobacter* species.

## DISCUSSION

This study has shown that the herbicide (Imazetapir) had no negative effect on the growth of nitrogen fixing bacterial population in the soil, that is, the percentage of survived bacteria after 70 days of incubation was 9.12%, and 11.66% for control. Fungicide and insecticide how-

ever did have a negative effect on bacterial population (Table 2); the percentage of the remaining survived nitrogen fixing bacteria was 0.012 and 0.0004% for Dimethoate and Bayleton 50, respectively. The number of nitrogen fixing bacteria reduced from $5.6 \times 10^6$ and $0.1 \times 10^6$ cell/g to $1 \times 10^3$ cell/ml when Dimethoate and Bayleton 50 were applied to the soil, respectively. When Imazetapir was applied, the number decreased from $34 \times 10^6$ to $3.1 \times 10^6$ cell/ml, while for control the population decreased from $36 \times 10^6$ to $4.2 \times 10^6$ cell/g. The results of this study are in agreement with previous studies, that is, the analysed strains of nitrogen fixing bacteria is resistant to herbicides and show no inhibition of growth observed when compared with the result of fungicide and insecticide application. Sudbakar et al. (2000) found that there were variable effects of pesticides on the growth of nitrogen fixing bacteria. *In vivo* study showed that amongst fungicides, Carbendazim reduced the bacterial

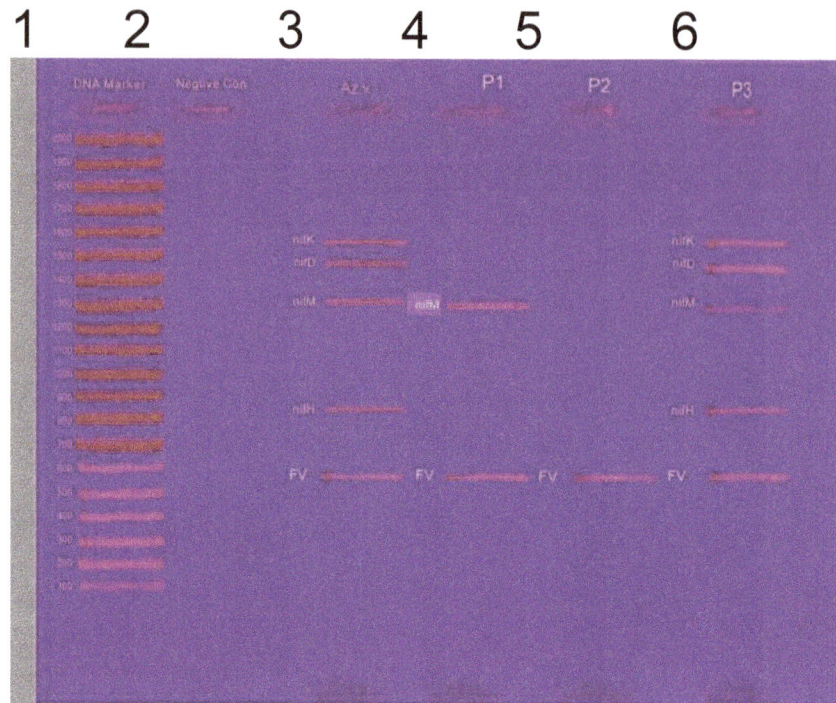

**Figure 2.** The PCR product of *Azotobacter vinelandii* isolated from the soil treated with pesticide as follows: Lane1, DNA marker; Lane 2, Negative control; Lane 3, Untreated *A. Vinelandii*; Lane 4, Treated with Dimethoate; Lane 5, Treated with Bayleton 50; Lane 6, Treated with Imazetapir.

population at all concentrations, but Dimethoate and wettable sulphur stimulated it and reduced it at higher concentration, and these results are supported by Gallori et al. (1991), Revellin et al. (1993), Taiwo and Oso (1997) and Dunfield et al. (2000) who reported that bacterial growth inhibition due to agrochemical, and they contain similar active ingredient that reduced the number of nitrogen fixing bacteria. Similar trend was reported by Martensson (1992) who observed that the fungicide treatment decreased the number of viable N2-fixing bacteria. Dimethoate decreased the growth of Rhizobium population (Castro et al., 1997).

The fixation of nitrogen was parallel to the population of both nitrogen fixing bacteria in soil treated with pesticide, Martinez et al. (1992) and Pozo et al. (1995) found that organo phosphorous insecticides profenofos and chloropyrifos reduced the number of aerobic nitrogen fixers and significantly decreased nitrogen fixation. Mubeen (2004) found negative effects of fungicides on nitrogen fixation.

Primer pair *nif*H1, *nif*H2, *nif*H3, *nif*V, *nif*U, FV genes of *A. chroococcum*, and *nif*H, *nif*K, *nif*D, *nif*M, and FV genes in *A. vinelandii* (Table 1) were used to amplify nif genes of *A. chroococcum* and *A. vinelandii* from three pots soil samples (Figures 1 and 2). The results show discrete bands of predicated size of approximately (140 to 1200 and 550 to 1550 bp) after being analyzed on conventional agaros gels. Lanes 3 and 4 are from Dimethoate and

Bayleton 50 treatment which reduced n*if*H, *nif*K, *nif*D, *nif*M gene in *A. vinelandii* (Figure 1). Also, lanes 1 and 2 are from Dimethoate and Bayleton 50 (Figure 2) when applied on the soil lost *nif*H1, *nif*H2, *nif*H3, *nif*V, *nif*U genes from *A. chroococcum*, while soil without pesticide lanes 5 and 3 are from *A. chroococcum* and *A. vinelandii* (Figures 1 and 2), respectively. For negative control, no bands were observed in lanes 1 and 2 as shown in Figures 1 and 2. These results are supported by other researchers (Cernakova, 1993; Mubeen, 2004).

It is concluded from this study that the pesticides have differential effect on the growth of nitrogen fixing bacteria, and their action vary at different sites. Indication has been observed that the pesticides (Dimethoate and Bayleton 50) which are under field condition possibly due to its high toxic nature reduced the population of these bacteria under field condition.

Their direct effect on nitrogen fixing bacteria was a decrease in the number of viable bacterial population and a high indirect effect was a reduction of nitrogen fixing genes, that is, *nif*H, *nif*K, *nif*U, *nif*M, *nif*H1, *nif*H2, *nif*H3, *nif*V genes of studied *Azotobacter* species and finally on amount of fixed nitrogen. While the other pesticide (Imazetapir) did not show any effects (Figures 1 and 2).

**REFERENCES**

Ayansina AD, Oso BA (2006). Effect of two commonly used herbicides on soil microflora of two different concentrations. Afr. J. Biochem. 5(2):129-132.

Bolognesi C (2003). Genotoxicity of pesticides: a review of human bionitoring studies. Mutat. Res. 543:251-272.

Castro S, Vinocur M, Permingiani M, Halle C, Taurian T, Fabra A (1997). Interaction of the fungicide Dimethoate and *Rhizobium* sp. in pure culture and under field conditions. Biol. Fertil. Soil 25:147-151.

Cernakova M (1993). Effect of insecticide nerametrine EK-15 on the activity of soil microorganisms. Folia Microbiologica, 38(4):331-334.

Dunfield KE, Siciliano S, Germida JJ (2000). The fungicide Thiram and Captan effects the phenotypic characteristics of *R. leguminosarium* strain clas determine by FAME and Biology analysis. Biol. Fertil. Soil 31:303-309.

Forbes BA, Sahm DF, Weissfeld AS (2002). Diagnostic Microbiology.Eleventh Ed. Mosby, Inc. USA.

Gallori E, Casalone E, Coella CM, Daly S, Polsimelli M (1991). 1,8 Naphthalic anhydride antidote enhance the toxic effect of Captan and Thiram fungicides on Azosprillium brasilense. Res. Microbiol. 142:1005-1007.

Helmut BF, Widmer W, Sigler V, Zeyer J (2004). New Molecular Screening Tools for Analysis of Free-Living Diazotrophs in Soil. Appl. Environ. Microbiol. 70:240-247.

Jayashreet VS, Karthick A, Kalaigandhi V (2007). Molecular study of *Azotobacter*nif H Gene by PCR. J. Nen. Microbil.152:871-884.

Johnsen K, Jacobsen CS, Torsovik V, Sorensen J (2001). Pesticides effect on bacterial diversity in agriculture soil- review. Biol. Fertil. Soil, Springer-Verlag. 33:443-453.

Kloepper JW, Lifshitz R, Zablotowicz RM (1989). Free--living bacteria inoculum for enhancing crop productivity. Trends Biotechnol. 7:39-44.

Martensson AM (1992). Effect of agrochemical and heavy metals on fast-growing rhizobia and their symbiosis with small seeded. Soil Biol. Biochem. 24 (5):435-445.

Martinez TMV, Salmeron V, GonzalezLopez J (1992). Effects of an organophosphorus insecticide, profenofos on agriculturalsoilmicroflora. Chemosphere, 24(1):71-80.

MrkovaakiN KL, Áaáiä N Mezei Sneÿana (2001). Primenamikro biološkog preparata u proizvodnji šeäernerepe. Zbornik instituta zaratar stvoipo vrtarstvo, 35:67-73.

Mubben, F. (2004). Biochemical and molecular approaches to study the effect of pesticides (Fungicide) on root-associated bacteria in wheat.PhD. thesis, Faculty of Science, Univ. Agric. Faisalabad Pakistan.

Pozo,C, Martinez TMV, Salmeron V, Rodelas B, Goles L, Opez, J (1995). Effect of chloropyrifos on soil microbialactivity. Environ. Toxicol. Chem. 14(2):187-192.

Rajeswari K, Kasthuri M (2009) Molecular characterization of *Azotobacterspp*.nif H gene isolated from marine source. Afr. J. Biotechnol. 8(24):6850-6855.

Revellin C, Leterme P, Catroux G (1993). Effect of some fungicide seed treatments on the survival of Bradyrhizobiumjaponicum and on the nodulation and yield of soya bean (*Glycine max* (L) *Merric*. Soil Fertil. 16:211-214.

Roweel DL (1996). Soil science: Methods and application. University of Reading, UK.

San-Tos A, Flores M (1995). Effects of glyphosate on nitrogen fixation of free-living heterotrophic bacteria(Abstract). Lett. Appl. Microbiol. 20(6):349-352.

Setubal JC, Santos PD, Goldman BS, Ertesvag H, Espin G, Rubio ML, Valla S, Almeida NF, Balasubramanian D, Cromes L, Curatti L, Du Z, Godsy E, Goodner B, Hellner-Burris K, Hernandez JA, Houmiel K, Imperial J, Kennedy C, Larson TJ, Latreille P, Ligon LS, Lu J, Mærk M, Miller NM, Norton SO, Carroll IP, Paulsen I, Raulfs EC, Roemer R, Rosser J, Segura D, Slater S, Shawn L, Stricklin Studholme DJ, Sun J, Viana CJ, Wallin, Wang B, Wheeler C, Zhu H, Dean DR, Dixon R, Wood D (2009). Genome sequence of *Azotobactervinelandii*, an obligate aerobe specialized to support diverse anaerobic metabolic processes. J. Bacteriol.191:4534-4545.

Steinberga V, Apsite A, Bicevskis J, Strikauska S, Viesturs U (1996). The effect of Azotobacterin on the crop yield and biologicalactivity of the soil. In: Wojtovich A, Stepkovska J, Szlagowska A (Ed): Proceedings of 2nd European Nitrogen Fixation Conference, pp. 191, Poznan.

Sudbakar P, Chattopadhyay GN, Gangwar SK, Ghosh JK, Saratchandra B (2000). Effect of common pesticides on nitrogen fixing bacteria of mullberry (*Monusalbal*). Indian J. Agric. Res. 34(4):211-216.

Taiwo LB, Oso BA (1997). The influence of some pesticides on soil microbial flora in relation to change in nutrient level, rock phosphate solublization and P-release under laboratory conditions. Agric. Ecosystem, Environ., 65:59-68.

Womble DD (2000). GCG: The Wisconsin package of sequence analysis programs methods. Mol. Biol. 132:3-22.

# 19

# Undesirable effects of drinking water chlorination by-products

## Khallef Messaouda, Merabet Rym and Benouareth Djamel Eddine

Department of Biology, Research laboratory, Biology, Water and Ecology, University of 8 Mai 1945, Guelma, Algeria.

The fundamental objective of water treatment is the protection of consumers from pathogenic microorganisms. Chlorination of drinking water is essential to prevent waterborne disease. However, chlorine reacts with organic matter present in surface waters to form various by-products suspected of being carcinogenic. In the last decade, several epidemiological studies have been conducted to determine the connection between exposure to these chlorination by-products and human health defects. The purpose of this paper is to evaluate the genotoxicity of drinking water of Annaba city. The study have been carried out in different points of water distribution and in the station of treatment, using two tests of determination of genotoxic risk by means of SOS chromotest (using the strain *Escherichia coli PQ37*). SOS chromotest showed genotoxic effect of the sample collected from the exit of treatment station.

**Key words:** Drinking water, genotoxicity, SOS chromotest, *Escherichia coli PQ37*.

## INTRODUCTION

Water is the most essential element to life on earth that is why it is a subject of attentive surveillance to prevent waterborne disease. The objectives of the surveillance of the quality of water destined for consumption are numerous and vary depending on means and process possibilities (Bouziani, 2000). Disinfection of drinking water has been widely practiced in drinking water treatment. It is essential to protect the public health and ensure water quality from the water treatment plant outlet to the consumer's tap (that is, during water distribution). Chlorine (Cl2) is used as the most common disinfectant due to its high efficiency to eliminate pathogens and protect human health against waterborne diseases (United States Environmental Protection Agency (US EPA), 2000; Liu et al., 2011). However, chlorine and other disinfectants react with natural organic matter and/or inorganic substances occurring in water to form various disinfection by-products such as trihalomethanes (THMs), haloacetic acids (HAAs) and other compounds (Legay et al., 2010). The presence of these compounds depends certainly on added quantities of chlorine but also on some organic matter and present organohalogenated by-products (AOX) in water, which can be of natural origin (humic and fulvic acids) or artificial one (residues of pesticides, phenols) (Zekkour and Beron, 2001; Potelon and Zysman, 1998).

The fundamental objective of the treatment of water is to protect the consumers of the pathogenic microorganisms and unpleasant or dangerous impurity for health (OMS, 1994; Jouany, 2000). Nevertheless, epidemiological studies suggested a possible link between the chloration, chlorinated by products and increase risk of several types of cancer. Since the first research works about disinfection by-products (Rook, 1974; Bellar et al.,

1974), trihalomethanes accepted a lot of attention because chloroforms have been shown as carcinogenic, further to work on laboratory animals. The United Sates Environmental Protection Agency (US EPA) (2000) reported that these THMs are human carcinogens, of which $CHCl_3$, $CHCl_2Br$ and $CHBr_3$ are carcinogen class B2 (human carcinogen) and $CHClBr_2$ is carcinogen class C (probable human carcinogen). The cancer of the genital tract and the gastro-intestinal cancer would be more frequent after consumption of chlorinated water (Rook, 1974; Urien, 1986; Monod, 1989; Lafferriere et al., 1999; Zekkour and Beron, 2001).

The objective of this paper was to investigate the effects of seasonal variation on physicochemical and genotoxic parameters of water quality and on the formation and species distribution of organohalogenated by-products of the water plant of Annaba city.

## MATERIALS AND METHODS

### Treatment of water

The treatment plant treats surface water coming from the dam on the Bou-Namoussa raw and water of the Bouchet Bridge and Hnichet, and water of the saline. Treatments performed on surface water are: prechlorination, coagulation floculation, decantation, filtration and sterilization.

### Sampling and sampling frequency

Four samplings (one sampling by season) have been performed from 2003 to 2004. The directed months were: April, June, September and January. Sampling was accomplished to study the variation of physicochemical parameters of raw water that may influence water quality during seasons; and those which are in direct contact with the production of organohalogenated compounds, proven harmful to human health. The studied raw water parameters were: pH, temperature, turbidity, colour, alcalinity, hardness *ISO 6059-1984 (F)*, organic matter *ISO 8467:1993 (F)*, ammonium *ISO 7150/1-1984 (F)* and chlorine request *ISO 7393-3:1990 (F)*.

### The research of the complete organohalogenated products (AOX)

The AOX is a parameter used for regulatory purposes for water quality. It represents the totality of chlorides and organically linked bromides, adsorbed on active charcoal. Volatile halogenated compounds in suspension are also assayed. The proportion of AOX of raw and treated water in the different sampling points is determined according to international norm *ISO 9562: 1989 (F)*, using the apparatus *coulomat 702cl*.

### Determination of genotoxic activity (the SOS chromotest)

SOS chromotest is a quantitative procedure, based on the measurement of two enzymatic activities ($\beta$-galactosidase and alkaline phosphatase) in liquid medium. It uses the SOS repair system of *Escherichia coli PQ37* strain, after 2 h of exposure. This assay is a simple, efficient and rapid test of genotoxicity that can beeasily adapted to the study of environmental water samples. The genotoxic activity of the concentration C is expressed by the

ratio $C = \beta/p$,

where $\beta$ represents the $\beta$-galactosidase activity and p the alkaline phosphatase activity. The induction factor for a compound at concentration C is defined as:

$$I(C) = R(C)/Ro$$

where $Ro$ is the genotoxic activity measured in the absence of this compound. A compound is considered genotoxic if the induction factor is higher than 1.5 according to Olivier and Marzin (1987).

## RESULTS AND DISCUSSION

For the system of water samples collected from the studied water plant, the effects of seasonal variation concentrations on water quality and on the formation and distribution of chlorinated by product have been investigated. Table 1 displays the results of the analysis of variance of two classification criteria (water and seasons). The results of the analysis of variance show the influence of seasons on the change of water quality. The content of organic matter, turbidity and the pH of drinking water vary from season to season.

The interpretation of these results points out that the seasonal variation significantly influences the temperature changing between hot season and cold season. This variation is highly significant and induces variation of turbidity that changes depending on the rate of pluviometry and the organic matter content of raw water. The change in the colour parameter which is in direct relation with the presence of humic and fulvic acids is also significant with seasons.

Turbidity, organic matter and colour introduce parameters linked to the presence of the potential forerunners of the chlorinated by-products and the temperature being key parameter in reaction chlorinates - organic forerunners. The same result was reported by Lafferriere et al. (1999). They demonstrated, by analyses of correlation, the effect of colour and temperature of raw water; as well as the effect of residual chlorine concentrations of treated water in the formation of trihalomethanes (a class of volatile chlorinated by-products).

The results reported in Figure 1 represent concentration of organohalogenated by-products (AOX) at different sampling sites. The analysis of bar charts representing the rates of organohalogenated by-products adsorbed on active charcoal apparently revealed that the autumn is the critical season of the apparition of organohalogenated by-products.

**Table 1.** ANOVA test of 2 controlled factors (values from P to IC = 95%).

| Parameter | Season | Water | Interaction (water × season) |
|---|---|---|---|
| pH | 0.836[NS] | 0.000*** | 0.009** |
| Temperature | 0.000*** | 0.858[NS] | 0.621[NS] |
| Turbidity | 0.004** | 0.001*** | 0.003** |
| Organic matter | 0.728[NS] | 0.001*** | 0.716[NS] |
| Colour | 0.086[NS] | 0.013* | 0.104[NS] |
| Alkalinity | 0.431[NS] | 0.579[NS] | 0.524[NS] |
| Hardness | 0.409[NS] | 0.646[NS] | 0.761[NS] |
| Conductivity | 0.192[NS] | 0.474[NS] | 0.330[NS] |
| Ammonium | - | - | - |
| chlorine request | / | / | / |

NS: no significant differences. - :The most part of data are less than the range (< 0.06), there is no statistical differences. / : insufficient data for statistical study.
P > 0.05, no statistical differences. P ≤ 0.05 *, significant differences. P ≤ 0.01**, significant increase. P ≤ 0.001***, very significant increase.

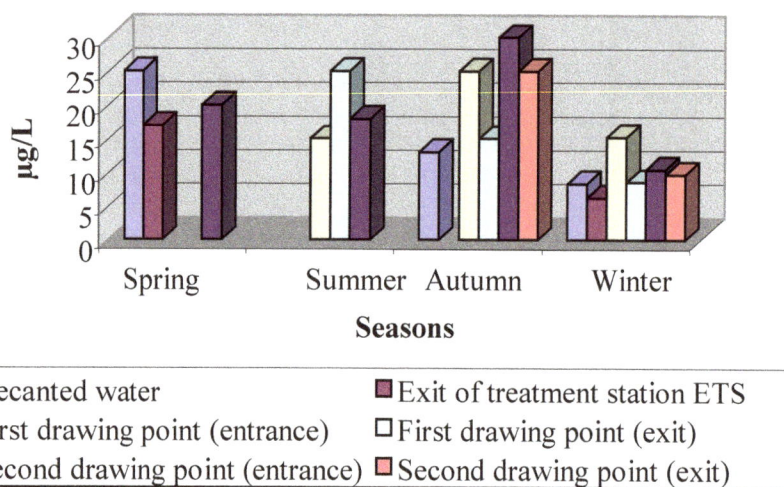

**Figure 1.** Spatiotemporal variation of AOX.

These results agree with those reported by Le et al. (1996). They found that contents of AOX in autumn, caused by the decomposition of organic matters notably after the fall of leaves, are 75 to 110% higher from those reported in spring.

## Standard SOS chromotest

Results of the present test are cited in Figure 2. The analysis of the interpretative bar charts of the mailman of induction I (C) showed that the sample Exit of Treatment Station (ETS) have a genotoxic activity superior to the other two. This proves that chlorinated by-products present in ETS are more genotoxic than those present in water and last point of sampling. Therefore, chlorinated by-products changes in the course of distribution are depending on the added doses of chlorine, to the stocking and to seasonal variation; as it leads to changes in physicochemical characteristics of water (pH, temperature, colour, turbidity and organic matter), which is consistent with the results of this work and previous works elaborated across the world to identify the adverse effect, especially genotoxic issues induced by compounds produced from the disinfection of waters intended for human consumption (Urien, 1986; Lebel et al., 1995; Williams et al., 1995, 1998; Chen and Weisel, 1998).

**(a) Winter**

**(b) Autumn**

**(c) Spring**

**(d) Summer**

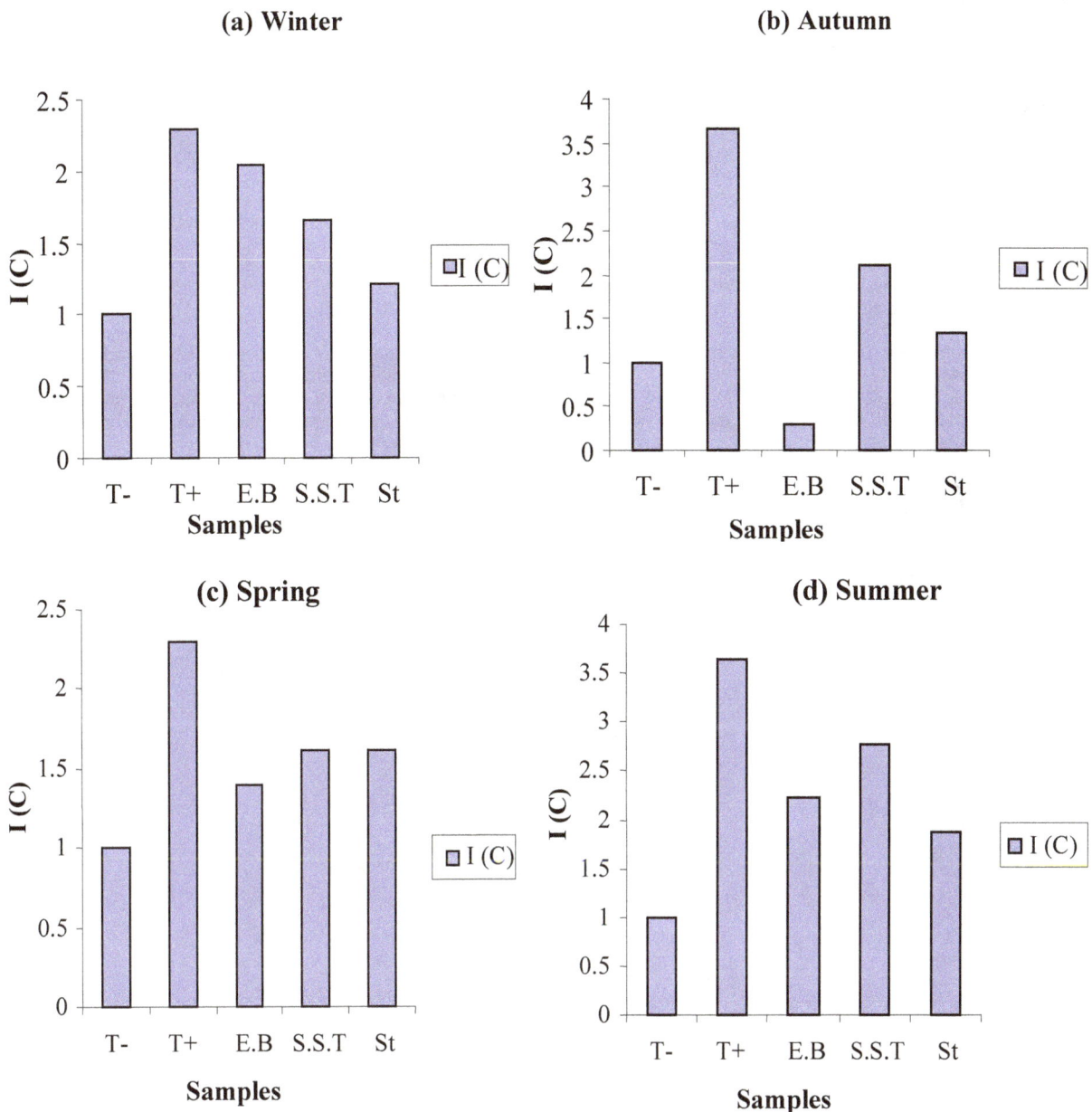

**Figure 2.** Results of standard SOS chromotest.

## Conclusion

The study of the genotoxic effects of drinking water, treated with chlorine, constitutes a new approach to assess the pollution of drinking water. The introduction of the genotoxic tests of water as a new quality criterion for drinking water, in addition to the already existent physicochemical and microbiological parameters, could be beneficial. Physicochemical identification, micro-biological quantification and study of the genotoxicity of

the potentially hazardous agents are important approaches to fix norms in order to minimize health risks. In other words, the group efforts aiming to restrict the production of the genotxic agents, notably by using less chlorine, must never put in danger the perfect disinfection of water intended for consumption.

**REFERENCES**

Bellard TA, Lichtenberg  JJ, Kroner RC (1974). The occurrence of organohalides in chlorinated drinking water. J. Am. Water Works Assoc. 66:703-706.

Bouziani M (2000). L'eau de la pénurie aux maladies. Editions IBN-KHALDOUN. p.247.

Chen WJ, Weisel CP (1998). Halogenated DBP, concentrations in a distribution system. J. Am. Water Works Assoc. 90(4):151-163.

Jouany JM (2000). L'ecotoxicology, hydrogéologie n°1 :31-41.

Lafferriere M, Levallois P, Gingras S (1999). La problématique des trihalométhanes dans les réseaux d'eau potable s'alimentant en eau de surface dans le bas Saint laurent. Vecteur environnement. Section Scientifique 38-43.

Le Curieux F, Giller S, Marzin D, Brice A, Erb F (1996). Utilisation de trois tests de génotoxicité pour l'étude de l'activité génotoxique de composés organohalogénés, d'acides fulviques chlorés et d'échantillons d'eau (non concentrés) en cours de traitement de potabilisation. Rev. Sci. Eau 1:75-95.

Lebel G, Benoit FM, Williams DT (1995). Etude d'un an sur les sous produits de désinfection. Santé Canada. 27.

Legay C, Rodriguz MJ, Sadiq R, Serodes JB, Levallois P, Proulx F (2011). Spatial variations of human health risk associated with exposure to chlorination by-products occurring in drinking water. J. Environ. Manag. 92:892-901.

Liu S, Zhu Z, Fan C, Qiu Y, Zhao J (2011). Seasonal variation effects on the formation of trihalomethane during chlorination of water from Yangtze River and associated cancer risk assessment. J. Environ. Sci. 23(9):1503-1511.

Monod I (1989). Mémento technique de l'eau tome 1 neuvième édition du CINQUANTENAIRE, 1200.

Olivier Ph, Marzin D (1987). Study of mutagenicity of 48 inorganic derivatives with SOS chromotest. Mut. Res.198:263-269.

Oms (1994). Directives de qualité pour l'eau de boisson (1994). Deuxième édition 1, Recommandations Genève, OMS, 202.

Potelon JL, Zysman K (1998). Le guide des analyses de l'eau potable, organochlorés (autres que pesticides). Ed. la lettre du cadre territorial. p. 253.

Rook JJ (1974). Formation of haloforms during chlorination of natural waters. Water Treat. Exam. 23:234.

Urien F (1986). Substances humiques naturelles de synthèse, thèse de docteur de troisième cycle, chimie et microbiologie de l'eau, 140.

Us Epa (2000). Microbial/ disinfection by-products federal advisory committee, stage2M-DBP agreement in principle.

Williams DT, Benoit FM, Lebel G (1998). Trends in levels of DBP. Environmetrics 9:555-563.

Williams DT, Lebel G, Benoit FM (1995). Etude nationale sur les sous produits de désinfection chlorés dans l'eau potable au Canada. Santé Canada. P 29.

Zekkour H, et BERON P (2001). Procédé innovateur de contrôle de la formation de composes organochlorés. Vecteur environnement, section scientifique 34(6):42-5.

# 20

# Metals bioavailability in the leachates from dumpsites in Zaria Metropolis, Nigeria

**Sani Uba[1] , Adamu Uzairu[1], Muhammad Sani Sallau[1], Hamza Abba[1] and Okunola Oluwole Joshua[2]**

[1]Department of Chemistry, Ahmadu Bello University, Zaria, Nigeria.
[2]National Research Institute for Chemical Technology, Zaria, Nigeria.

**Landfill leachates pose a significant threat to both surface water and groundwater especially the wells adjacent to landfills. The study investigated the bioavailability of zinc (Zn), copper (Cu), lead (Pb), cadmium (Cd) and mercury (Hg) from leachates of ten huge dumpsites across the metropolitan city of Zaria. The trends in the mean concentrations of the metals (mg/L) among the fractions were; Zn: total > mobile > particulate > dissolved; Pb: total > mobile > particulate > dissolved; Cd: mobile > dissolved > total > particulate; Hg: particulate > mobile > total > dissolved, respectively. All the concentrations of the metal ions were above the world Health Organization (WHO) (2006) and United States Environmental Protection Agency (USEPA) (2000) tolerable limits across the sites, with the exception of lead at the control site which was not detected. The order of the metals bioavailability was; Cd > Hg > Zn > Pb > Cu, with more than 49% found in the bioavailable phase. Thus, the underground waters within the vicinity of the dumpsites were greatly at the risk of being polluted by these toxic metals and subsequently affecting the inhabitants who use the water for drinking and other domestic activities untreated, through the food chain transfer. The health implications associated with the toxic metals include an irreversible damage to nervous system, gastric and intestinal disorder, heart disease, liver, brain damage, mental retardation and teratogenic effects.**

**Key words:** Fractionation, heavy metals, leachates, dumpsites, Zaria.

## INTRODUCTION

Inadequate municipal and industrial waste collection and disposal creates a range of environmental problems in Zaria metropolis wHere considerable amount of waste ends up in open dumps or drainage system, threatening both surface water and groundwater quality. This provides a breeding ground for disease carrying pests, create health hazards, pollute the air, soil and sometimes groundwater and surface water, as well as deteriorating the beauty of the area (Botkin and Keller, 1998). The situation is high in Zaria metropolis where most households could not make use of garbage collection containers; lack of most solid wastes services in crowded, low-income neighbors is a major contributor to the high mobility and mortality among the urban poor.

The incoming wastes in the city originate mainly from households and industries (Benedine et al., 2011).

The open dumpsites are well known to release large amounts of hazardous and otherwise deleterious chemicals to nearby groundwater, surface water, soil and to the air via leachates and landfill gases. It is known that such releases contain a wide range variety of potential carcinogens and potentially toxic chemicals that represent a threat to public health (Fredlee et al., 2003). Leachates have been implicated as environmental pollutants such as air, soil, plants, surface and ground water pollution. Sufficient number of individuals near dumpsites would experienced an average increased cancer risk, at least 1 in 1000 (Fredlee et al., 2003).

As many studies have shown, municipal refuse may increase heavy metals concentrations in soils and underground water (Carlson et al., 1976; Albores et al., 2000; Okoronkwo et al., 2005a, b) which may have effects on the host soils crops and human health (Smith et al., 1996; Nyle and Ray 1999). Thus, the environmental impacts of leachates emanating from dumpsites are greatly influenced by their heavy metals contents. However, while total heavy metals contents is a critical measure in assessing risk of a refuse dumpsite, it does not provide a predictive insight on the bioavailability, mobility and fate of the heavy metals contaminants (Albores et al., 2000). Thus, it is the chemical form or species of the heavy metals that is an important factor in assessing their impacts on the environment as it controls their bioavailability and mobility (Norvell, 1984).

Past investigations on the metals impact of municipal refuse dumpsites leachates in Nigeria were concerned only with total heavy metals determination (Amadi et al., 2011; Aiyesanmi et al., 2011). The objective of this study therefore was to investigate the chemical fractionation of cadmium (Cd), copper (Cu), lead (Pb), mecury (Hg), and zinc (Zn) in dumpsites leachates of Zaria metropolis so as to assess their potential mobility, bioavailability and fate. Thus, the human health and ecological risks associated with the refuse dumpsites to the underground water (hand-dug wells) will be assessed.

## MATERIALS AND METHODS

### Quality assurance

All reagents used were of analytical grade, de–ionized water was used for the preparation of the standard solutions, all the glassware and polythene sample bottles were washed with liquid soap, rinsed with water, soaked in 10% $HNO_3$ for 24 h, cleaned thoroughly with double distilled de- ionized water and dried. The analytical results obtained were validated with spiked samples. The percentage recoveries of the metals and the analytical precision was confirmed with the triplicates throughout the study (Todorovi et al., 2008). Procedural blanks, reagents blanks and preparation of standard solutions were carried under clean laboratory environment.

### Description of the study area

Zaria metropolis is located at latitude 11° 07' N and longitude 07° 42' E, and is presently one of the most important cities in Northern Nigeria (Uba et al., 2008) (Figure 1). It has problems of environmental sanitation such as improper disposal of refuse near residential areas resulting in the contamination of the underground water via leachates emanating from the dumpsites since most of the wells near the dumpsites were poorly covered and opened (Figure 2). It has a total area of 300 km$^2$ and constitutes four major settlements; Zaria City, Tudun Wada, Sabon Gari and Samaru (Zaria at a glance, 2013). It has a tropical continental climate with a pronounced dry season, lasting up to seven months (October to May). During the dry season, a cool period is usually experienced between November and February. This emanates from the influence of the North-eastern winds (the Harmattan) which control the tropical continental air mass coming from the Sahara. This weather prevails over most parts of the country. The North-East (NE) winds are characterized by hazy to dusty conditions and low temperatures, as low as 10°C at night. In the afternoon, up to 40°C is sometimes recorded. The humidity also drops to less than 15% in December/January.

Zaria experiences a brief period of hot but dry weather in March and April, followed by a progressive incursion of tropical maritime air mass from the Atlantic Ocean which displaces the NE (Harmattan) winds. During this short period, the mean daily maximum temperatures are fairly stable, and they range from 38 to 42°C. After that, the South Westerly Monsoon winds laden with moisture bring the rain in thunderstorms and squalls with heavy fall of high intensities. The rainy season lasts from May to September/ October with long-term annual rainfall of 1,040 mm in about 90 rain days (Zaria at a glance, 2013). The relatively deep tropical ferruginous soils and climate conditions of Zaria are suitable and sustain a good cover of savanna woodland (Northern Guinea Savanna), with a variety of grasses woody shrubs and short trees. Ten huge dumpsites were selected which covered the metropolitan city, and a control/uncontaminated site was selected 300 m away from the Kusfa dumpsite which is a new settlement without any dumping activities as summarized in Table 1.

### Samples collections

Leachate samples were collected from ten dumpsites and a control site from June to August, 2011 in the rainy season from randomly selected leachate drains at the sites. The samples were collected in the well labeled clean polythene bottles that were rinsed with the leachates prior to sample collection. The samples for elemental analysis were collected in 1 L polyethylene bottles while those for mercury analysis were collected in glass bottles (American Public Health Association (APHA), 2005).

### Samples pre-treatment

Samples for mercury analysis were preserved in 1 ml concentrated $H_2SO_4$ and 1 ml 5% $K_2Cr_2O_7$ solution for every 100 ml samples. The samples for elemental analysis were preserved in 2 ml concentrated $HNO_3$ (Aiyesanmi et al., 2011; APHA, 2005).

### Chemical fractionation of heavy metals in the dumpsite-leachates

The chemical fractionations of the samples were carried out on the

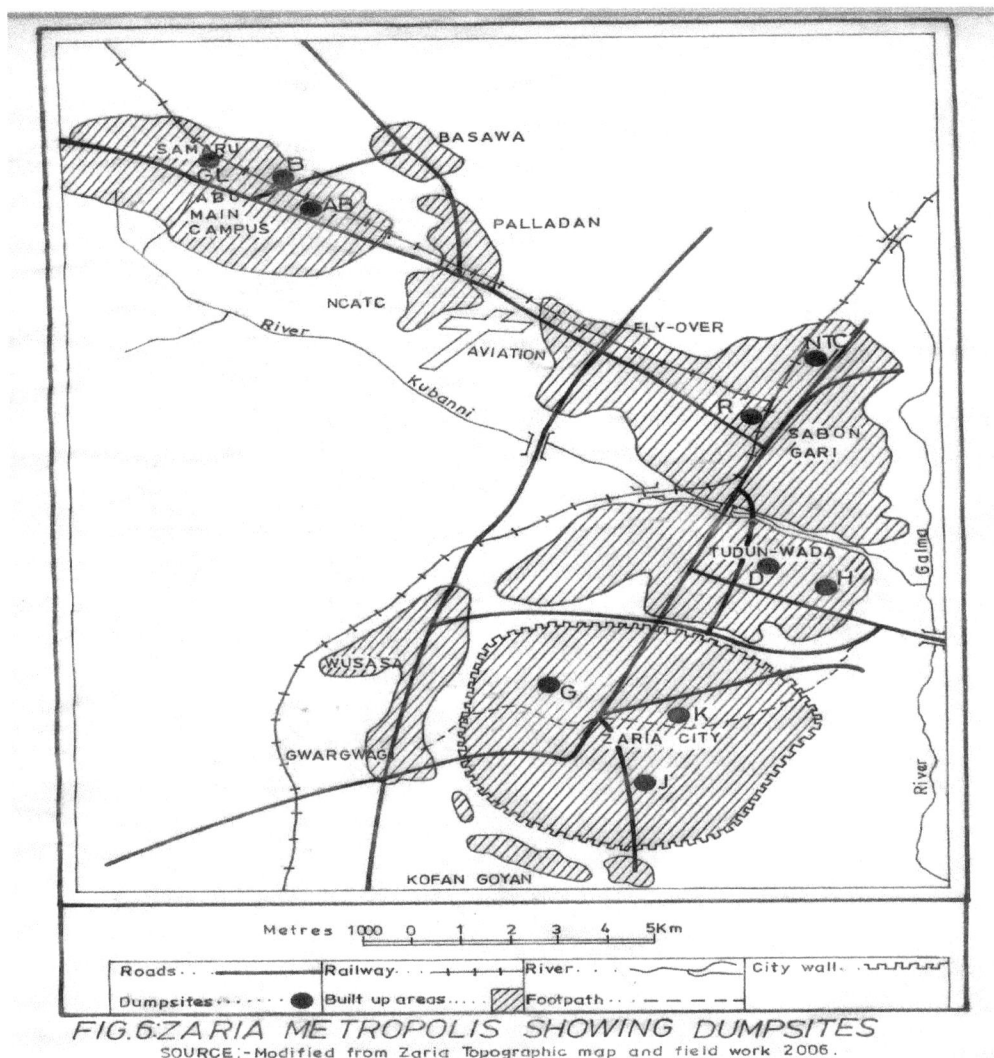

**Figure 1.** Zaria map showing different sampling points.

principle proposed by Bäckström et al., 2003; Wakawa et al., 2008). Basically, there are three steps that were involved, while fraction (IV) was taken as the difference between fractions (III) and (I) and the fractionation was carried out as follows;

**Fraction I (dissolve phase):** 50 ml of the leachate samples were decanted from the sampling vessel and filtered through 0.5 μm teflon filters and the solution was then acidified with 5 ml 2% HNO$_3$ and made up to 25 ml with de-ionized water (Bäckström et al., 2003).

**Fraction II (mobile phase):** 50 ml of the samples were decanted and acidified with 2% HNO$_3$ (5 ml). These were followed by filtration through 0.5 μm teflon filters after 24 h. The solutions were then made up to 25 ml with de-ionized water (Bäckström et al., 2003).

**Fraction III (total fraction):** 5 ml of 2% HNO$_3$ were added to the 50 ml of the samples and the solution were stirred vigorously to suspend all the particulate matter and then filtered through 0.5 μm

teflon filter after 24 h (Bäckström et al., 2003).

**Fraction IV (particulate fraction):** Particulate fraction was the difference between total residual fraction and dissolve phase (BäckstrÖm et al; 2003).

**Samples analysis**

The digests of the samples were analysed for Zn, Pb, Cu, Cd and Hg using AAS-650 (varian double beam) and the validation of the procedure for metal determination was conducted by spiking samples with multi-element standard solutions containing 5 mg/L of the metals with the exception of Cd where 4 mg/L of the spiked sample was used. Spiked samples were used under the same experimental conditions used for the procedural blanks as samples the acceptable recoveries (>98.4%) from the spiking experiment had validated the experimental procedure (Table 2).

**Figure 2.** One of the dumpsites in Zaria metropolis showing one of the sampling point (RA).

### Statistical analysis

The data were expressed as means ± standard deviation. To show whether there is significant difference between the mean concentrations of the metals across the sites, one way analysis of variance (ANOVA) was used and the Pearsons moment correlation (r) was used to establish the degree of relationship among the fractions of the analysed metal ions across the sites using a statistical software package for social science (SPSS version 16).

## RESULTS

### Chemical fractionation of metals in the leachates (dumpsite-leachates)

Tables 3 to 7 showed the percentages of the bioavailable of zinc, lead, copper cadmium and mercury in the fractionated leachates across the sites in addition to the concentrations of various fractions (mg/L). The highest

total extractable fraction of zinc (Table 3) was recorded at the control site, with 98.674% in the bio-available phase. The order of the bioavailable fractions of zinc across the sites followed the trend: CTR > SA > AJ > JK > PR > SH > BG > DD > KU > RA > NTC. Table 9 showed the results for the analysis of variance (ANOVA), and a significant difference of the mean concentrations of the metal were recorded across the sites at $p < 0.05$, with the exception of cadmium which was not significant at 95% confidence.

Furthermore, Table 4 showed the elevated levels of lead recorded across the sites, with the extractable fractions predominantly found in the bio-available phase (>65%) except the control site which was below the detection limit (BDL). The highest bio-available lead was recorded at BG-dumpsite leachates. The order of mobility Table 7 showed the concentrations of mercury in the fractionated leachate samples in which 29.23 (CTR) to 100% (SH, SA, PR and DD) were in the bioavailable

**Table 1.** Description of the sampling points (dumpsites) and various activities being discharged at each dumpsite.

| Sampling site | Sampling site | Activities | No./wells | House holds |
|---|---|---|---|---|
| NTC | National tobacco company | Tobacco wastes, residential waste and workshops | 2 | Household and shops |
| DD | Dandaji | Household wastes | 2 | Households |
| SH | Shafi | Households wastes, brick molding and automobile | 2 | Households and shops |
| RA | Railway station | Carpentry, wood workshops mechanic workshops, residential and wood wastes | 2 | Market and households |
| JK | Jeka da Kwarinka | Household | 3 | Households |
| PR | Prince road | School and households wastes | 2 | Schools & households |
| SA | Samaru | Households wastes | 3 | Households |
| KU | Kusfa | Households wastes | 3 | Households |
| BG | Babban Gwari | Residential waste | 2 | Households |
| AJ | Alkali Jae | Households waste | 4 | Households |
| CTR | Control | No dumping activity | 3 | Households |

**Table 2.** Fractions descriptions and WHO limits.

| Symbol | Description | Unit | Tolerable limit (WHO, 2006) |
|---|---|---|---|
| I | Dissolved fraction | mg/L | Depends on the metal |
| II | Mobile fraction | mg/L | Depends on the metal |
| III | Total fraction | mg/L | Depends on the metal |
| IV | Particulate fraction | mg/L | Depends on the metal |
| Zn | Zinc | mg/L | 5 |
| Pb | Lead | mg/L | 0.001 |
| Cu | Copper | mg/L | 1.5 |
| Cd | Cadmium | mg/L | 0.003 |
| Hg | mercury | mg/L | 0.001 |
| Bioavailable | Equal to the sum of dissolved and mobile phases | % | - |

phase. The bioavailability trend across the sites was: PR > SH > SA > DD > RA > JK > AJ > BG > NTC > KU > CTR. It was observed that the levels obtained were well above the WHO (2006) limits both across the sites and among the fractions. The distribution trend among the fraction was: particulate > mobile > total > dissolved. Similarly,

one way ANOVA showed a significant difference among the fractions at $p < 0.05$ as shown in Table 9. The metal was positively correlated with the lead however, negative correlation of the metal ion was recorded with copper, cadmium and zinc, revealing an inverse relationship. The high concentrations recorded at the sites may not be

unconnected with dumpsites constituents where cadmium containing waste formed part of the constituents and the total fraction was significantly not different at $p < 0.05$. Table 8 showed the degree of association of the metal ions, the variables showed significant positive correlations with each other and with different metal ions with

**Table 3.** Concentrations (mean ± SD) mg/L of zinc in the fractionated dumpsites leachates.

| Fraction | Sites | | | | | | | | | | | |
|---|---|---|---|---|---|---|---|---|---|---|---|---|
| | AJ | BG | CTR | DD | JK | KU | SA | SH | RA | PR | NT | STD |
| I | 0.30±0.02 | 0.16±0.01 | 0.04±0.06 | 0.29±0.02 | 1.16±0.08 | 0.39±0.03 | 2.71±0.02 | 0.47±0.03 | 0.15±0.03 | 0.12±0.09 | 0.04±0.03 | 5.00 |
| II | 0.13±0.09 | 0.13±0.06 | 0.06±0.04 | 0.17±0.01 | 1.19±0.08 | 0.18±0.01 | 0.20±0.01 | 4.64±0.03 | 0.10±0.07 | 3.71±0.03 | 0.10±0.02 | 5.00 |
| III | 0.34±0.001 | 0.46±0.03 | 0.35±0.03 | 0.53±0.04 | 1.53±0.01 | 4.12±0.03 | 2.72±0.02 | 2.82±0.02 | 0.83±0.05 | 1.01±0.07 | 1.26±0.09 | 5.00 |
| IV | 0.04±0.03 | 0.31±0.02 | 0.01±0.05 | 0.24±0.02 | 0.37±0.03 | 3.73±0.02 | 0.2±0.02 | 2.35±0.02 | 0.80±0.06 | 0.89±0.06 | 1.23±0.09 | 5.00 |
| Bioavailable (%) | 95.63 | 70.77 | 98.67 | 80.52 | 91.310 | 55.680 | 96.650 | 77.150 | 57.580 | 84.490 | 53.390 | - |

**Table 4.** Concentrations (mean ± SD) mg/L of lead in the fractionated dumpsites leachates.

| Fraction | Sites | | | | | | | | | | | |
|---|---|---|---|---|---|---|---|---|---|---|---|---|
| | AJ | BG | CT | DD | JK | KU | SA | SH | RA | PR | NT | STD |
| I | BDL | 0.35±0.03 | BDL | BDL | BDL | 0.25±0.02 | BDL | BDL | 0.36±0.03 | 0.10±0.07 | BDL | 0.001 |
| II | 0.33±0.02 | 0.32±0.02 | BDL | 0.33±0.02 | 0.32±0.02 | 0.35±0.03 | 0.27±0.02 | 0.36±0.0 | 0.34±0.02 | 0.36±0.03 | 0.33±0.02 | 0.001 |
| III | 0.38±0.03 | 0.38±0.03 | BDL | 0.30±0.02 | 0.33±0.02 | 0.39±0.03 | 0.33±0.02 | 0.38±0.02 | 0.39±0.03 | 0.33±0.02 | 0.32±0.02 | 0.001 |
| IV | 0.38±0.03 | 0.03±0.02 | BDL | 0.30±0.02 | 0.33±0.02 | 0.13±0.09 | 0.33±0.02 | 0.02±0.01 | 0.39±0.03 | 0.23±0.02 | 0.32±0.02 | 0.001 |
| Bioavailable (%) | 65.260 | 97.580 | BDL | 67.650 | 66.390 | 88.140 | 64.240 | 97.210 | 73.690 | 77.500 | 66.940 | - |

**Table 5.** Concentrations (mean ± SD) mg/L of cadmium in the fractionated dumpsites leachates.

| Fraction | Sites | | | | | | | | | | | |
|---|---|---|---|---|---|---|---|---|---|---|---|---|
| | AJ | BG | CTR | DD | JK | KU | SA | SH | RA | PR | NT | STD |
| I | 0.04±0.02 | 0.06±0.4 | 0.002±0.001 | 0.05±0.03 | 0.055±0.3 | 0.07±0.04 | 0.05±0.040 | 0.07±0.05 | 0.06±0.004 | 0.05±0.00 | 0.05±0.03 | 0.003 |
| II | 0.07±0.05 | 0.06±0.04 | 0.02±0.01 | 0.06±0.04 | 0.07±0.05 | 0.06±0.04 | 0.06±0.03 | 0.05±0.04 | 0.06±0.033 | 0.138±0.02 | 0.02±0.01 | 0.003 |
| III | BDL | 0.07±0.05 | 0.03±0.02 | 0.06±0.04 | 0.06±0.04 | 0.03±0.02 | 0.07±0.05 | 0.05±0.04 | 0.02±0.01 | 0.03±0.02 | 0.03±0.02 | 0.003 |
| IV | BDL | 0.01±0.05 | BDL | 0.02±0.01 | 0.01±0.03 | BDL | 0.02±0.09 | BDL | BDL | BDL | BDL | 0.003 |
| Bioavailable (%) | 100.00 | 96.45 | 100.00 | 90.68 | 97.43 | 100.00 | 89.94 | 100.00 | 100.00 | 100.00 | 100.00 | - |

few exceptions. For example, the dissolved fractions of zinc (ZnLI) was positively correlated and significant with the particulate fractions (0.433). Other variables that were positively correlated with the fraction of zinc were PbLII, PbLIII, PbLIV, CdLI, CdLIII, CdLIV and HgLI respectively. ZnLI was highly positively correlated to HgLIV (particulate phase) but negatively correlated to HgLII. Similarly, the mobile fraction (ZnLII) of zinc was strongly positively correlated to CdLIV, CuLIV and HgLI fractions with r-values of 0.885, 0.687 and 0.557, respectively. Other mild positive correlations were observed with the PbLII, PbLII, PbLIII, CdLI, CdLII, CdLIII, CuLIII and CuLIV. Furthermore, the ZnLII (mobile fraction) of

**Table 6.** Concentrations (mean ± SD) mg/L of copper in the fractionated dumpsites leachates.

| Fraction | Sites | | | | | | | | | | | STD |
|---|---|---|---|---|---|---|---|---|---|---|---|---|
| | AJ | BG | CTR | DD | JK | KU | SA | SH | RA | PR | NT | |
| I | 0.091±0.07 | 0.05±0.03 | BDL | 1.11±0.08 | 0.06±0.04 | 0.07±0.05 | 0.08±0.06 | 0.03±0.02 | 0.03±0.02 | 0.001±0.01 | 0.07±0.05 | 1.500 |
| II | 0.003±0.001 | 0.03±0.02 | BDL | 0.08±0.01 | 0.12±0.01 | 0.08±0.05 | 0.03±0.002 | 0.07±0.01 | 0.07±0.01 | 0.09±0.01 | 1.39±0.01 | 1.500 |
| III | 0.002±0.001 | 0.21±0.02 | BDL | 1.60±0.01 | 0.33±0.02 | 0.10±0.01 | 0.07±0.01 | 0.4±0.03 | 0.49±0.04 | 0.59±0.04 | 0.08±0.06 | 1.500 |
| IV | 0.001±0.09 | 0.16±0.01 | BDL | 0.48±0.03 | 0.27±0.01 | 0.04±0.03 | BDL | 0.36±0.03 | 0.47±0.03 | 0.59±0.04 | 0.006±0.001 | 1.500 |
| Bioavailable (%) | 63.920 | 58.440 | 0.000 | 100.000 | 100.000 | 79.050 | 50.000 | 49.910 | 100.000 | 78.470 | 63.430 | - |

**Table 7.** Concentrations (mean ± SD) mg/L of mercury in the fractionated dumpsites leachates

| Fraction | Sites | | | | | | | | | | | STD |
|---|---|---|---|---|---|---|---|---|---|---|---|---|
| | AJ | BG | CTR | DD | JK | KU | SA | SH | RA | PR | NT | |
| I | 1.05±0.08 | 0.59±0.42 | BDL | 1.94±0.14 | 0.92±0.07 | 0.65±0.05 | 2.16±0.02 | 5.13±0.36 | 1.10±0.78 | 3.37±0.02 | 1.20±0.09 | 0.001 |
| II | 3.71±0.03 | 0.80±0.01 | 0.69±0.02 | 0.84±0.005 | 1.07±0.08 | 0.85±0.01 | 0.69±0.05 | 0.74±0.01 | 2.17±0.01 | 1.03±0.01 | 1.94±0.01 | 0.001 |
| III | 2.72±0.01 | 1.15±0.08 | BDL | 0.92±0.07 | 1.24±0.08 | 14.96±0.11 | 1.71±0.11 | 0.72±0.05 | 1.45±0.01 | 1.62±0.01 | 2.74±0.02 | 0.001 |
| IV | 0.75±0.01 | 0.57±0.004 | 1.67±0.01 | BDL | 0.32±0.02 | 14.31±0.10 | BDL | BDL | 0.34±0.002 | BDL | 1.54±0.01 | 0.001 |
| Bioavailable (%) | 90.850 | 81.860 | 29.230 | 100.000 | 90.900 | 53.490 | 100.000 | 100.000 | 93.260 | 100.000 | 79.380 | - |

mercury was strongly positively correlated to HgLIII and HgLIV (at r = 0.696 and 0.656) while CdLI, CdLII, CdLIII, HgLI, ZnLIV, PbLI and PbLII were negatively correlated as show in Table 8. The particulate fraction of zinc (ZnLIV) had the highest significant positive correlation with HgLIII and HgLIV with r-values of 0.791 and 0.781 at p < 0.05. Other positive correlations of dissolved zinc (ZnLI) were observed with the PbLIV, CdLI, CdLIV and HgLI respectively.

As shown in Table 8, zinc was strongly positively correlated to dissolved, mobile, total and particulate fractions while the mobile phase of zinc was positively correlated to the particulate fraction (r = 0.312). Furthermore, all the variables (frac-

tions) were positively correlated, the highest r-value was observed on correlating mobile and total lead (r = 0.948) at p < 0.05 as shown in Table 7.

Similarly, cadmium fractions were positively correlated with the exception of mobile and particulate fractions which showed negative correlation. Copper fractions, were strongly positively correlated with the dissolved and particulate, particulate and total fractions (r = 0.869 and 0.748), respectively. However, negative correlations were recorded among the fractions of mercury across the sites with the exception of particulate and total fractions which were strongly positively correlated (0.974) at p < 0.05 (Table 8).

## DISCUSSION

On comparing the results obtained for zinc with the standard limits (USEPA, 2000; WHO, 2006), sites KU, SA, SH and PR were contaminated (concentration > 5 mg/L). Thus, the zinc in the analysed leachate samples was readily bio-available to the environment contaminating especially, the underground water due to leachates percolation. Zinc pollution is known to induce vomiting, dehydration, abdominal pain, dizziness and lack of muscular co-ordination (WHO, 1999). Overall, the mobile fractions had the highest concentrations of the total extractable Zinc across the sites. The concentrations recorded were

**Table 8.** Correlation matrices for the fractionated metals in the leachates.

| Fractions | ZnLI | ZnLII | ZnLIII | ZnLIV | PbLI | PbLII | PbLIII | PbLIV |
|---|---|---|---|---|---|---|---|---|
| ZnLII | -0.067 | | | | | | | |
| ZnLIII | 0.433* | 0.245 | | | | | | |
| ZnLIV | -0.159 | 0.311 | 0.261 | | | | | |
| PbLI | -0.290 | -0.217 | 0.372 | 0.254 | | | | |
| PbLII | 0.006 | 0.312 | 0.379 | 0.379 | 0.254 | | | |
| PbLIII | 0.143 | 0.182 | -0.148 | -0.148 | 0.379 | 0.948** | | |
| PbLIV | 0.296 | -0.310 | 0.383 | 0.383 | 0.148 | 0.409 | 0.379 | |
| CdLI | 0.180 | 0.220 | 0.553 | 0.550 | 0.383 | 0.841 | 0.869** | 0.145 |
| CdLII | 0.068 | 0.386* | 0.008 | 0.003** | 0.151 | 0.493** | 0.388 | 0.210 |
| CdLIII | 0.361 | -0.186 | 0.170 | -0.060 | 0.116 | -0.149 | -0.103 | -0.134 |
| CdLIV | 0.687 | 0.687** | -0.276 | 0.038 | -0.384 | -0.213 | 0.002 | 0.040 |
| CuLI | -0.056 | -0.0202 | -0.217 | -0.190 | -0.230 | 0.118 | -0.022 | 0.209 |
| CuLII | -0.204 | -0.138 | -0.021 | 0.120 | -0.205 | 0.145 | 0.04 | 0.236 |
| CuLIII | -0.167 | 0.147 | -0.231 | -0.122 | -0.055 | 0.289 | 0.094 | 0.178 |
| CuLIV | -0.251 | 0.557** | -0.187 | -0.009 | 0.183 | 0.426* | 0.242 | 0.108 |
| HgLI | 0.138 | 0.885** | 0.308 | 0.269 | -0.294 | 0.433 | 0.331 | -0.112 |
| HgLII | -0.249 | -0.261 | -0.371 | -0.215 | -0.19 | 0.254 | 0.293 | 0.596** |
| HgLIII | -0.044 | -0.194 | 0.696** | 0.791** | 0.333 | 0.254 | 0.308 | -0.062 |
| HgLIV | -0.118 | -0.223 | 0.656** | 0.781** | 0.331 | 0.063 | 0.118 | -0.233 |

| Fractions | CdLI | CdLII | CdLIII | CdLIV | CuLI | CuLII | CuLIII | CuLIV |
|---|---|---|---|---|---|---|---|---|
| CdLII | 0.176 | | | | | | | |
| CdLIII | 0.43 | -0.119 | | | | | | |
| CdLIV | 0.154 | 0.019 | 0.515 | | | | | |
| CuLI | 0.89 | -0.270 | 0.297 | 0.601 | | | | |
| CuLII | 0.128 | -0.395 | -0.185 | -0.198 | -0.590 | | | |
| CuLIII | 0.218 | 0.288 | 0.128 | 0.429* | 0.869* | -0.154 | | |
| CuLIV | 0.280 | 0.625** | -0.176 | 0.003 | 0.326 | -0.246 | 0.748 | |
| HgLI | 0.379 | 0.379 | -0.077 | 0.046 | 0.047 | -0.720 | 0.302 | 0.526 |
| HgLII | -0.123 | -0.017 | -0.381 | -0.365 | -0.128 | 0.192 | -0.225 | -0.174 |
| HgLIII | 0.347 | 0.030 | 0.060 | -0.211 | -0.107 | 0.021 | -0.225 | -0.301 |
| HgLIV | 0.219 | -0.089 | 0.087 | -0.259 | -0.121 | -0.01 | -0.251 | -0.345 |

| Fractions | HgLI | HgLII | HgLIII | HgLIV |
|---|---|---|---|---|
| HgLI | | | | |
| HgLII | -0.207 | 1 | | |
| HgLIII | -0.232 | -0.023 | 1 | |
| HgLIV | -0.309 | -0.130 | 0.974 | 1 |

*Correlation significant at $p < 0.05$, **correlation significant at $p < 0.01$.

higher than the values of 0.37 to 0.65 mg/L reported by Aiyesanmi et al. (2011) in Benin City for the total elemental analysis of leachates. The difference might be attributed to the different composition of the analysed dumpsites.

The lead concentrations recorded suggests that there was a common source of pollution by the metal ions as significant difference among the fractions was observed at $p < 0.05$. When the concentrations (total extractable) across the sites were compared with those of the international standard (USEPA, 2000; WHO, 1999) they all exceeded the tolerable limit of 0.05 mg/L with the exception of the fractions at the control site. The total extractable fractions were higher than the range of 0.05

**Table 9.** ANOVA for leachates across the sites.

| Fractions | Parameter | Sum of squares | df | Mean square | F | P |
|---|---|---|---|---|---|---|
| ZnL1 | Between groups | 11.609 | 10 | 1.161 | | |
| | Within groups | 0.023 | 11 | 0.002 | 559.924 | 0.000 |
| | Total | 11.632 | 21 | | | |
| ZnLII | Between groups | 49.561 | 10 | 4.956 | | |
| | Within groups | 0.090 | 11 | 0.008 | 604.118 | 0.000 |
| | Total | 49.651 | 21 | | | |
| ZnLIII | Between groups | 28.885 | 10 | 2.888 | | |
| | Within groups | 0.095 | 11 | 0.009 | 335.592 | 0.000 |
| | Total | 28.979 | 21 | | | |
| ZnLIV | Between groups | 24.853 | 10 | 2.485 | | |
| | Within groups | 0.056 | 11 | 0.005 | 491.905 | 0.000 |
| | Total | 24.909 | 21 | | | |
| PbLI | Between groups | 0.416 | 10 | 0.042 | | |
| | Within groups | 0.001 | 11 | 0.000 | 573.362 | 0.000 |
| | Total | 0.416 | 21 | | | |
| PbLII | Between groups | 0.196 | 10 | 0.020 | | |
| | Within groups | 0.003 | 11 | 0.000 | 80.510 | 0.000 |
| | Total | 0.198 | 21 | | | |
| PbLIII | Between groups | 0.228 | 10 | 0.023 | | |
| | Within groups | 0.003 | 11 | 0.000 | 82.211 | 0.000 |
| | Total | 0.231 | 21 | | | |
| PbLIV | Between groups | 0.423 | 10 | 0.042 | | |
| | Within groups | 0.002 | 11 | 0.000 | 242.765 | 0.000 |
| | Total | 0.425 | 21 | | | |
| CdL1 | Between groups | 0.006 | 10 | 0.001 | | |
| | Within groups | 0.000 | 11 | 0.000 | 93.537 | 0.000 |
| | Total | 0.006 | 21 | | | |
| CdLII | Between groups | 0.018 | 10 | 0.002 | | |
| | Within groups | 0.000 | 11 | 0.000 | 157.957 | 0.000 |
| | Total | 0.018 | 21 | | | |
| CdLIII | Between groups | 0.003 | 10 | 0.000 | | |
| | Within groups | 0.005 | 11 | 0.000 | 0.701 | 0.708 |
| | Total | 0.009 | 21 | | | |
| CdLIV | Between groups | 0.001 | 10 | 0.000 | | |
| | Within groups | 0.000 | 11 | 0.000 | 575.287 | 0.000 |
| | Total | 0.001 | 21 | | | |

**Table 9.** Contd.

| | | | | | | |
|---|---|---|---|---|---|---|
| CuLI | Between groups | 1.931 | 10 | 0.193 | | |
| | Within groups | 0.003 | 11 | 0.000 | 686.461 | 0.000 |
| | Total | 1.934 | 21 | | | |
| CuLII | Between groups | 3.030 | 10 | 0.303 | | |
| | Within groups | 0.005 | 11 | 0.000 | 689.227 | 0.000 |
| | Total | 3.035 | 21 | | | |
| CuLIII | Between groups | 3.952 | 10 | 0.395 | | |
| | within groups | 0.009 | 11 | 0.001 | 508.609 | 0.000 |
| | Total | 3.960 | 21 | | | |
| CuLIV | Between groups | 0.977 | 10 | 0.098 | | |
| | Within groups | 0.003 | 11 | 0.000 | 419.862 | 0.000 |
| | Total | 0.980 | 21 | | | |
| HgLI | Between groups | 40.242 | 10 | 4.024 | | |
| | Within groups | 0.126 | 11 | 0.011 | 351.747 | 0.000 |
| | Total | 40.368 | 21 | | | |
| HgLII | Between groups | 16.376 | 10 | 1.638 | | |
| | Within groups | 0.069 | 11 | 0.006 | 262.326 | 0.000 |
| | Total | 16.445 | 21 | | | |
| HgLIII | Between groups | 322.132 | 10 | 32.213 | | |
| | Within groups | 0.614 | 11 | 0.056 | 577.196 | 0.000 |
| | Total | 322.746 | 21 | | | |
| HgLIV | Between groups | 328.814 | 10 | 32.881 | | |
| | Within groups | 0.517 | 11 | 0.047 | 699.358 | 0.000 |
| | Total | 329.331 | 21 | | | |

ANOVA run at $p < 0.05$, if p is $< 0.05$, there was a significant difference among the fractions with the variable at 95% confidence otherwise, there was not.

to 0.12 mg/L reported by Manpanda et al. (2007) in Zimbabwe and lower than 0.35 to 0.97 mg/L reported by Ahlberg et al. (2006) in Sweden, respectively. It was also noted that if significant quantity of lead was leached into the groundwater, cytogenetic alteration such as kidney and brain damage or birth defects results especially when ingested through the food chain or drinking water (Ademoroti et al., 1996; Aiyesanmi et al., 2011).

The extractable fractions of cadmium were compared with the WHO (2006) standard limits of 0.003 and 0.001 mg/L (WHO, 2006; USEPA, 2003) respectively, overall, the results showed higher values with few exceptions. The recorded concentrations in this study were below the ranges of 0.02 ± 0.01 to 0.24 ± 0.31 mg/L and 3. 62 ± 0.01 to 8.15 mg/L reported by Aiyesanmi et al (2011) in

Benin City and Ahlberg et al. (2006) in Sweden, respectively. Analysis of variance (ANOVA) showed a significant difference (at $p < 0.05$) both among the fractions and across the sites. In addition, there was a positive correlation between the lead and Zinc (Pb to Zn) across the sites suggesting a common source of pollution. Cadmium is toxic when inhaled even in trace amount in dust/particulates during incineration/burning at dumpsite because of its carcinogenicity (Aiyesanmi et al., 2011). It is also known that it is very hazardous and of no use to biological processes (Watanabe et al., 2008).

The levels of copper recorded in this study were lower than > 1.5 mg/L reported by Ikem et al. (2002) in Lagos. The distribution pattern among the fractions was Cu: total > particulate > mobile > dissolved. Copper in the blood exist in two forms: bound to ceruplasmin (85 to 95%) and the rest 'freely' loosely bound to albumin. The free copper is toxic as it generates reactive oxygen species such as superoxide, hydrogen peroxide and the hydroxyl radical. These damages proteins and DNA (Brew et al., 2010). The levels of mercury recorded were significantly high, quiet above the WHO tolerable limit across the sites, and significant amount was found in the bioavailable fraction, thus the metal was readily leachable into the nearby open wells resulting to serious health problems such as chromosomal segregation, disruption and inhibition of cell division.

## CONCLUSION

The leachates samples were heavily polluted by zinc, copper, cadmium and mercury including those at the control site. However, lead was not detected at the control in all the fractions of the samples. Furthermore, significant amounts of the fractionated metals were found in the mobile phase showing a threat to the open wells within the vicinity of the dumpsites. Overall, more than 49% of the analysed toxic metals were found in the bioavailable fractions (dissolved + mobile fractions) resulting to serious health problems such as typhoid fever, cholera and other water borne related diseases to the residents who relied heavily on the untreated well waters for drinking and other domestic activities due to erratic and inadequate water supply in the city.

## ACKNOWLEDGEMENTS

The authors wish to acknowledge the Petroleum Trust Development Fund (PTDF) and Ahmadu Bello University Staff Development for their financial support. Furthermore, the staff of the Multi – user Science Research Laboratory and the entire staff of Chemistry Department, Ahmadu Bello University, Zaria were acknowledged for their support and analytical assistance.

## REFERENCES

Ahlberg G, Gustafssion O, Wedel P (2006). Leading of metals from sewage sludge during one year and their relationship to particle size. J. Environ. Pollut. 144 (2):545-553.

Aiyesanmi AF, Imoisi OB (2011). Understanding leading behaviors of landfill leachate in Benin city, Edo State, Nigeria through dumpsites monitoring. British J. Environ. Climate Change 1(4):190-200.

Albores AF, Peret – Cid BI, Gomes EF, Lopez EF (2000). Comparison between sequential extraction procedures and single extraction procedures and for metal partitioning in Sewage Sludge Samples. Analyst 125:1353-357.

Amadi AN (2011).Assessing the effects of Aladimma dumpsites on soil and groundwater using water quality index and factor analysis. Australian J. Basic Appl. Sci. 5 (11):763-770.

APHA (2005).Standard methods for the examination of water and waste, 20th ed. APHA –AWWA– WPCF, Washington, D.C USA.

BäckstrÖrm M, Nilson U, Karsten BA. Stelan K (2003). Speciation of heavy metals in road run off and roadside total deposition. Water, Air and Soil pollut.147:343-66.

Benedine A, Robert TA, Abbas II (2011). The impact of spatial distribution of solid waste dumps on infrastructure in samara zaria, Kaduna State, Nigeria using geographic information system (GIS). Res. J. Inf. Tech. 3(3):113-117

Brew GJ (2010). Copper toxicity in the general population .Clin.Neurophysiol. 121(4):459-60. Doi10.1016j.clinph.2009.12.015

Carlson CW (1976). Land application of waste material, soil conservation. Soil Am.Ankeny, IOWA, R. pp 3-7.

Mapanda F, Nyamadzawo G, Nyamangara J, Wuta M (2007). Effects of discharging acid mine drainage evaporation pond on chemical quality of the surrounding soil and water. J. Phy. Chem. Earth. 32:1366-1375

Norvell WA (1984). Comparison of chelating agents, metals in diverse soil materials. Soil Sci. Am.J.48:1285-1292.

Nyle CB, Ray RN (1999).The nature and properties of soils.12th ed. USA.pp743-785.

Okoronkwo NE, Igwe JC, Onwucheiewe EC (2005). Risk and health implication of polluted soil for crop production. Afr. J. Biotechnol. 4(13):1520-1524.

Okoronkwo NE, Odemelon SA, Ano OA (2005). Levels of toxic elements in soils of abandoned waste dumpsite. Afr. J. Biotechnol. 5 (13):1241-1244.

Smith CJ, Thomas P, Cook FJ (1996). Accumulation of Cr, Pb, Cu, Ni, Zn and Cd in Soil following Irrigation with untreated urban effluents in Australia Environ. Pollut. 94 (3):317-323.

Todorovi P, Djordveji D, Anthony JS (2008). Lead distribution in water and its association with sediment constituents of Barje lake (Lesuoval, Yugoslavia). J. Serb. Chem. Soc. 66 (1):697-708.

United States Environmental protection Agency (USEPA, 2003). National water quality Inventory. http://www. Epa. Gov.

USEPA (2000). United States environmental protection agency, national water quality Inventory, http: //www,epa.gov/3056/2000 report.

Wakawa RJ, Uzairu A, Kagbu JA, Balarabe ML (2008). Impact assessement of effluents discharged on physicochemical parameters and some heavy metals concentration in surface water of river Challawa, Kano, Nigeria. Afr. J. Pure Appl. Chem. 2(10):100-106.

WHO (2006).Guideline for drinking water quality, 2nd edition Recommendation. World Health Organization. Geneva 1:30-113.

# 21

# Occurrence of some pesticides in Bhoj wetland Bhopal and their effect on phytoplankton community: An ecological perspective

## Ayaz A. Naik[1] and Wanganeo A.[2]

[1]Biological Oceanography Division, National Institute of Oceanography, Dona Paula, Goa-403004, India.
[2]Department of Environmental Science and Limnology B.U Bhopal-462026, M.P India.

**The present communication deal with the analysis of water samples from Bhoj wetland to detect three pesticides (chlorpyrifos, monocrotophos and endosulfan) and to assess their individual toxicity to the phytoplankton community. Higher concentration of most of the pesticides was found to be in the pre-monsoon (March to May) 0.9 µg/l (monocrotophos) and post-monsoon (September to December) 1 µg/l (monocrotophos and chlorpyrifos) period. Among the aquatic organisms, phytoplankton communities are the key targets for the pesticides because of their ecophysiological similarities with terrestrial plants. A standard 96 h static algal bioassay was followed to determine pesticide effects on the population growth rate of phytoplankton. At higher concentrations of all the pesticides elicited a significant effect on population growth rate by maximum inhibition of the cell division, but toxicity would not be expected at typical environmental concentrations. The population growth rate $EC_{50}$ average values determined for Chlorophyceae, Cyanophyceae and Bacillariophyceae varied in the range of 16.1 to 32.3 µg/l for chlorpyrifos, 8.6 to 14.3 µg/l for monocrotophos and 4.2 to 15 µg/l for endosulfan, respectively. Therefore, decrease in phytoplankton populations resulting from pesticide exposure could occur at higher concentrations in aquatic systems where pesticides are present in mixture. Detrimental effects on phytoplankton population growth rate could impact nutrient cycling rates and food availability to higher trophic levels. Characterizing the toxicity of chemical mixtures likely to be encountered in the environment may benefit the pesticide registration and regulation processes.**

**Key words:** Pesticides, monsoon, phytoplankton, toxicity.

## INTRODUCTION

Agricultural area has the potential to pollute the aquatic ecosystem via the popular use of pesticides. Several of pesticides are being used in India both in agriculture and public health sectors. Although the uses of pesticides have resulted in increased food production and other benefits, it has raised concerns about potential adverse effects on environment and human health. Since the pesticides are lipid soluble in nature, cumulative accumulation of low concentrations of these in the body fat of mammals might pose potential hazards in the long run (Metcalf, 1997). Some of the pesticides like di-chloro di-phenyl tri-chloroethane (DDT) and di-chloro di-phenyl di-chloroethane (DDD) are also of significant concern because of their chiral structure (Ali et al., 2002). The greatest potential for unintended adverse effects of pesticides is through the contamination of the hydrologic systems, which supports aquatic life and related food chains and is used for drinking water, irrigation, recreation and many more purposes. The presence of pesticides in the environment even at nano/pico level can cause severe damage to human health (Ali et al., 2008). In aquatic bodies these pesticides affect many non target organisms like fish and birds due to biomagnifications through food chain (Kaushik et al., 2010). The persistence of these substances in aquatic ecosystem has special significance as they are picked up by aquatic organisms like plankton, and in the process pesticide residues enter in the food chain (Sarkar et al., 2003). These are bioaccumulative, relatively stable and carcinogenic (Jain and Imran, 1997), so needs close monitoring. Pesticides get introduced into natural aquatic system by various means incidentally during manufacturing and through surface water runoff from the agricultural land after their application. In addition, some pesticides are deliberately introduced into aquatic system to kill undesirable pests such as weeds, algae and vectors.

Plankton forms an important component of an aquatic ecosystem, reflects the condition in that environment and controll the flow of energy besides production, as such their well being is of utmost importance viz-a-viz aquatic water body metabolism (Wetzel, 1975; Michael, 1985). When these chemical substances come in contact with the aquatic biota, the later changes their metabolic activity and thus affect the aquatic environment adversely (Nimmo and Mcwen, 1994). Algae are useful indicators of potential pollution as they respond by stimulation, inhibition or both to all toxicants affecting water quality (Rajendran and Venugopalan, 1983). The variability in their responses usually depends on the toxicant concentrations, duration of exposure and the test species subjected to study. Characterized by their differential sensitivity to chemical compounds, algae have been recommended as test species in detecting the occurrence of pesticides in aquatic environments (Hersh and Crumpton, 1987) and the current scientific understanding concerning the phytotoxic effects of several kinds of contaminants is based mostly on the studies of a few green algae. Algal assays are considered as a supplement to chemical analysis in the assessment of

pollution and have been proved to be more sensitive than *Daphnia* tests for monitoring toxicities in aquatic environments (Baun et al., 1998). Considering their importance in an aquatic ecosystem, some species of algae present in the water body were considered in the present investigation to assess the impact of pesticides viz., chlorpyrifos, monocrotophos and endosulfan commonly used in the agricultural catchment area. In spite of persistent nature and consequent interactions of pesticides with biosphere, relatively few studies have been conducted on the effect of pesticides on freshwaters.

The Bhoj Wetland situated in the Western side of the city has a water stretch from west to east and basically bounded between the latitude 23° 13' to 23° 16' and longitude 77° 18' to 77° 24' was selected for the present study (Figure 1). The water body is a wetland of National importance and is one of the sixteen wetlands in the country that have been so far identified for conservation and management. It is the main source of potable water, besides, being a habitat for various migratory birds and a source of revenue by way of tourism. A lot of deterioration has been observed in its water quality with the passage of time. This is mainly because of various anthropogenic activities taking place in its widely spread catchment area. Keeping this in view, the present study was undertaken to assess residual pesticide concentration in water and study their toxicity on the plankton population.

## MATERIALS AND METHODS

Water samples were collected from two different sites (30 cm below the surface) during the Premonsoon (March to May), Monsoon (June to August) and the Post monsoon (September to December) period in brown glass bottles of 1 liter capacity and were transported to the laboratory. For analysis of pesticides, 1 liter of the unfiltered water sample from each location was extracted by liquid-liquid extraction in a separatory funnel using n-hexane. The combined solvent extracts were demoisturised using anhydrous granular sodium sulphate and concentrated in a rotary evaporator to a final volume of 2 ml. All the samples were analyzed for different pesticides viz. endosulfan, monocrotophos and chlorpyrifos on Antek-2000 gas chromatograph equipped with Ni[63] ECD. For all extraction, GC grade n-hexane (Spectrochem, India; 99%) was used. The pesticides standards of 99.9% purity were procured from Sigma-Aldrich, USA. All the analysis was carried out in duplicate and the recoveries of the individual pesticides were determined through spiked sample method. Recovery correction factor were applied to the final results.

### Sample collection and culture

Fresh water phytoplankton samples were directly collected into one liter plastic bottles from Bhoj Wetland Bhopal. Environmental factors such as water temperature and pH of the sampling station was measured at the spots (date will be published elsewhere). Thereafter recommended nutrient media (Aquacentre ltd. USA) was added to the bottles and were immediately transferred to plant

**Figure 1.** Satellite map of Bhoj wetland showing sampling stations.

growth chamber (Vision, VS-3D, Korea) with the temperature range of 20 to 25°C and then it was kept for three days. From the incubated samples, 1 ml of the sample was transferred to the Sedgwick rafter cell for the observation of the desirable abundance and community changes of phytoplankton under and inverted microscope.

### Acute toxicity test

A standard 96 h static algal bioassay (ASTM, 1996) was followed to determine the potential toxicity of three pesticides (endosulfan, monocrotophos and chlorpyrifos) on phytoplankton. Known volume of multispecies algal cultures were exposed to a series of concentrations of test material and an untreated control was maintained to measure the normal growth of the algae incubated in plant growth chamber (Vision, VS-3D, Korea) with the temperature of 25°C, light intensity 180 $\mu E\ m^{-1}\ s^{-1}$ and light: dark cycle of 12:12, respectively. Algal toxicity may be expressed in terms of a broad range of responses, encompassing those that are inhibitory (50% reduction in cell number at specified time interval) and cell death observed in Sedgwick rafter cell. The effects of test materials were evaluated by measuring the growth of phytoplankton in treated and untreated cultures. Cell density was assessed in control as well as treatment containers at an interval of 24 h up to 96 h using Sedgwick-rafter cell following the method given in American Public Health Association (APHA) (1995). Based on the cell count, data was analyzed for the determination of $EC_{50}$, 96 h and limiting percentage/percentage inhibition in the growth relative to control systems (Mary and Rubber, 1983).

### Limiting percentage/percentage inhibition:

$I = \mu_c - \mu_{tox} / \mu_c \times 100$

Where I is the inhibition, $\mu_c$ is the average specific growth rate of control and $\mu_{tox\ is}$ the average specific growth rate of phytoplankton in the test series.

### RESULTS AND DISCUSSION

The present study revealed the occurrence of different pesticides viz. chlorpyrifos, monocrotophos and endo-sulfan in the surface water of the Bhoj Wetland, Bhopal in different seasons, summarized in the Table 1. The average water temperature and pH in the system varied in the range between 20 to 25°C and 7 to 8, respectively (Data communicated elsewhere). Higher concentration of most of the pesticides was found during the pre-monsoon and post-monsoon period. In the present study, the concentration of endosulfan in the surface water collected from Site-A (under the influence of agriculture) detected the lowest concentration (0.01 μg/l) during post-monsoon whereas; highest concentration (0.12 μg/l) was detected during monsoon. Minimum concentration (0.7 μg/l) of monocrotophos was detected during monsoon and the maximum concentration (1 μg/l) was detected during post-monsoon at Site-A. Concentration of chlorpyrifos detected during the present investigation at the same site ranged between 0.5 μg/l (monsoon) to 1 μg/l (post-monsoon), whereas, the concentration of these pesticides were found to be below detection limit at Site-B during the study period. The contamination of water by pesticides at Site-A may be due to the magnitude of the

**Table 1.** Residual pesticide concentration in surface water of Bhoj Wetland, Bhopal.

| Period of sampling | Site A | | | Site B |
|---|---|---|---|---|
| | Endosulfan (µg/l) | Monocrotophos (µg/l) | Chlorpyrifos (µg/l) | |
| Pre monsoon | 0.073 | 0.9 | 0.79 | *Nil |
| Monsoon | 0.12 | 0.7 | 0.5 | *Nil |
| Post monsoon | 0.01 | 1 | 1 | *Nil |
| Average | 0.07 | 0.87 | 0.76 | *Nil |
| EPA (1994) criteria for aquatic organisms | 0.22 | 1 | 20 | - |

*Pesticide residual concentration was below detection limit

agriculture fields associated with the pesticide use whereas, absence of pesticides or the below detection limit of pesticides at Site-B may be due to the dilution factor and the non agricultural catchment. The presence of pesticides in the water body can be attributed to the maximum usage of these chemical substances in the agricultural catchment which finds their way to the waterbody during rainy period and during the period of their applications also.

While comparing the concentrations of endosulfan, monocrotophos and chlorpyrifos with the Environmental Protection Agency (EPA) (1994) water quality criteria for fresh water aquatic organisms, it appears that all the values were found to be within the prescribed limit. Among other investigators who contributed in the same field are Lalah et al. (2003) who reported low concentration of endosulfan and other pesticides in Tana and Sabaki Rivers in Kenya. Bakre et al. (1990) found maximum pesticides in water from Mahala reservoir with the onset of rains which was attributed to the sub soil movement of water. The levels of pesticides in the surface water systems in India as well as outside India depends on many factors like drainage basin, flow rate, particulate matter content, depth of water body etc. Several studies were conducted on pesticide contamination in lakes and reservoirs from other parts of the world (Kucklick et al., 1994; Tanabe et al., 1983). However, some investigators have assessed organochlorine residues in water reservoirs in India, that is, Mahala water reservoir, Jamuna water (Agarwal et al., 1956), Ganga water (Nayak et al., 1995; Sinha, 1991) and rural pond water (Dua et al., 1996).

Hosokawa et al. (1995) reported the presence of pesticide residues in rivers receiving discharge from agricultural activities. Jensen et al. (1969) reported ultra trace levels of pesticides at places like North Pole snow which was far from the site of their application. Nayak et al. (1995) reported many water samples from Ganga river contaminated with pesticides which exceeded the safe limit. Dua et al. (1996) assessed the organochlorine

insecticide residue in water from five lakes of Nanital (UP). Various other workers who also contributed in the field of environmental research include the works of Ahad et al. (2006), Singh et al. (2006), Moore et al. (2007), Singh et al. (2005), Sarkar et al. (2003) and Abbassy et al. (2003). Pesticides not only do their intended job, but may also adversely affect non targeted beneficial species. Contamination of surface waters by pesticides has been reported to have direct toxic effects on populations of phytoplankton. Pesticides can affect the structure and function of aquatic communities by changing the species composition of an algal community. The toxicity of some pesticides on algae is higher than toxicity reported by several authors on organisms like zooplankton, filter-feeding invertebrates and fishes (Kreutzweiser et al., 1998; Kreutzweiser and Faber, 1999). This is an important fact because in the freshwater ecosystem, algae are important primary producers in the food chain, with phytoplankton providing food for a diverse community of invertebrates and fishes. Depending on pesticide toxicity, aquatic contamination could result in a die-off of many algal species, thus decreasing the food source (Gomez et al., 2004). During the present 96 h acute toxicity tests of different pesticide chemicals to phytoplankton population, the later were affected in a different manner, indicating significant decline in population growth at higher concentrations. Concentration of chlorpyrifos tested in the range of 4.3 to 129 µg/l recorded a considerable inhibitory effect in growth of Chlorophyceae, Cyanophyceae and Bacillariophyceae groups by 87.1, 85.6 and 91.3%, respectively at concentration of 129 µg/l as compared to the control sample (Figure 2). Monocrotophos at the concentrations in the range of 0.19 to 28.5 µg/l recorded inhibition in growth by 79.8, 80.2 and 69.3% at concentration of 28.5 µg/l in the Chlorophyceae, Cyanophyceae and Bacillariophyceae. groups, respectively in comparison to the untreated control sample (Figure 3).

The present investigation report is in complete agreement with the findings of Megharaj et al. (1986) who also

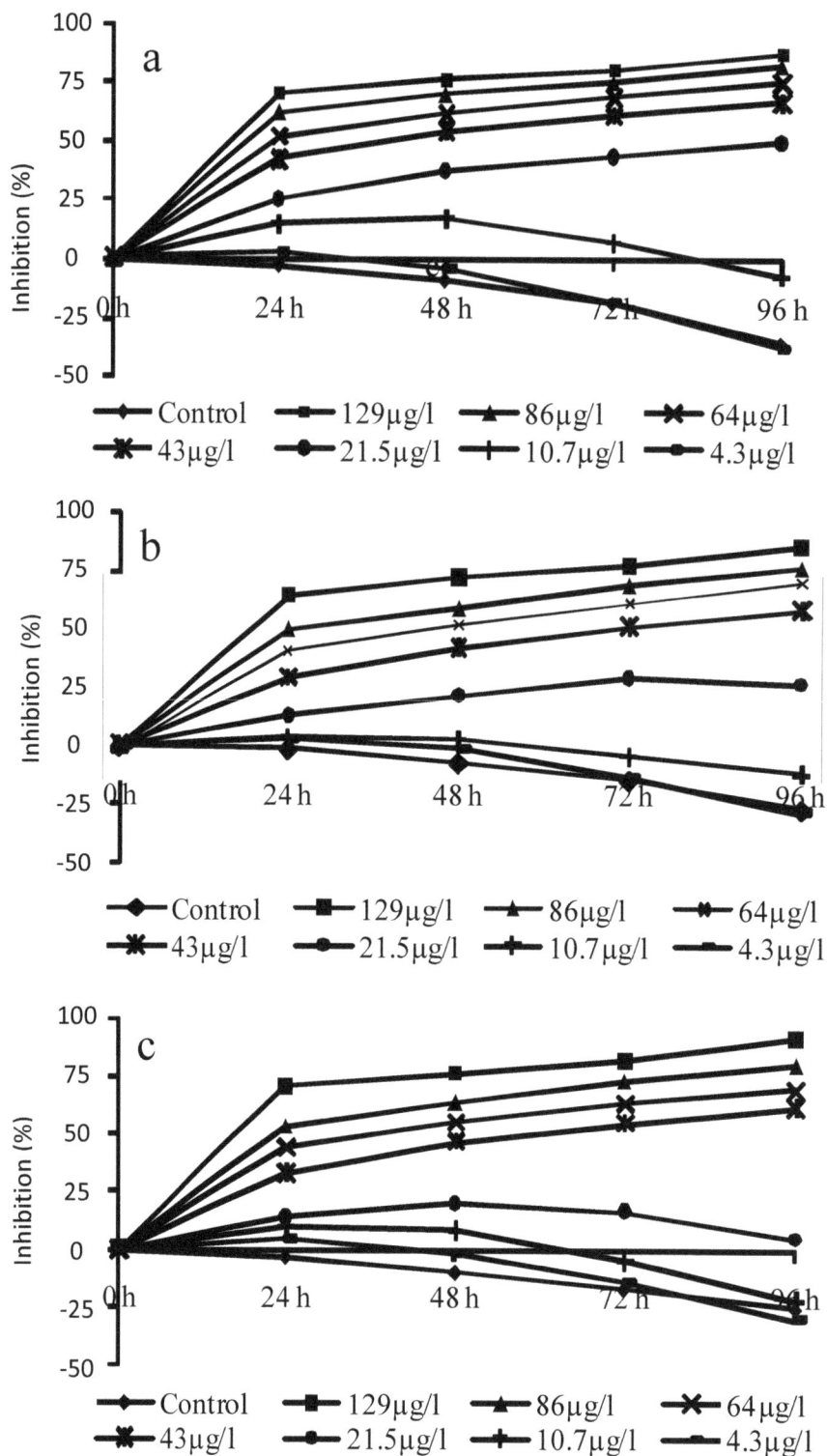

**Figure 2.** Growth inhibitory effect of chloropyrifos on (a) Chlorophyceae, (b) Cyanophyceae and (c) Bacillariophyceae.

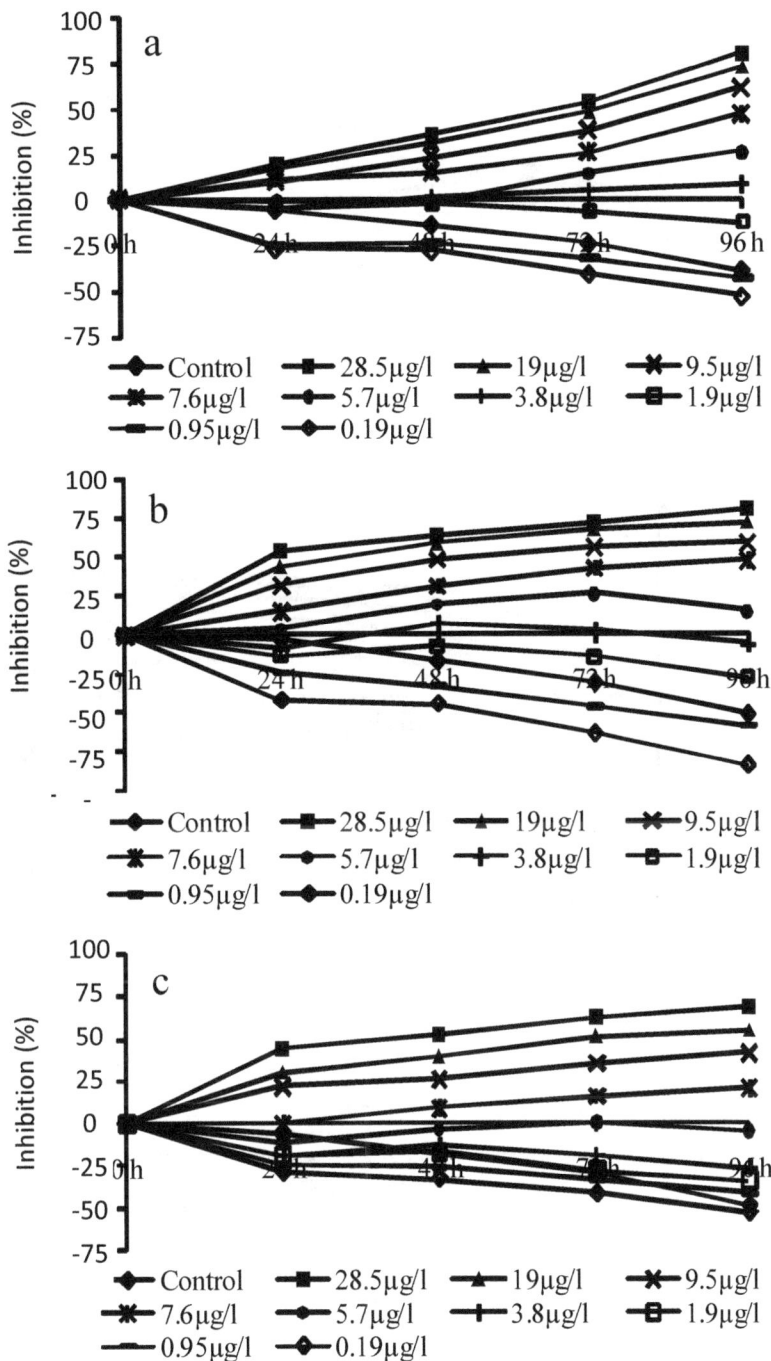

**Figure 3.** Growth inhibitory effect of monocrotophos on (a) Chlorophyceae, (b) Cyanophyceae and (c) Bacillariophyceae.

recorded higher concentration of monocrotophos inhibiting the growth of green algae and at lower concentration recorded an increase in growth rate of algae. Similarly, concentrations of endosulfan in the range of 0.06 to 18 µg/l recorded growth inhibition by 70.7, 72.7 and 56.8% at higher concentration (18 µg/l) in

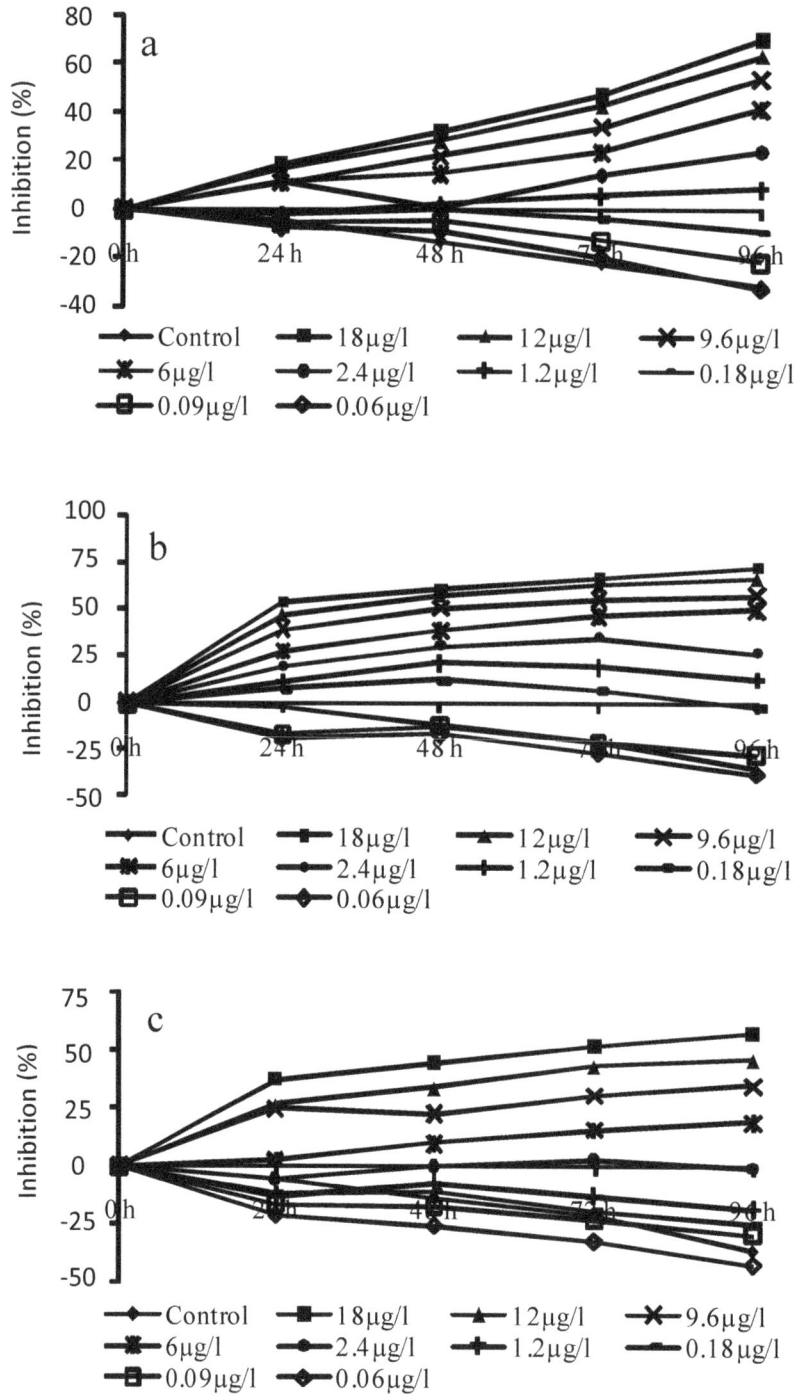

**Figure 4.** Growth inhibitory effect of Endosulfan on (a) Chlorophyceae, (b) Cyanophyceae and (c) Bacillariophyceae.

the group, Chlorophyceae, Cyanophyceae and Bacillariophyceae, respectively (Figure 4). The tested pesticides recorded stimulatory growth effect at lower concentrations while increasing the concentration and

time of exposure maximum inhibitory effect in the growth of algal groups was observed at higher concentrations during the 96 h acute toxicity testing procedure.

Rajendran and Venugopalan (1983) are of the opinion

that the highly toxic nature of organochlorine insecticides on algae might be due to their uptake metabolism involving surface adsorption followed by very rapid absorption across the cell wall. Higher concentration of endosulfan also suppressed growth in *Chlorella ellipsoidea* (Asma and Mathew, 1995). Goebel et al. (1982) noticed the impact of endosulfan in higher concentrations on algae. Kasai and Hanazato (1995) investigated that Volvocales and Cryptophyceae were clearly reduced by herbicides application. Khalil and Mostafa (1986) recorded inhibition in the growth of green alga *Phormidium* sp. at higher concentration of insecticides. Maule and Wright (1984) also reported the inhibitory effect of some pesticides to green algae. A similar growth pattern had been reported in *Chlorella protothecoides* by Saroja and Bose (1982). Tandon et al. (1988) also observed the inhibitory effect of endosulfan on *Aulosira* sp. Further, Kumar et al. (2008) observed reduction in growth of *Aulosira fertilissima* and *Nostoc muscorum* at higher concentration of endosulfan. Kapoor and Arora (2000) while working on blue green algae reported the level of tolerance of BHC (an insecticide) found to be very low. Tubea et al. (1981) however, reported increase in growth inhibition in *Lyngbya* sp. as herbicides concentration increased. A good number of investigations point out significant growth reduction of unicellular algae when treated at different concentrations of organochlorine insecticides like DDT, heptachlor and lindane (Menzel et al., 1970; Powers et al., 1975; Subramaniam et al., 1979; Ramachandran et al., 1980; Sahu et al., 1992; Penalva and Fernandez, 1992; Fliedner and Klein, 1996; Krishnaswamy, 1997).

As discussed earlier, the present study indicates that lower concentrations of pesticides have a stimulatory effect on algal growth. This may probably be attributed to the fact that the pesticides selected in the present investigation are mostly phosphate based and may have acted as growth enhancers at lower concentrations. Menzel et al. (1970) also reported the enhanced growth and cell division at lower concentrations of DDT and heptachlor. Peterson and Batley (1993) have also emphasized the possibility of a nutrient effect at lower concentrations of endosulfan where it may act as a carbon source for plankton growth. Further, lower concentrations of endosulfan recorded in an increase in the cell density of Bacillariophyceae when exposed to furadan (Kar and Singh, 1978).

EC$_{50}$ values determined by certain investigators with respect to different algae prove to be advantageous for evaluating the toxicity of various organophosphates (Walsh and Alexander, 1980; Ibrahim, 1983; Kent and Currie, 1995; Chen et al., 1997; Zou et al., 1998). In the present investigation, effective concentrations (concentration responsible for 50% inhibition in the cell density of the phytoplankton) were recorded based on the

sensitivity of different classes to different pesticides. Sensitivity not only varies among toxicants but also among taxonomic groups and species within taxa. No species is always most sensitive or always the least sensitive (Wang and Freemark, 1995).

Chlorophyceae, Cyanophyceae and Bacillariophyceae in the present study recorded average EC$_{50}$ value for chlorpyrifos in the range of 16.1 to 32.3 µg/l, monocrotophos in the range of 8.6 to 14.3 µg/l and endosulfan in the range of 4.2 to 15 µg/l, respectively (Figure 5). Chlorophyceae was found to be more sensitive to chlorpyrifos than Cyanophyceae. Tang et al. (1997) also recorded some Chlorophycean species to be more susceptible to atrazine than diatoms. Based on the toxicity of these pesticides to different phytoplankton communities, endosulfan was found to be comparatively highly toxic whereas, chlorpyrifos was found to be least toxic amongst the tested pesticides.

## Conclusion

It has been observed that higher concentration of various pesticides used in the experimental work affected phytoplankton community as a whole. The tests of acute toxicity with pesticides confirmed the high toxicity of the chemical to selected species of aquatic community. The results proved Chlorophyceae to be the most sensitive test group among the phytoplankton. It is necessary to protect natural surface waters from accidental escape of pesticides from treated areas. Assuming that the application of pesticides is done in an appropriate way, the pesticide should degrade quickly and never be present in surface waters at lethal concentrations. Toxicity is a biological response and this needs to be taken into account in formulating realistic guidelines on the acceptable upper limits on pesticide contamination of the environment.

Toxicity testing has a clear role in safeguarding environmental quality but a considered selection of the testing methods is essential for obtaining relevant results. Toxicity of pesticides is highly dependent on the duration, frequency, intensity of exposure, and the susceptibility of the target organism which is influenced by age, sex, fitness and genetic variation. Though a new generation of bioassays has been developed and is used in ecotoxicological practice, "standard" toxicity tests on living organisms are still essential and more relevant in the evaluation of the toxicity of compounds for higher animals, including humans.

## ACKNOWLEDGEMENTS

The authors are thankful to the Director ITRC Lucknow

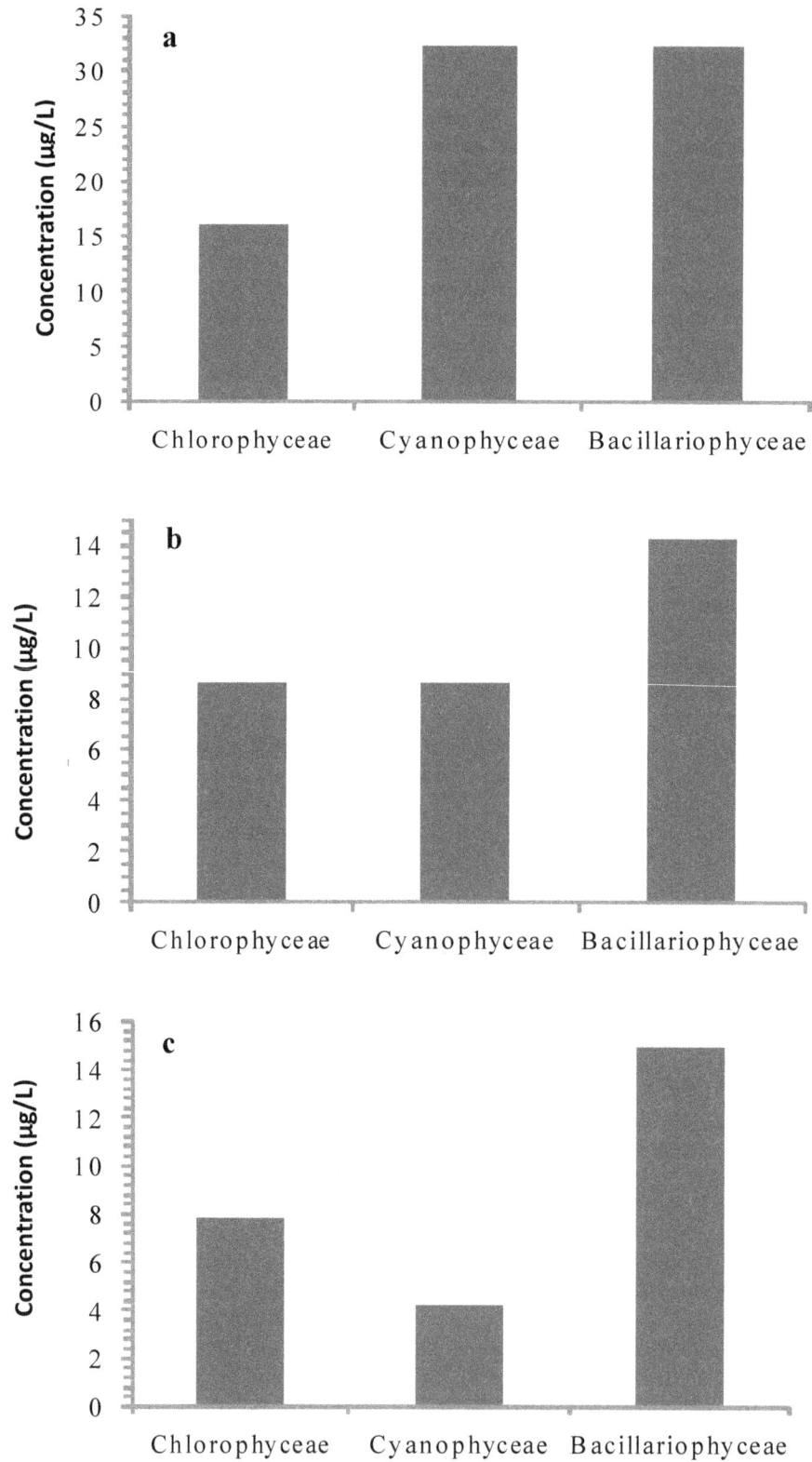

**Figure 5**. Population growth rate effectivec concentration (EC-50) of (a) Chloropyrifos, (b) Monocrotophos and (c) Endosulfan for phytoplankton

for providing support for the analysis of the samples. The Vice Chancellor of Barkatullah University Bhopal is greatly acknowledged for providing the Laboratory facilities and authors are also grateful to the reviewers and editor for structuring the manuscript.

## Conflicts of interest

No competing interests exist.

## REFERENCES

Abbassy MS, Ibrahim HZ, Abdel-Kader HM (2003). Persistent Organochlo -rine Pollutants in the Aquatic Ecosystem of Lake Manzala, Egypt. Bull. Environ. Contam. Toxicol. 70:1158-1164.

Agarwal HC, Mittal PK, Menon KB, Pillai MKK (1956). DDT residues in the river Jamuna in Delhi, India. Water Air Soil Pollut. 28:89-104.

Ahad K, Mohammad A, Mehboob F, Sattar A, Ahmad I (2006). Pesticide Residues in Rawal Lake, Islamabad, Pakistan Bull. Environ. Contam. Toxicol. 76:463-470.

Ali I, Aboul-Enein HY (2002). Determination of chiral ratio of O, p-DDT and p-DDD pesticides on polysaccharides chiral stationary phases by HPLC under reversed-phase mode. Environ. Toxicol.17 (4) 329-333.

APHA (1995). Standard methods for the examination of water and wastewater. APHA, AWWA and WPCF. New York.

Asma VM, Mathew KJ (1995). Impact assessment of biocides on microalgae: A study in-vitro. C.M.F.R.I. Spl. Publ. 61:112116.

ASTM "American Society for testing materials" (1996). Annual Book of ASTM Standards, Vol.11.05, West Conshohocken. PA-1, 402.

Bakre PP, Mishra V, Bhatnagar P (1990). Organochlorine residue in water from the Mahala water reservoir, Jaipur, India. Environ. Pollut. 63:275-281.

Baun A, Bussarawit N, Nyholm N (1998). Screening of pesticide toxicity in surface water from an agricultural area at Phuket Island (Thailand). Environ. Pollut. 102 (2-3):185-190.

Chen B, Chen M, Wu Z (1997). Study on the toxicity of fenvalerate and tetramethrin to ocean algae and shell fish. J. Fish. Sci. China. 4 (2):51-55.

Dua VK, Kumari R, Sharma VP (1996). HCH and DDT contamination of rural ponds of India. Bull. Environ. Contam. Toxicol. 57:568-574.

EPA (1994). Federal Radiation Protection Guidance for Exposure of the General Public. Proposed Recommendations. U.S. Environmental Protection Agency, Washington, D.C., Federal Register, 59 (246): 66414.

Fliedner A, Klein W (1996). Effects on lindane on the planktonic community in freshwater microcosms. Ecotoxicol. Environ. Saf. 33(3): 228-235.

Goebel H, Gorbach S, Knauf W, Rimpu RH, Huttenbach H (1982). Properties, effects, residues and analytics of the insecticide Endosulfan. Residue Rev. 83:1-174.

Gomez De Barreda Ferraz, D, Sabater CJ, Carrasco M (2004). Effect of propanil, tebufenozide and mefenacet on growth of four freshwater species of phytoplankton: A microplate bioassay. Chemosphere. 56: 315.

Hersh CM, Crumpton WG (1987). Determination of growth rate depression of some algae by atrazine. Bull. Environ. Contam. Toxicol. 39: 1041-1048.

Ibrahim EA (1983). Effect of some common pesticides on growth and metabolism of the unicellular algae. Skeletonema costatum, Amphiroa paludosa and Phaeodactylum tricornutam. Aquat. Toxicol. 3:1-14.

Jain CK, Imran A (1997). Determination of Pesticides in Water, Sediments and Soils by Gas Chromatography, Int. J. Environ. Anal.

Chem.68:1, 83-101, DOI: 10.1080/03067319708030482.

Jensen S, Johnels AG, Oilson M, Otterlind G (1969). DDT and PCB in marine animals from Swedish waters. Nature. 224:247-253.

Kapoor K, Arora L (2000). Comparative studies on the effect of peticides on nitrogen fixing cyanobacterium Cylindrospermum majus, Kutz. Ind. J. Env. Sci. 4 (1):89-96.

Kar S, Singh PK (1978). Toxicity of carbofuran to blue-green alga Nostoc muscorum. Bull. Environm. Contam. Toxicol. 20: 707-714.

Kasai F, Hanazato T (1995). Effects of the triazine herbicide simetryn on freshwater plankton communities in experimental ponds. Environ. Pollut. 89 (2):197-202.

Kaushik A, Sharma HR, Jain S, Dawra J, Kaushik CP (2010). Pesticide pollution of river Ghaggar in Haryana, India. Environmental monitoring and assessment. 160(1-4), 61-69.

Kent RA, Currie D (1995). Predicting algal sensitivity to a pesticide stress. Environ. Toxicol. Chem. 14 (6):983-991.

Khalil Z, Mostafa IY (1986). Interactions of pesticides with fresh-water algae: Effect of methomyl and its possible degradation by Phormidium fragile. J. Environ. Sci. Health Part B Pestic. Food. Contam. Agric. Wastes. 21(4):289-302.

Kreutzweiser DP, Faber MJ (1999). Ordination of zooplankton community data to detect pesticide effects in pond enclosures. Arch. Environ. Contam. Toxicol. 36: 392.

Kreutzweiser DP, Gunn JM, Thompson DG, Pollard HG, Faber MJ (1998). Zooplankton community responses to a novel forest insecticide, tebufenozide (RH-5992), in littoral lake enclosures. Can. J. Fish Aquat. Sci. 55(3):639-648.

Krishnaswamy CS (1997). Biflagellated unicellular algae as detectors of pesticide contamination in water. Proceedings of the 5th symposium on our environment (Lee HK, Wong MK ed.). 44 (1-3):487-502.

Kucklick JR, Bidleman TF, MeConnel LL,Walla MD, Ivanov GP (1994). Organochlorine in water and biota of lake Baikal, Siberia. Environ. Sci. Technol. 28:31-37.

Kumar S, Habib K, Fatma T (2008). Endosulfan induced biochemical changes in nitrogen-fixing cyanobacteria. Sci. Total Environ. 403(1-3):130-138.

Lalah JO, Yugi PO, Jumba IO, Wandiga SO (2003). Organochlorine Pesticide Residues in Tana and Sabaki Rivers in Kenya. Bull. Environ. Contam. Toxicol. 71:298-307.

Mary M, Ernest R (1983). Effects of Pesticides on Pure and Mixed Species Cultures of Salt Marsh Pool Algae. Bull. Environ. Contam. Toxocol.30, 464-472.

Maule A, Wright SJL (1984). Herbicide effects on the population growth of some green algae and cyanobacteria. J. Appl. Bacteriol. 57(2):369-379.

Megharaj M, Venkateswarlu K, Rao AS (1986). Growth response of four species of soil algae to monocrotophos and quinalphos. Environ. Pollut. Ser. A Ecol. Biol. 42(I):15-22.

Menzel DW, Anderson J, Randtke A (1970). Marine phytoplankton vary in their response to chlorinated hydrocarbons. Science. 167:1724-1726.

Metcalf RL (1997). Pesticides in aquatic environment In: Khan MAQ(ed) Pesticides in environment, Plenum Press, New York, p127.

Michael RG (1985). Use of rotifers and cladocerans as potential bioindicators of Indian fresh water ecosystem. Symp. Biomonitoring State Environ. pp. 82-83.

Moore MT, Lizotte RE, Knight SS, Smith S, Cooper CM (2007). Assessment of pesticide contamination in three Mississippi Delta oxbow lakes using Hyalella azteca. Chemosphere, Volume 67, Issue 11, pp. 2184-2191.

Nayak AK, Raha R, Das AK (1995). Organochlorine pesticide residue in middle stream of the Ganga river, India. Bull. Environ. Contam. Toxicol. 54:68-75.

Nimmo DR, Mcwen LC (1994). Pesticides. Handbook of Ecotoxicology, Vol. 2, (Ed.) P. Calow. 155-203. Cambridge, MA. Pub. Blackwell Scientific.

Penalva S, Fernandez F (1992). Lindane effects on the growth, size and composition of two marine unicellular algae, Monochrysis lutheri

(Droop) and *Phaedactylum tricornutum* (Bohlin). Proceedings of the FAO-UNEP-IOC-Workshop on the Biological effects of pollutants on marine organisms. 69:235-243.

Peterson SM, Batley GE (1993). The fate of endosulfan in aquatic ecosystems. Environ. Pollut. 82:143-152.

Powers CD, Rowland RG, Wurster CF (1975). The toxicity of DDE to marine dinoglagellate. Environ. Pollut. 9:253-262.

Rajendran N, Venugopalan VK (1983). Effects of pesticides on phytoplankton production. Mahasagar. 16 (2):193-197.

Ramachandran S, Rajendran N, Venugopalan VK (1980). Effect of pesticides DDT, Dimethoate and Sevin on the growth of marine diatom *Coscinodiscus concinnus* in culture. Mahasagar. 13 (3):235-238.

Sahu J, Das MK, Adhikary SP (1992). Reaction of blue-green algae of rice field soils to pesticide application. Trop. Agric. 69 (4):362-364.

Sarkar UK, Basheer VS, Singh AK, Srivastava SM (2003). Organochlorne pesticide residues in water and fish samples: First report from rivers and streams of Kumaon Himalayan region, India. Bull. Environ. Contam. Toxicol. 70:485-493

Saroja G, Bose S (1982). Effects of methyl parathion on the growth, cell size, pigment and protein content of *Chlorella protothecoides*. Environ. Pollut. 27A:297-308.

Singh KP, Malik A, Mohan D, Takroo R (2005). Distribution of Persistent Organochlorine Pesticide Residues in Gomti River, India. Bull. Environ. Contam. Toxicol. 74:146-154.

Singh SK, Raha P, Banerjee H (2006). Banned Organochlorine Cyclodiene Pesticide in Ground Water in Varanasi, India. Bull. Environ. Contam. Toxicol. 76:935-941.

Sinha AK (1991). A comprehensive study of Ganga and its dependent. In. Krishnamuru, CR, Bilgrami KS, Dass TM, Mathur RP (eds). The Ganga –a scientific study, Northern Book Centre, New Delhi P 125-140.

Subramaniam BR, Lingaraja T, Venugopalan VK (1979). Effect of low concentration of DDT on the growth and production of marine diatom *Skeletonema costatum* (Grv). Curr. Sci. 48 (5):226-228.

Tanabe S, Hidaka H, Tatsukawa R (1983). PCBs and chlorinated Hydrocarbon pesticide in Antarctic atmosphere and hydrosphere. Chemosphere. 12:277-288.

Tandon RS, Lal R, Rao VVSN (1988). Interaction of endosulfan and malathion with blue-green algae Anabaena and *Aulosira fertilissima*. Environ. Pollut. 52:1-9.

Tang JX, Hoagland KD, Siegfried BD (1997). Differential Toxicity of Atrazine to Selected Freshwater Algae. Bull. Environ. Contam. Toxicol. 59: 631-637.

Tubea B, Hawxby K, Mehta R (1981). The effects of nutrient, pH and herbicide levels on algal growth. Hydrobiologia. 79 (3):221-228.

Walsh GE, Alexander SV (1980). A marine algal bioassay method: Results with pesticides and industrial wastes. Water Air Soil Pollut. 13: 45-55.

Wang W, Freemark K (1995). The use of plants for Environmental monitoring and assessment. Ecotoxicol. Environ. Saf. 30:289-301.

Wetzel RG (1975). Limnology. Pub.W. B. Sundars Co., Philadelphia.

Zou Li, Cheng G, Li Y, Lin K (1998). Study on the toxicity of 11 kinds of organic phosphorus pesticides to marine microalgae. Mar. Environ. Sci. 17 (3):29-34.

# 22

# Organic pollution from the Songhua River induces NIH 3T3 cell transformation: Potential risks for human health

**Jia-Ren Liu[1, 2]\*, Hong-Wei Dong[1], Xuan-Le Tang[1], Jia Yu[1], Xiao-Hui Han[1], Bing-Qing Chen[1], Chang-Hao Sun[1] and Bao-Feng Yang[3]**

[1]Public Health College, Harbin Medical University, 157 BaoJian Road, NanGang District, Harbin, P. R. China 150086.
[2]Harvard Medical School, 300 Longwood Avenue, Boston, MA, 02115-5737, USA.
[3]Department of Pharmacology, Harbin Medical University, Harbin, Heilongjiang, P. R. China 150081.

**Epidemiological investigation has shown that organic contamination of the Songhua River is a risk factor for tumor development among residents who live nearby. A mutagenesis is induced by organic contamination using short-term genotoxic bio-assays. To further investigate the risk of carcinogenic potential to human health, the NIH3T3 cell line was used to examine the induction of transformation by diethyl ether extracts of water samples taken from the Songhua River in the summer of 1994 and the winter of 1995. The results indicated that the malignant transformed foci were induced by diethyl ether extracts. Cellular transformation frequencies showed a dose response. Malignant cells possessed typical characteristics in cell growth and in the cellular anchorage dependent test while the control cells did not. Thus, this study demonstrates diethyl ether extracts of water samples could induce cell transformation of NIH3T3 cells. Evidence was provided the possible relationship between organic pollution and carcinogenic potential.**

**Key words:** Diethyl ether extracts, NIH3T3 cells, cell transformation, malignant cells, the Songhua River.

## INTRODUCTION

The Songhua River is the third largest river in the People's Republic of China, with a catchment area of about 556,800 km$^2$. The Songhua River crosses three provinces including Heilongjiang, Jilin and Inner Mongolia and consists of the NenJiang River, the Second Songhua River, and the Songhua River. The Songhua River is the major freshwater source for industry and agriculture, as well as the source of drinking water for 62.25 millions

residents living along it (from a 2005 government report). It is known that the Songhua River has been heavily polluted by waste water from industry and domestic sewage. Xu et al. (1990) reported that 152 organic compounds were detected in the Songhua River by methods of gas chromatography (GC), Gas chromatography/mass spectrometry (GC/MS), high pressure liquid chromatography (HPLC) and total ion chromatography (TIC). Of these compounds, 19% were polycyclic aromatic hydrocarbon (PAHs), 14% were chloro-compounds, 13% were aromatic compounds and 54% were other compounds. An epidemiological investigation indicated that organic contamination of the Songhua River is a risk factor for tumor development among the residents living along it. Organic pollutants from the Songhua River were tested for mutagenicity using the *Salmonella typhimurium* assay and diethylether

---

\*Corresponding author. E-mail: Jiarenliu@yahoo.com, JiarL@ems.hrbmu.edu.cn.

**Abbreviations: MNNG;** N-methyl-n'-nitro-n- nitrosoguanidine, **DMSO;** dimethyl sulphoxide, **EDTA;** ethylenediamine tetraacetic acid.

extracts of 1.7 or 3.5 L water equivalent / plate were positive with TA98 (-S9) and TA1538 (-S9) (Zhu et al., 1985). Yang's study also reported that the frequency of chromosome aberrations (CA) and sister chromatid exchanges (SCE) of blood lymphocytes in residents living along the Songhua River were higher than that of the control group (Yang et al., 1991). However, there is no further direct evidence to elucidate the high tumor development of residents who lived along with the Songhua River except for genotoxicity assays that only respond to genotoxic agents.

*In vitro* cell transformation tests using NIH3T3 cells can simulate the process of animal two-stage carcinogenesis by treating the cells under different growth conditions, and therefore potentially detecting not only initiating activity, but also promoting activity of chemicals. Cell transformation is the induction of certain phenotypic alterations in cultured cells that are characteristic of tumorigenic cells. These phenotypic alterations can be induced by exposing mammalian cells to carcinogens. Transformed cells that have acquired the characteristics of malignant cells have the ability to induce tumors in susceptible animals (Berwald and Sachs, 1963; Berwald and Sachs, 1965). *In vitro* transformed cells exhibit morphological changes related to neoplasia. The phenomenon of morphological cell transformation involves changes in behavior and growth control of cultured cells, such as alteration of cell morphology, disorganized pattern of colony growth, and acquisition of anchorage-independent growth (Weber et al., 1976; Combes et al., 1999). Transformed cells then become able to grow in semi-solid agar (anchorage-independent growth), produce autocrine growth factors and can evolve to tumorigenicity when injected into appropriate hosts. Although partly data in conclusion of cell transformation were mentioned in previous study (Liu et al., 2007), the detail information about cell transformation in NIH 3T3 cells should be stated because there is no further direct evidence to elucidate the high tumor development of residents who lived along with the Songhua River except for genotoxic assays that only respond to genotoxic agents. Thus, the objectives of this study were to determine the transformational effect of organic extracts taken from the Songhua River and the risk of carcinogenic potential to human health.

## MATERIALS AND METHODS

### Reagents and chemicals

The chemicals were obtained from the following sources: N-methyl-n'-nitro-n- nitrosoguanidine (MNNG), and dimethyl sulfoxide (DMSO) from the Sigma Chemical Company (St. Louis, MO); Trypsin, ethylenediamine tetraacetic acid (EDTA), Giemsa, agar (low melting point), and Trypan Blue Stain from Tian Xiang Ren Chemical Company (Beijing, China). Methanol, acetone, diethyl ether, and anhydrous sodium sulfate were analytical grade and obtained from Harbin Chemical and Bio-Reagent Co. (Harbin, China).

### Sampling collections and extraction

During the summer of 1994 and the winter of 1995, we collected 820 L and 990 L water samples directly taken from the intake of the most distant water treatment plant of Harbin, located 15 km above Harbin. The pH value in each sample was 7.30. For the extraction of pollutants, GDX-102 (40 - 60 bead mesh) resin (Tianjin Chemical Co., China) was washed by swirling and decanting 4 times with 10 resin-volumes of acetone for 12 h each time, followed by absolute methanol for another 48 h in the same volume and distilled water 3 times. The extract was stored in distilled water at 4°C (Junk et al., 1974; Wang et al., 1983). Glass columns, 30 mm (inside diameter) × 350 mm, were filled with distilled water before the addition of sufficient washed GDX-102 resin (100 g dry weight) to give a bed height of 15 - 20 cm. Flow was regulated with a 3-way nylon stopcock. Distilled water (about 250 ml) was passed through the GDX-102 resin before using these columns.

Water samples (80 - 100 L) taken from the Songhua River were passed through each column. Water was dropped onto glass wool, protecting the resin from splatter, at 8-10ml/min. Finally, each column was washed with 3 resin-volumes of distilled water 3 times prior to diethyl ether extraction. Three volumes (resin column) of diethyl ether were added to the column. The resin in the column was swirled with glass rod during each exaction until the final ether eluted solution was colorless. Distilled lab grade water was used as the control. The effluent was dehydrated with anhydrous sodium sulfate and evaporated within a KD concentrator until nearly dry. The residue was dissolved in diethyl ether (< 1ml) and kept in a -30°C refrigerator for cell transformation assays. The diethyl ether in the samples was evaporated and the residues were dissolved in DMSO prior to the assay. The cell transformation assay was performed when the samples were done in October 1995.

### Cells, media, and culture conditions

NIH3T3 cell line, original from NIH Swiss mouse embryo, was bought from Cancer Research Institute of Beijing (China). NIH3T3 cells were cultured in RPMI 1640 medium (Gibco Co. Los Angeles, USA) supplemented with 10% FBS (fetal bovine serum, FBS, Gibco Co.) at 37°C in a humidified incubator under 95% air and 5% $CO_2$. Cells were passaged before confluence by a mixture of ethylenediamine tetraacetic acid (EDTA) and trypsin, usually at 3 day intervals. In order to obtain constant transformation results, many frozen stock ampoules of the cells were not passaged more then 4 times after recovering cells from liquid nitrogen. One ampoule was thawed and used in each individual cell toxicity and transformation experiment.

### Test procedures

#### Cytotoxicity assay

A preliminary cytotoxicity assay was carried out for determination of test concentrations. For the evaluation of growth inhibition and cytotoxic effect of initiating chemicals, NIH3T3 cells were seeded at a density of $5.0 \times 10^3$ cells per 25 ml culture bottle (20 $cm^2$ of bottle surface, four bottles per dose). The glass culture bottle was clear, did not have any blotches, and had a smooth surface to show that cells attached to the bottom of the glass bottle. Twenty-four hours later, the cultures were exposed to diethyl ether extracts of water samples and the medium containing the diethyl ether extracts was discarded after 24 h exposure. The cells in the bottles were washed three times with PBS (phosphate buffered saline, pH 6.8). Then the culture medium was replaced with fresh RPMI 1640 medium with 10% FBS. The same procedures were used for a positive (MNNG 1.5 µg/ml), negative (DMSO, 0.05 % v/v) and blank controls. After

**Table 1.** Cytotoxicity in NIH 3T3 cells treated with MNNG and diethyl ether extracts of water samples taken from the Songhua River in the summer of 1994 and the winter of 1995 (n = 4).

| Season | Doses (L water/ml) | Number of seeded cell | No. of colonies Mean ± SD | Absolute CFE (%) | Relative CFE (%) |
|--------|------|------|------|------|------|
| Winter | 0.198 | 5000 | 1050 ± 300 | 21.0** | 41.9 |
|        | 0.099 | 5000 | 1600 ± 452 | 32.0** | 63.9 |
|        | 0.050 | 5000 | 2230 ± 402 | 44.6 | 89.1 |
| Summer | 0.205 | 5000 | 800 ± 202 | 16.0** | 31.9 |
|        | 0.102 | 5000 | 1025 ± 201 | 20.5** | 40.9 |
|        | 0.051 | 5000 | 1395 ± 355 | 27.9** | 55.7 |
|        | 0.026 | 5000 | 1710 ± 669 | 34.2* | 68.2 |
| MNNG | 1.5 µg/ml | 5000 | 1940 ± 595 | 38.8 | 77.5 |
| DMSO | 0.05 % (v/v) | 5000 | 2505 ± 574 | 50.1 | 100.0 |
| Blank | 0 | 5000 | 2410 ± 375 | 48.2 | 96.2 |

CFE: colony-forming efficiency; * $P < 0.05$, ** $P < 0.01$, compared to the negative control group (DMSO) using $\chi^2$ test.

$$\text{Absolute CFE } (\%) = \frac{\text{Number of colonies formed}}{\text{Number of cells seeded}} \times 100\%$$

$$\text{Relative CFE } (\%) = \frac{\text{Number of colonies on treatment}}{\text{Number of colonies on negative control}} \times 100\%$$

one week, the cells were fixed with methanol and stained with Giemsa. Cell colonies, cells which originated from one cell and contained more than 20 cells per colony were counted under the microscope.

The colony-forming efficiency (CFE) was calculated as follows:

### Transformation assay

The assay procedure for transformation experiments was described in detail in previous reports (Huang, 1985; Dunkel et al., 1991). Exponentially growing NIH3T3 cells were plated at a density of 3.0 × 10³ cells per 25 ml glass bottle in RPMI 1640 medium with 10% FBS. Fifteen bottles were used for each dose. After 24 h incubation, different levels of the diethyl ether extracts were added as the initiation treatment for 24 h at 37 °C in 95% air and 5% $CO_2$. Then, the medium containing the diethyl ether extracts was discarded. The cells were washed three times with PBS. Then the culture medium was replaced with 3 ml RPMI 1640 fresh medium with 10% FBS at three day intervals until 21 days after sample treatment (Huang, 1985). Both a negative and the positive controls were used in this experiment. Cells in most bottles were fixed with methanol at 21 days after treatment, stained with Giemsa and scored for transformed foci. Other bottles which were not fixed by methanol were used to determine the cell growth curve and semi-solid agar culture. A rubber scraper ring (10 mm diameter) was used to separate transformed foci from surface of bottles with trypsin / EDTA mixture. Verification of the transformed morphology was made with an invert microscope at ×10, ×20, and × 40 magnifications.

*(a) Criteria for transformation:* Scoring of transformed foci was characterized by the following morphological criteria, which discriminate transformed foci by four morphological characteristics: (1) foci of more than 2 mm in diameter, (2) deep basophilic staining, (3) piling up of cells forming a dense multi-layer, and (4) random orientation of cells at the edge of foci, overlapping nuclei, and basophilia. The criteria were considered suspect when either of criteria (1) without (4) or (3) without (1) and (4) was met (Huang,

1985; Dunkel et al., 1991).

The transformation frequency (TF) per bottle was calculated as follows:

$$\text{TF/bottle } (\%) = \frac{\text{Average transformed foci per bottle}}{\text{Average survival foci per bottle}} \times 100$$

*(b) Transformed cell growth (Huang, 1985):* For each extract and control, 1 × 10⁵ cells from transformed foci, including suspect foci, from the cellular transformation test were seeded in 6-well plates. Beginning the next day, three wells for each extract or control were selected at random and the number of living cells was determined by Trypan Blue Stain. The mean for each set of 3 wells was determined and the curves of cell growth were plotted.

*(c) Cellular anchorage dependent test (Ruhaut, 1979):* Briefly, 6-well plates were prepared with a 1.5 ml underlay of 0.6% soft agar in RPMI 1640 containing 10% FBS. 1 × 10⁴ cells in 1.5 ml of 0.3% soft agar (low melt) were placed on top in RPMI 1640 containing 10% FBS. The plates were incubated at 37 °C in 95% air and 5% $CO_2$. Cell colonies were observed after 3 - 4 weeks. The number of colonies was counted and the result was reported as mean ± standard deviation.

### Data analysis and statistical evaluations

All analyses were conducted using the SPSS for Windows statistical package, version 11.0 (SPSS Inc., Chicago, IL, USA). For each experimental condition (that is, the same cell number in the cytotoxicity assay and anchorage dependence test or per bottle in the transformation assay), foci data were expressed as mean ± SD. The statistical analysis of experimental values in the cell transformation test was done using the $\chi^2$ test with a one-sided Fisher's Exact Test (Armitage, 1971). A $P$-value of ≤ 0.05 was considered statistically significant.

A positive result was recorded if there was a significant increase in the number of foci ($P < 0.05$), with one of the following: (1) a dose-response relationship, (2) 2 or 3 transformed foci in two successive concentrations, or (3) 3 or more transformed foci in a concentration (Huang, 1985; Dunkel et al., 1991).

**Table 2.** Activity of transformation in NIH3T3 cells induced by MNNG and diethyl ether extracts of water samples taken from Songhua River in the summer of 1994 and the winter of 1995.

| Season | No. of bottles fixed | Doses (L water/ml) | Total no. of cell foci | | | Cell foci/bottle | | | TF/bottle (%) |
|---|---|---|---|---|---|---|---|---|---|
| | | | TCF | SCF | SVF | TCF | SCF | SVF | |
| Winter | 8 | 0.198 | 16 | 8 | 179 | 2.0 ± 1.3** | 1.0 ± 0.8 | 22.4 ± 11.3** | 8.9** |
| | 9 | 0.099 | 15 | 4 | 452 | 1.7 ± 1.0** | 0.4 ± 0.5 | 50.2 ± 21.8 | 3.3** |
| | 8 | 0.050 | 1 | 3 | 507 | 0.1 ± 0.4 | 0.4 ± 0.5 | 63.4 ± 26.8 | 0.2 |
| Summer | 11 | 0.205 | 14 | 6 | 431 | 1.3 ± 1.1** | 0.6 ± 0.7 | 39.2 ± 14.7** | 3.3** |
| | 10 | 0.102 | 16 | 22 | 800 | 1.6 ± 1.2** | 2.2 ± 1.6** | 80.0 ± 24.1 | 2.0** |
| | 8 | 0.051 | 7 | 10 | 523 | 0.9 ± 0.8 | 1.3 ± 1.3* | 65.4 ± 15.1 | 1.3** |
| | 9 | 0.026 | 0 | 5 | 635 | 0 | 0.6 ± 0.7 | 70.6 ± 26.7 | 0 |
| MNNG | 9 | 1.50 | 22 | 5 | 305 | 2.4 ± 1.7** | 0.6 ± 0.9 | 33.9 ± 10.8** | 7.2** |
| DMSO | 9 | 0.05% (v/v) | 0 | 3 | 600 | 0 | 0.3 ± 0.5 | 66.8 ± 22.5 | 0 |
| Blank | 8 | 0 | 0 | 0 | 665 | 0 | 0 | 83.1 ± 20.4 | 0 |

*$P < 0.05$, **$P < 0.01$, compared to the negative control group (DMSO) using $\chi^2$ test. MNNG: N-methyl-n'-nitro-n-nitrosoguanidine; TCF: transformed cell foci; SCF: suspect cell foci; SVF: survival cell foci; Total TF: frequencies of (transformed + suspect) cell foci.

# RESULTS

## Cytotoxicity

The results of cytotoxicity are as shown in Table 1. Diethyl ether extracts of water samples taken from the Songhua River were toxic to NIH 3T3 cells and the frequency of cellular colonies was reduced as the concentration increased. Both the absolute CEF and relative CEF showed a dose-response relationship. The maximal doses for each sample were selected according to the cytotoxicity assay.

## Transformation test

Based upon the cytotoxicity results, all concentrations of samples were used in the transformation test. As shown in Table 2, the diethyl ether extracts from water samples taken from the Songhua River showed a positive response in the transformation test. The TCF per bottle and suspect cell foci (SCF) per bottle showed a dose-response relationship. A significant low frequency of formation of the survival cell foci (SVF) per bottle was observed in 0.198 L water / ml of winter sample, 0.205 L water /ml of summer sample when compared to the negative control group ($P < 0.01$ and $P < 0.05$). Dose-dependent relationships were also observed in the frequency of cellular transformation (TF) per bottle as the concentrations of samples increased. There was no evidence of transformation in the negative control group. As noted in Table 2, the suspect transformed foci were noted in all samples, including the negative control groups except for the blank control. STF would further determine whether there have malignant cells in the foci, especially, in the negative control. In contrast to the normal cellular pattern of NIH3T3 cells (Figure 1A), Figure 1B shows at lower power (×40) evidence of cell transformation (invasive and piling up properties)

and Figure 1C, at higher power (×100), shows evidence of cellular transformation (fibroblastic, criss-crossing, densely stained cells, random orientation at edges of foci and basophilia).

## Malignant cellular transformed foci measurement

To further characterize the transformed cells from transformed foci, cell growth curves and colony formation assays in semi-solid agar were performed. The cell growth curves showed that the MNNG and the diethyl ether extract of the winter sample did not differ in growth compared to the blank control group within 3 days. After 4 days, the transformed cells grew rapidly while the growth of cells in the blank control group was slow (Figures 2A and 2B). The cells from transformed foci of the winter sample grew quickly and the shape of cell growth curves was nearly the same after 5 days. The cell growth curve in the summer

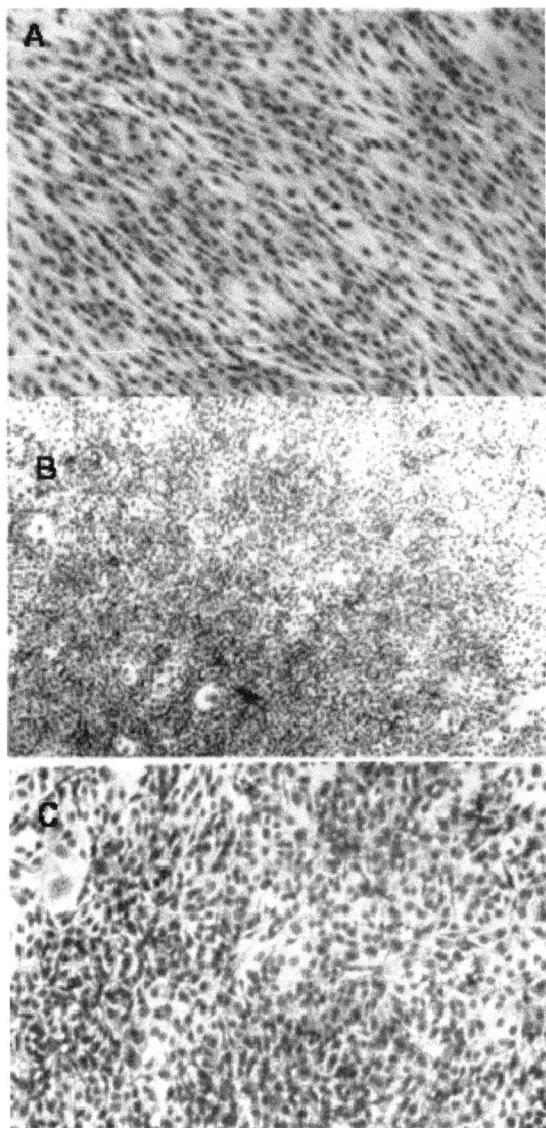

Figure 1. Morphological difference in the cell foci. (A) Non-transformed NIH 3T3 cells showing contact inhibition and growth in monolayer (×100). (B) Typical transformed cell foci showing invasive and piling up properties (×40). (C) Typical transformed foci showing fibroblastic, criss-crossing, densely stained cells, random orientation at edges of foci, and basophilia (×100).

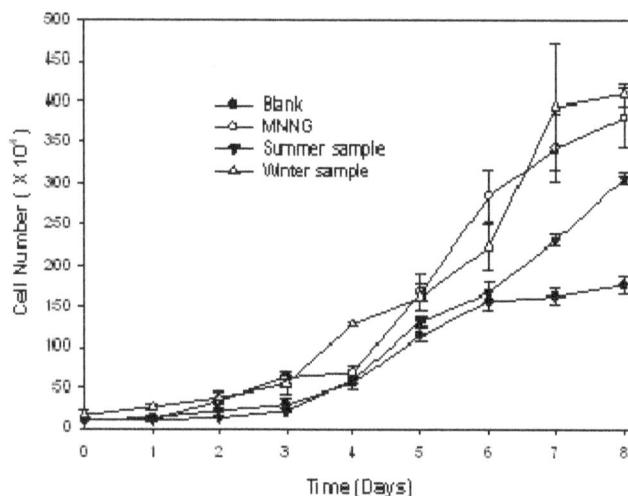

Figure 2. The cell growth curves of NIH 3T3 cells from the experimental and control groups. To determine the rate of cell growth from the experimental and control groups, the cell growth curves were plotted. The cell growth curve of the diethyl ether extract of the winter sample and the positive control did not differ in growth compared to the blank control group within 3 days. After 4 days, the transformed cells had grown rapidly while the growth of cells in the blank control group was slow. The cells from transformed foci of the winter sample grew quickly and the shape of cell growth curves was nearly the same as the positive control group after 5 days. The cell growth curve in the summer sample was only different from the blank control group after 6 days.

## DISCUSSION

Surface water pollution is a very serious public health and aquatic ecosystem problem, especially in rivers that are used as a source of drinking water Ohe, et al. 2004, Liu, et al, 2007, 2009 Ohe, et al. (2004) reported that pollutants from surface waters in Europe, Asia, South America, etc. were determined to be mutagenic / genotoxic using a variety of bioassays which demonstrated that these environmental mixtures contain many toxicants that may have the risk of carcinogenic potential. Umbuzeiro et al. (2001) reported that compiled data obtained during the last 20 years from more than a thousand samples and found that TA98 was more sensitive than TA100 and 79% of the mutagenicity was detected by this strain, regardless of the presence of S9-mix. However, in vitro mammalian cells also can further predict carcinogenic potential. Quantitative cell culture systems provide procedures for detecting potential carcinogens in a relatively short period of time compared with the years required to complete in vivo tests (Dunkel et al., 1991). Mammalian cells in culture are particularly appropriate in quantitative cell culture systems (Saffiotti, 1983). Quantitative dose-response relationships have been demonstrated with established mammalian cell systems in fibroblasts derived from mouse embryo cells. With most of these systems, the dose-response relationship for

sample was only different from the blank control group after 6 days. In the Colony formation assay, the plates were incubated for 3 - 4 weeks and the number and size of colonies were analyzed. The transformed cells formed colonies while no colonies were observed in the control groups including blank control and DMSO groups (Table 3 and Figure 3). These results confirm that transformed cells from all diethyl ether extracts lost anchorage dependence and grew in an anchorage independent manner, whereas the normal cells did not.

**Table 3.** Colony formation of transformed cells picked up from suspect foci of bottle in semi-solid agar.

| Groups | Total no. of foci picked up | No. of colonies/1,000 cells |
| --- | --- | --- |
| | | Mean ± SD |
| Transformed foci from experimental groups | 15 | 53 ± 39** |
| MNNG | 18 | 133 ± 48** |
| DMSO | 5 | 0 |
| Blank | 5 | 0 |

**P < 0.01, compared to the negative control group (DMSO).

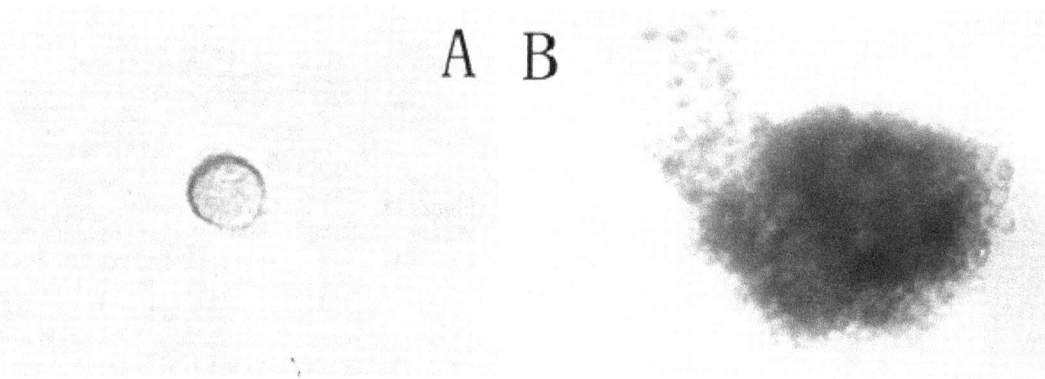

**Figure 3.** Growth of NIH3T3 cells transformed in semi-solid agar. To determine the cells' ability for anchorage independent growth, about $1 \times 10^4$ cells were suspended in 1.5ml of 0.3% top agar containing PRMI 1640 plus 10% FBS and overlaid onto 1.5ml of bottom agar (0.6% agar in RPMI 1640 containing 10 % FBS) in a 6-well plate. After incubation for 3 - 4 weeks at 37°C, we found that the cells that had transformed had acquired the ability to grow in soft agar and colony formation (B, ×100). In contrast, the non-transformed cells did not produce any anchorage independent colonies (A, ×400).

for transformation by an organic chemical carcinogen is consistent, indicating a direct cause- and -effect relationship between carcinogen and transformation. The use of mammalian cell culture (*in vitro*) transformation systems is an important technique for short-term testing for potential carcinogens, since the mechanisms involved in such systems may be similar to the processes of neoplastic transformation *in vivo* (IARC, 1980, Barrett, 1985). Hu et al. (2004) reported that arsenic at the low doses of 0.1 and 0.5 µmol/L for 110 days could induce the morphological transformation of NIH3T3 cells. The transformed cells had the characteristics of malignant cells, not only growing in semi-solid agar, but also still surviving over 50 passages.

The Songhua River is located at the junction of the temperate and cold-temperate zones. The region has a long, cold winter, a torrid, rainy summer; and a dry, windy spring. Early August was chosen for the summer sample because it had more rain and early January was chosen for the winter season because the level of water in the Songhua River was at its lowest. The Songhua River is a major freshwater source for industry and agriculture in the area. However, the increasing population, industry and agricultural activities around the Songhua River lead to the introduction of contaminants and the possible pollution of the river. Zhu et al. (1985); Zhang et al. (1988)

and Zhu et al. reported that compiling mutagenicity tests of diethyl ether extracts of fish and water samples taken from the Songhua River and epidemiological investigations on cancer mortalities for residents of Harbin and Zhaoyuan county along with the Songhua River revealed the following results: (1) Mutagenicity tests showed that diethyl ether extracts from fish and water samples are mutagenic. Treatment of 1.7 L water equivalent / plate and 6.0 g fish equivalent / plate in the *Ames* test indicated a positive response in a dose-dependent manner; (2) a case-control study of gastric and liver cancers showed that the cancer mortalities were significantly higher for residents who drank water or ate fish from the Songhua River. The cancer mortalities were significantly higher for residents of Harbin, who drank water of the Songhua River, than for those residents drinking ground water. The cancer mortalities were significantly higher for residents of Zhaoyuan county near the Songhua River, as they ate more fish; (3) The certain relationship between cancer mortalities and organic contamination of the Songhua River was noted; (4) Further laboratory evidence was needed to explain this relationship. Yang's study (Yang et al., 1991) also investigated the frequencies of CA and SCE of blood lymphocytes from 59 residents who lived near the Songhua River over 5 years and 52 residents as a control group. The results of CA in residents living near the Songhua River were higher than that in the control residents but the difference was not significant. However, the frequency of SCE in these same residents was significantly higher than in the control group. This indicated that water pollution including mercury and methyl-mercury from the Songhua River is a main factor for increasing frequencies of CA and SCE in the blood lymphocytes of residents who lived along the river.

In this study, the TF per bottle for the blank control group is zero and 3 SCF for the DMSO control group; however, there were three foci in the DMSO control group which were suspect in that they possessed anchorage dependence in the semi-solid agar culture test. This result in the negative control is very astonishing to us, because a high transformation frequency was obtained in the cell line by searching related articles. Rubin (2005) reported that spontaneous neoplastic transformation develops within days in the NIH3T3 cell line through differential inhibition of their proliferation under contact inhibition. Contact inhibition and reduction in serum concentration select for the same cellular phenotype that increases saturation density and generates transformed foci. This allows the selection of cells that are better able to proliferate and reduces spontaneous transformation under restricted conditions such as low seeding cell density, low passage number, the same serum concentration in the media as well as short experimental period. Rubin et al. (1989), in another study, showed that NIH3T3 cells could produce transformed foci spontaneously if kept in the confluent state for more than 10 days. Cells maintained in continuous exponential multiplication in the sub confluent state by transfer every 2 - 3 days in medium with 10% calf serum failed to develop the capacity to produce foci in 2% calf serum, but those transferred in the same way in 2% calf serum or in 10% fetal bovine serum, which is a less potent growth stimulant, did develop that capacity to an increasing degree over time. The formation of foci depends on the type and concentration of bovine serum used in the medium and on the passage history of the cells. The number of transformed cells increased sharply with the time that a culture remained in the confluent state. In our study, concentrations of diethyl ether extracts of water samples taken from the Songhua River in the summer of 1994 and the winter of 1995 could induce typical transformed foci in NIH3T3 cells. TF per bottle showed a dose-response relationship. The TF also showed a dose-dependent manner as our previously study (Liu et al., 2007). Transformed NIH3T3 cells lost contact inhibition, and the speed rate of their growth increased. Normal cells reach contact inhibition and growth rate slows after 4 days. The cell growth curve in the DMSO control group were also determined in the previously study (Liu et al., 2007). The shape of cell growth curve in the DMSO control group is nearly the same as the blank control group. Although the cell growth curve from diethyl ether extract of water sample in the

summer was a bit different in comparison with the blank control group, the number of transformed cells in cellular foci was less than those of the diethyl ether extract of water sample in the winter sample and the positive control group. The latter was proved by the anchorage dependent test. Thus, the cell growth speeds are different from each other. NIH3T3 cells are anchorage dependent, this is, and they must have a solid surface to adhere to in order to grow. Transformed NIH3T3 cells lose anchorage dependence and grow into cluster of cells in semi-solid agar.

No single method has been universally accepted for the statistical evaluation of chemical responses in cell transformation assays. The distribution of transformed foci is generally symmetric across culture vessels for all treatment conditions. However, some researchers have noted abnormal distributions, and relatively large numbers of foci occur at random in both control and carcinogen-treated cultures (Rundell, 1984). The current literature describes two main ways to evaluate the frequency of transformed foci. Frequencies can be evaluated using non-parametric $x^2$ or one-tailed Fisher's Exact tests (Armitage, 1971). Alternatively, parametric methods can be employed following suitable transformation of the data. After examining several mathematical transformations, this study employed parametric analyses following logarithmic (that is, log10) transformation Matthews, 1986; Rundell and Guntakatta, 1983). ANOVA (analysis of variance) was carried out on the log10-transformed frequency values, and differences between responses were investigated using a modified student's t-test.

We have measured the cell transformation of diethyl ester extracts of water samples from the Songhua River, a major source of drinking water in the North Eastern region of China. This study is unique in that it used a cell transformation assay to demonstrate that the water samples possess the risk of potential carcinogenesis. Cell morphology changed dramatically as reflected by transformed foci. The cells from transformed foci have the characteristics of malignant cells. These results indicate that organic extracts from the Songhua River samples have induced cell transformation of NIH3T3 cells and provide further an evidence of a relationship between organic pollution and carcinogenic potential. Based on these results, we recommend that future analyses of organic pollutants be both qualitative and quantitative and a further epidemiological study of the population is needed.

## ACKNOWLEDGMENTS

This research was supported by grants from Dr. E. J. Love from the department of Community Health Sciences, Faculty of Medicine, University of Calgary, Canada. We thank Shu-Yuan Zhao and Lin Guan for the general collection and extraction of the water.

## REFERENCES

Armitage P (1971). Statistical Methods in Medical Research. Blackwell Sci. Publ. Oxford. 135-138.

Berwald Y, Sachs L (1963). In Vitro Cell Transformation with Chemical Carcinogens. Nature. 200: 1182-4.

Berwald Y, Sachs L (1965). In vitro transformation of normal cells to tumor cells by carcinogenic hydrocarbons. J. Natl. Cancer. Inst. 35(4): 641-61.

Combes R, Balls M, Curren R, Fischbach M, Fusenig N, Kirkland D, Lasne C, Landolph J, LeBoeuf R, Marquardt H, McCormick J, Müller L, Rivedal E, Sabbioni E, Tanaka N, Vasseur NHY (1999). Cell transformation assays as predictors of human carcinogenicity. ATLA. 27: 745-767.

Dunkel VC, Rogers C, Swierenga SH, Brillinger RL, Gilman JP, Nestmann ER (1991). Recommended protocols based on a survey of current practice in genotoxicity testing laboratories: III. Cell transformation in C3H/10T1/2 mouse embryo cell, BALB/c 3T3 mouse fibroblast and Syrian hamster embryo cell cultures. Mutat Res. 246(2): 285-300.

Matthews EJ (1986). Assessment of chemical carcinogen-induced transforming activity using BALB/c-3T3 cells. J. Tissue Culture Methods. 10:157-164.

Hu Y, Jin XM, Wang GQ, Snow ET (2004). NIH3T3 cells transformation induced by low dose arsenic in vitro. Chinese J. Xinjiang Med. Univ. 27:1-5.

Huang XS, Chen XR (1985). Experimental methods of mutation and deformity for environmental substances. Sci. Tech. Publ. China: Hangzhou, Zejiang, 173-194. Intl. Agency for Research on Cancer (IARC) (1980). Monographs on the evaluation of the carcinogenic risk of chemicals to humans. Intl. Agency for Research on Cancer: Lyon,

Barrett JC (1985). Cell culture models of multistep carcinogenesis. Intl. Agency for Research on Cancer: Lyon. 181-202.

Rundell JO (1984). In vitro transformation assays using mouse embryo cell lines: BALB/c-3T3 ceils. Humana Press: New Jersey. 279-285.

Junk GA, Richard JJ, Grieser MD, Witiak D, Witiak JL, Arguello MD, Vick R, Svec HJ, Fritz JS, Calder GV (1974). Use of macroreticular resins in the analysis of water for trace organic contaminants. J Chromatogr. 99(0): 745-62.

Liu JR, Dong HW, Tang XL, Sun XR, Han XH, Chen BQ, Sun CH, Yang BF(2009) Genotoxicity of water from the Songhua River, China, in 1994-1995 and 2002-2003: Potential risks for human health. Environ Pollut. 157(2): 357-64.

Liu JR, Pang YX, Tang XL, Dong HW, Chen BQ, Sun CH (2007). Genotoxic activity of organic contamination of the Songhua River in the north-eastern region of the People's Republic of China. Mutat Res. 634(1-2): 81-92.

Ohe T, Watanabe T, Wakabayashi K (2004). Mutagens in surface waters: A review. Mutat Res. 567(2-3): 109-49.

Rubin H (2005). Degrees and kinds of selection in spontaneous neoplastic transformation: an operational analysis. Proc. Natl. Acad. Sci. USA. 102(26): 9276-81.

Rubin H, Xu K (1989). Evidence for the progressive and adaptive nature of spontaneous transformation in the NIH 3T3 cell line. Proc. Natl. Acad. Sci. USA. 86(6): 1860-4.

Rundell J, Guntakatta M, EJ, M (1983). Criterion development for the application of BALB/c-3T3 cells to routine testing for chemical carcinogenic potential. Plenum Publ. New York. 302-327.

Saffiotti U (1983). Evaluation of mixed exposure to carcinogens and correlations of in vivo and in vitro systems. Environ Health Perspect. 47:319-24.

Ruhaut R (1979). An overview of the problem of thresholds for chemical carcinogens. IARC Sci. Publ. (25):191-202.

Umbuzeiro GA, Roubicek DA, Sanchez PS, Sato MI (2001). The Salmonella mutagenicity assay in a surface water quality monitoring program based on a 20-year survey. Mutat Res. 491(1-2): 119-26.

Wang JL, Yun LJ, Wang HZ, Sun B, Cai HD (1983). National products of macroporous resin absorbed organic compounds from water and its applications. China Envin Sci. & Techn. 3:1-3.

Weber MJ, Hale AH, Yau TM, Buckman T, Johnson M, Brady TM, LaRossa DD (1976). Transport changes associated with growth control and malignant transformation. J. Cell Physiol. 89(4): 711-21.

Xu GY, Gao YF, He ZS, Jia YQ (1990). Investigation of organic pollutants in the Songhua River basin. Chinese J. Envin. Sci. 11: 29-31.

Yang BC, Sun XL, Chen Q, Zheng WG, Lu WB, XL, W (1991). Investigation of the effect of polluted water in Songhua River on the chromosome aberrations and SCE of inhabitants. Chines Zhonghua Yu Fang Yi Xue Za Zhi. 25: 296-297.

Zhu ZG (1984). Study on the relationship between morality of tumor in a city and water pollution. Chinese J. of Environ. Health. 1: 3-4.

Zhang YE, Guo DL, Tao KS, Zhu ZG (1988). Research on the relationship between tumor incidence and water pollution. Chinese J. Harbin. Med. Uni. 22: 26-28.

Zhu ZG, Gan HF, Guo DL, Xie X, Pang YX, Wang XZ, Zhang YE, Shi LT, Qiu XH, Xu ChX, Wang DC, Ma TH, Si JL, Sun J (1985). Study on carcinogenic potentiality of organic contamination of the Songhua River. China J. Environ. Sci. 5: 7-12.

Zhu ZG, Gan HF, Guo DL, Xie X, Pang YX, Wang XZ, Zhang YE, Shi LT, Qiu XH, Xu ChX, Wang DC, Ma TH, Si JL, Sun J (1985). Study on carcinogenic potentiality of organic contamination of the Songhua River. China J. Environ. Sci. 5: 7-12.

# 23

# Fluoride and thyroid function in children in two villages in China

**Quanyong Xiang[1]\*, Liansheng Chen [1], Youxin Liang[2], Ming Wu[1] and Bingheng Chen[2]**

[1]Jiangsu Province Center for Disease Control and Prevention, 172 Jiangsu Road, Nanjing 21009, P. R. China.
[2]School of Public Health, Fudan University (formerly Shanghai Medical University), 138 Yixueyuan Rd., Shanghai (200032), China.

**Eighty two children, aged 8 - 13 years old, from Wamiao village (severe endemic fluorosis area), from Xinhuai village (nonendemic fluorosis area) were 88 (as a control group), were recruited in this study. The prevalence of dental fluorosis (DF) were 85.37% (Wamiao) and 6.82% (Xinhuai) in two village's children respectively; drinking water fluoride (F⁻) in children's household shallow well from 0.62 - 4.00 mg/L in Wamiao and 0.23 - 0.76 mg/L in Xinhuai; serum total triiodothyronine (TT3), total thyronine (TT4), thyroid-stimulating hormone (TSH) were 1.47 ± 0.28 and 1.47 ± 0.33 ng/mL, 9.67 ± 1.76 and 9.22 ± 2.54 µg/dL, 3.88 ± 2.15 and 2.54 ± 2.07 µIU/mL in two villages children respectively. The prevalence of DF, drinking water F⁻, serum TSH in Wamiao village was significantly higher than that in Xinhuai village. As the children in Wamiao village were divided into different subgroups according to their severity of DF, serum TT3 and TSH showed significant difference in different groups. The results in this study confirmed that the high F⁻ exposure can caused functional abnormalities of thyroid, and the different severity degree of DF may be relation to significant deviation in the serum levels of thyroid hormone.**

**Key words:** Thyroid function, fluoride, dental fluorosis.

## INTRODUCTION

Professor John Grevers had sent specimens of mottled teeth, found the identical condition in the teeth of people with goiter in Utrecht; Grevers also obtained laboratory evidence that there was a clear association of his clinical cases of goiter with mottled enamel; Goldemberg even became convinced that the goiters were, in fact, caused by excessive intake of fluoride (F⁻) (Schuld, 2005).

Some animal studies indicated that the rats had thyromegaly, increasing or reducing of thyroid weight, reducing of follicle and atrophy of follicular epithelium of thyroid when the rats were exposed to F⁻ for 6 ~ 12 months (Editorial, 1976). The research results also indicated that F⁻ can affect the hormone secretion of the thyroid (Chuanhua et al., 1998; Desun et al., 1994; Fuzun et al., 2001; H Wang et al., 2009; Hu Aiwu et al., 2007; Juvenal et al., 1978; Liu Guoyan et al., 2008; Xiuan Z et al., 2006). Yaming et al. gave the results that excessive long-term intake of F⁻, with or without adequate intake, are a significant risk factor for the development of thyroid dysfunction (Yaming et al., 2005). A few future studies on human showed different results. Xiaoli et al. (1999) reported that serum thyronine (T4) reduced significantly, but the triiodothyronine (T3); thyroid-stimulating hormone (TSH) increased significantly, in the 8 - 12 years old children in endemic fluorosis areas in China. T3 and T4 concentrations in the serum of the patients with endemic fluorosis were significantly below the Normal reference value (Guimin et al., 2001). Whereas the study by Mingyin F (Mingyin et al., 1994) and other researchers (Baum et al., 1981; Eichner et al., 1981) indicated that the high F⁻ intake does not have effects on thyroid-function. In order to get a better understanding between the F⁻ intake and the children's thyroid function, this study investigated the TT3 (total T3), TT4 (total T4), and TSH in the serum of the children in endemic fluorosis and

---

*Corresponding author. E-mail: quanyongxiang@yahoo.com.cn.

**Abbreviations: DF,** Dental fluorosis; **F⁻,** fluoride; **T3,** 3,5,3'-triiodothyronine; **TT3,** total 3,5,3'- triiodothyronine; **FT3,** free 3,5,3'- triiodothyronine; **T4,** thyronine; **TT4,** total thyronine; **FT4,** free thyronine; **TSH,** thyroid-stimulating hormone; **IDD,** iodine deficiency disorder.

**Table 1.** F⁻ in drinking water in children's household shallow well in two villages.

| Village | No. of samples | F⁻ (mg/L) (M±S) | Range (mg/L) |
|---|---|---|---|
| Wamiao | 82 | 2.36 ± 0.70 | 0.62 ~ 4.00 |
| Xinhuai | 88 | 0.36 ± 0.10 | 0.23 ~ 0.76 |

Note: t = -26.47, p = 0.000.

**Table 2.** Prevalence of DF in children in two villages.

| Village | No. of samples | No. of DF | Prevalence of DF (%) |
|---|---|---|---|
| Wamiao | 82 | 70 | 85.37 |
| Xinhuai | 88 | 6 | 6.82 |

Note: $\chi^2$ = 105.94, p = 0.000.

**Table 3.** TT3 concentration in serum in the children in two villages (ng/mL).

| Village | No. of samples | TT3 (M±S) | Range |
|---|---|---|---|
| Wamiao | 62 | 1.47 ± 0.28 | 1.01 ~ 2.10 |
| Xinhuai | 68 | 1.47 ± 0.33 | 0.66 ~ 2.20 |

Note: t = 0.855, p = 0.394.

**Table 4.** TT4 concentration in serum in the children in two villages (μg/dL).

| Village | No. of samples | TT4 (M ± S) | Range |
|---|---|---|---|
| Wamiao | 58 | 9.67 ± 1.76 | 5.98 ~ 15.09 |
| Xinhuai | 61 | 9.22 ± 2.54 | 5.22 ~ 15.41 |

Note: t = 1.111, p = 0.269.

non-endemic fluorosis areas, the dental fluorosis (DF) and the drinking water F⁻ concentration in children's household shallow well were also analyzed.

**MATERIALS AND METHODS**

The investigation of age, gender, and DF was conducted from February, 2003 - June, 2003 in Wamiao village (a severe endemic fluorosis area) and Xinhuai village (a non-endemic fluorosis), and the samples of water and blood were also collected during this time. The basic information of these two villages was formerly reported (QY Xiang et al., 2004).

Fasting venous blood samples (2 - 2.5 mL) were collected and preserved in clean plastic centrifuge tubes, which were immediately centrifuged for 10 min at 3000 rpm. Serum was quickly removed to other clean plastic tubes and kept in a refrigerator at -40℃. The TT3, TT4 and TSH were measured at the time of September, 2006, with the Test Kit, which were bought from Hainan Huamei Medicine Co. LTD, manufactured by BioCheck, Inc. (Foster City, CA94404, USA).

The drinking water samples, which were collected from the household shallow wells in each child's family, were kept in clean plastic bottles and analyzed within two week. F⁻ in drinking water was measured with an F⁻ ion selective electrode according to the National Standard of China (National Standard of P.R. China, 1999). A dentist and a specialist in endemic fluorosis control and prevention examined the children for dental fluorosis with a mouth mirror, forceps, and a probe under natural light. Dean's classification was used for diagnosing dental fluorosis. The six grades of Dean's classification scale for dental fluorosis are: none (normal enamel) (the score marked for 0), suspected or questionable (0.5), very mild (1), mild (2), moderate (3), and severe (4) (Chinese Ministry of Health 1991). Statistical analysis of the prevalence of dental fluorosis was made according to the rates of DF%.

Data were analyzed using SPSS Software. Before the investigation, the informed consent must be signed by the children's parents. We have complied with all requirements of International Regulations for the human investigation.

**RESULTS**

There were 170 children in this study, 82 in Wamiao village (46 male and 36 female), 88 in Xinhuai village (52 male and 36 female). The average age was 11.00 ± 1.44 in Wamiao village and 10.84 ± 1.67 in Xinhuai village.

The F⁻ concentration in drinking water and the prevalence of DF in two villages were shown in Table 1 - 2. The results indicated that the drinking water F⁻ and the prevalence of DF in children in Wamiao village were significantly higher than that in Xinhuai village.

The TT3 and TT4 concentrations in children's serum in two villages were not have significant difference. But the TSH concentration in the children's serum in Wamiao village was higher than that in Xinhuai village, there was a significant difference between two villages. The details were shown in Table 3 - 5.

The means of TSH/TT3 and TSH/TT4 in Wamiao village were significant higher than that in Xinhuai village, but the means of TT3/TT4 in Wamiao village was

**Table 5.** TSH concentration in serum in the children in two villages (μIU/mL).

| Village | No. of samples | TSH (M±S) | Range |
|---|---|---|---|
| Wamiao | 62 | 3.88 ± 2.15 | 0.19 ~ 8.82 |
| Xinhuai | 67 | 2.54 ± 2.07 | 0.71 ~ 9.37 |

Note: t = 3.604, p = 0.000.

**Table 6.** Compared the difference in the values of TT3/TT4, TSH/TT3, TSH/TT4 between the two villages.

| Village | TT3/TT4 | | TSH/TT3 | | TSH/TT4 | |
|---|---|---|---|---|---|---|
| | No.* | Mean ± S | No. | Mean ± S | No. | Mean ± S |
| Wamiao | 40 | 0.151 ± 0.037 | 55 | 2.735 ± 1.485 | 41 | 0.416 ± 0.218 |
| Xinhuai | 41 | 0.170 ± 0.044[#] | 66 | 1.932 ± 1.813[#] | 42 | 0.284 ± 0.191[#] |

Note: * the number of subjects. # compared with Wamiao village. TT3/TT4: t = 2.028 p = 0.046; TSH/TT3: t = 2.628 p = 0.010 ; TSH/TT4: t = 2.925 p = 0.004.

**Table 7.** The correlation between the serum TT3, TT4, TSH in the children's serum and drinking water F⁻ in two villages.

| Village | TT3 and F⁻ | | TT4 and F⁻ | | TSH and F⁻ | |
|---|---|---|---|---|---|---|
| | PC | p | PC | p | PC | p |
| Wamiao | 0.087 | 0.502 | 0.057 | 0.672 | 0.023 | 0.858 |
| Xinhuai | 0.108 | 0.381 | -0.167 | 0.198 | -0.112 | 0.381 |

Note: F⁻ (drinking water fluoride). PC (Pearson correlation).

**Table 8.** The relationship between the DF score and the serum TT3, TT4, TSH in the children in Wamiao village.

| DF (group) | TT3 (ng/mL) | | TT4 (μg/dL) | | TSH (μIU/mL) | |
|---|---|---|---|---|---|---|
| | No. of samples | TT3 (M ± S) | No. of samples | TT4 (M ± S) | No. of samples | TSH (M ± S) |
| 1 | 11 | 1.19 ± 0.18 | 7 | 9.33 ± 1.92 | 8 | 3.31 ± 1.26 |
| 2 | 14 | 1.38 ± 0.28 | 15 | 9.20 ± 1.00 | 14 | 5.15 ± 2.68 |
| 3 | 22 | 1.54 ± 0.21 | 19 | 10.24 ± 2.18 | 25 | 3.50 ± 1.71 |
| 4 | 12 | 1.49 ± 0.34 | 12 | 9.46 ± 1.88 | 12 | 3.61 ± 2.50 |
| 5 | 3 | 1.40 ± 0.38 | 5 | 9.86 ± 1.04 | 3 | 3.74 ± 2.06 |

Note: TT3: 1 and 3, t = -4.680 p < 0.000; 1 and 4, t = -2.57 p = 0.018. TSH: 1 and 2, t = -2.183 p = 0.041.

significant lower than that in Xinhuai village. The details were shown in Table 6.

Each child's serum concentration of TT3, TT4, and TSH was compared with the drinking water F⁻ concentration in their household shallow well in two villages, there were not significant relationships between the TT3, TT4, TSH and the drinking water F⁻. See Table 7.

As shown in Table 8, In Wamiao village, there were not significant trend between the dental fluorosis score

(severity of DF) and the serum TT3, TT4, TSH concentration (the children with 0 and 0.5 DF score were divided into group 1, 1 score was group 2, 2 score was group 3, 3 score was group 4, 4 score was group 5). But the results of the comparison between each group indicated that the TT3 concentration has a significant difference between group 1 and group 3, group 1 and group 4; there were also significant differences between the group 1 and 2 in the TSH concentration. In Xinhuai

**Table 9.** The relationship between the DF score and the serum TT3, TT4, TSH in the children in Xinhuai village.

| DF (score) | TT3 (ng/mL) | | TT4 (µg/dL) | | TSH (µIU/mL) | |
|---|---|---|---|---|---|---|
| | No. of samples | TT3 (M ± S) | No. of samples | TT4 (M ± S) | No. of samples | TSH (M ± S) |
| 1 | 63 | 1.48 ± 0.33 | 56 | 9.34 ± 2.58 | 62 | 2.53 ± 2.09 |
| 2 | 5 | 1.30 ± 0.21 | 5 | 7.82 ± 1.55 | 5 | 2.72 ± 1.98 |

**Table 10.** The TT3, TT4, and TSH results by age and gender in Wamiao village.

| | Age | TT3 | | | | TT4 | | | | TSH | | | |
|---|---|---|---|---|---|---|---|---|---|---|---|---|---|
| | | No. | High | Normal | low | No. | High | Normal | Low | No. | High | Normal | Low |
| Male | 8 | 2 | 0 | 2 | 0 | 1 | 0 | 1 | 0 | 2 | 1 | 1 | 0 |
| | 9 | 4 | 0 | 4 | 0 | 3 | 0 | 3 | 0 | 4 | 0 | 4 | 0 |
| | 10 | 5 | 0 | 5 | 0 | 4 | 1 | 3 | 0 | 5 | 0 | 5 | 0 |
| | 11 | 5 | 1 | 4 | 0 | 6 | 0 | 6 | 0 | 4 | 2 | 2 | 0 |
| | 12 | 10 | 0 | 10 | 0 | 11 | 1 | 10 | 0 | 9 | 2 | 7 | 0 |
| | 13 | 10 | 0 | 10 | 0 | 10 | 0 | 10 | 0 | 10 | 2 | 8 | 0 |
| **Total** | | 36 | 1 | 35 | 0 | 35 | 2 | 33 | 0 | 34 | 7 | 27 | 0 |
| Female | 8 | 0 | 0 | 0 | 0 | 2 | 1 | 1 | 0 | 2 | 0 | 2 | 0 |
| | 9 | 2 | 0 | 2 | 0 | 2 | 0 | 2 | 0 | 2 | 1 | 1 | 0 |
| | 10 | 4 | 0 | 4 | 0 | 3 | 0 | 3 | 0 | 5 | 1 | 4 | 0 |
| | 11 | 8 | 0 | 8 | 0 | 8 | 1 | 7 | 0 | 10 | 2 | 8 | 0 |
| | 12 | 9 | 0 | 9 | 0 | 6 | 0 | 6 | 0 | 7 | 1 | 6 | 0 |
| | 13 | 3 | 0 | 3 | 0 | 2 | 0 | 2 | 0 | 2 | 0 | 2 | 0 |
| **Total** | | 26 | 0 | 26 | 0 | 23 | 2 | 21 | 0 | 28 | 5 | 23 | 0 |

village there was not a significant difference between the two groups in TT3, TT4, and TSH (see Table 9).

In China as clinical diagnosis biomarkers, the reference normal values of TT3, TT4, TSH were 0.80 ~ 2.00 ng/mL, 5.00 ~ 13.00 µg/dL, 0.34 ~ 5.60 µIU/mL. As shown in Tables 10 - 11, there was 1 subject with high serum TT3 concentration in each village; and 4 in Wamiao, 5 in Xinhuai with high serum TT4; and 12 in Wamiao, 9 in Xinhuai with high serum TSH. There was not significant difference between the gender and total in the ratio of high TT3, TT4, and TSH in two villages indicated by the results of Chi-Square Tests.

## DISCUSSION

In Wamiao and Xinhuai village primary school, the less economic development conditions are not allowed the school to provide food and drinking water for students and teachers; they must back to home to have meals for breakfast, lunch, and dinner; a few students may be bring boiling water in plastic or glass bottle from family to school. In our formerly investigation, the daily total intake of drinking water in two villages' children was: 156.0 ml crude water and 1085.1 ml boiled water included average 2 mL/child/day tea. The rate of crude water was only

14.38% for the total daily intake of drinking water. Each family has a household shallow well in the yard; the average use age of the shallow well was 5.16 ±3.12 years in Xinhuai village and 8.27 ±3.02 years in Wamiao village. Students in these two villages hardly ever move from one residential site to another, and hardly drink the market sell water. So the F⁻ exposure history of the subjects was relatively clear and the F⁻ in drinking water was the main source of the F⁻ intake (QY Xiang, et al 2004; QY Xiang et al. 2005). Therefore, it was relatively easy to explore the exact relationships between the drinking water F⁻ and the serum concentration of TT3, TT4, and TSH.

In 1990, the Government of China has an official commitment to the world that the hazard of IDD (iodine deficiency disorder) will be removed in China before 2000. So, in China, all the countrymen will eat the qualified iodize salt from 1994. In Wamiao and Xinhuai village the mean levels of iodine in children's urinary were over 280.70±87.16 µg/L and 300.96 ± 92.88 µg/L, and there was not significant difference in childrens' urinary iodine concentration in two villages (Q Xiang et al., 2003). Neither village was identified as being in an area of endemic iodine deficiency area according to The Manual of Prevention and Treatment of Endemic Iodine Deficiency published by Chinese Ministry of Health

**Table 11.** The TT3, TT4, and TSH results by age and gender in Xinhuai village.

| | Age | TT3 | | | | TT4 | | | | TSH | | | |
|---|---|---|---|---|---|---|---|---|---|---|---|---|---|
| | | No. | High | Normal | Low | No. | High | Normal | Low | No. | High | Normal | Low |
| Male | 8 | 6 | 0 | 6 | 0 | 5 | 1 | 4 | 0 | 6 | 2 | 4 | 0 |
| | 9 | 3 | 0 | 3 | 0 | 3 | 1 | 2 | 0 | 3 | 0 | 3 | 0 |
| | 10 | 5 | 0 | 5 | 0 | 3 | 0 | 3 | 0 | 5 | 1 | 4 | 0 |
| | 11 | 9 | 0 | 9 | 0 | 10 | 1 | 9 | 0 | 8 | 0 | 8 | 0 |
| | 12 | 11 | 1 | 10 | 0 | 11 | 1 | 10 | 0 | 11 | 0 | 11 | 0 |
| | 13 | 4 | 0 | 4 | 0 | 5 | 1 | 4 | 0 | 5 | 0 | 5 | 0 |
| **Total** | | 38 | 1 | 37 | 0 | 37 | 5 | 32 | 0 | 38 | 3 | 35 | 0 |
| Female | 8 | 4 | 0 | 4 | 0 | 0 | 0 | 0 | 0 | 4 | 2 | 1 | 1 |
| | 9 | 4 | 0 | 4 | 0 | 1 | 0 | 1 | 0 | 4 | 2 | 2 | 0 |
| | 10 | 5 | 0 | 5 | 0 | 3 | 0 | 3 | 0 | 4 | 0 | 4 | 0 |
| | 11 | 1 | 0 | 1 | 0 | 1 | 0 | 1 | 0 | 1 | 1 | 0 | 0 |
| | 12 | 9 | 0 | 9 | 0 | 10 | 0 | 10 | 0 | 9 | 1 | 8 | 0 |
| | 13 | 7 | 0 | 7 | 0 | 9 | 0 | 9 | 0 | 7 | 0 | 7 | 0 |
| **Total** | | 30 | 0 | 30 | 0 | 24 | 0 | 24 | 0 | 29 | 6 | 22 | 1 |

(Chinese Ministry of Health, 1989). Thus urinary iodine levels do not appear to affect the differences in TT3, TT4, and TSH in children between the two villages.

In this study, there were not significant relationships between the drinking water F- and the serum concentration of TT3, TT4, and TSH. There was significant difference between the two villages in serum TSH, TT3/TT4, TSH/TT3, and TSH/TT4; the serum TSH, TSH/TT3, and TSH/TT4 were significantly higher in Wamiao village than that in Xinhuai village, but the values of TT3/TT4 was on the contrary which was founded in the study of H Wang in rats (H Wang et al., 2009). In the study of Susheela AK (Susheela et al., 2005) and his co-worker, the results indicated that in Two-thirds of the sample children have elevated TSH in the sample group with high drinking water fluoride, but none are below normal; in this study, as shown in Tables 10 - 11, there were 19.35% in Wamiao and 13.42% in Xinhuai with higher serum TSH than normal, and none are below normal. Serum TSH level in children in endemic fluorosis area was significant higher than that in non endemic fluorosis area in the study of (Xiaoli et al., 1999). With the increasing F- in the diet, the serum TSH level was correspondingly increased in the young pigs and rats (Liu Guoyan et al. 2008; Xiuan et al. 2006). Those results were consistent with this study. But there were opposed results in the study of (Xiaowei et al., 1994). There were not significant differences between the residents of endemic fluorosis areas and non-endemic fluorosis areas in serum TSH in the study of Mingyin et al (Mingyin et al., 1994).

The results in this study indicated that there were not significant difference between two villages in serum TT3 and TT4 in the children. But in the study of Xiaowei (Xiaowei et al. 1994) indicated that serum TT3 and FT3

(free T3) were significant increased, serum FT4 (free T4) was significant reduced in adults in the higher F- drinking water area compared with the control area. Serum T4 was significant lower, T3 was significant higher in the children in the endemic fluorosis area compared with the non endemic fluorosis area in the report of Xiaoli (Xiaoli et al., 1999).

It has long been suggested that DF is associated with IDD and thyroid dysfunction (Susheela et al. 2005). The results reported in Table 7 revealed significant deviations in the serum levels in TT3 and TSH in the children in Wamiao village between the different DF groups. The serum TT3 in Group 3 (with mild dental fluorosis) and 4 (with moderate dental fluorosis) were significant higher compared with group 1 (with normal and suspected dental fluorosis). The serum TSH in group 2 (with very mild dental fluorosis) was significant higher than that in group 1.

F- and iodine are belong to chlorine group element, and F- is more active than iodine. The mechanism of F- causing the functional disorder of appendix cerebri - thyroid may be: 1. F- can compete with iodine and influence the absorption and condensing of iodine in thyroid; 2. F- can influence the biologic activity of functional enzyme system in thyroid; 3. F- can influence the feedback mechanism of hypothalamus and adenohypophysis of appendix cerebri and control the secretion of thyroid directly (Xiaowei et al., 1994).

The statement in the guest editorial of Andreas Schuld said that: "In fact, DF is a developmental disorder-originating from aberrant thyroid hormone metabolism. It is well established that DF can only occur as a result of excessive fluoride exposure during crucial times of development Thyroid hormone (TH) deficiency leads to delayed tooth eruption. The more F- ingested, the longer it takes for the tooth to erupt. The later in life maturation of enamel is completed, the greater is the severity of DF. At the same time, other risk factors known to influence DF are identical to those observed in thyroid dysfunction. Thus, while DF gets more severe at higher altitudes, the same is generally true for iodine deficiency" (Schuld, 2005).

The results in this study indicated or suggested that the high $F^-$ exposure can caused the thyroid functional abnormalities, and the different severity degree of DF may be relation to deviation in the serum levels of thyroid hormone. There were different results in the past reports about these. So the exact relationship among fluoride, DF, and appendix cerebri-thyroid function need to further study.

## ACKNOWLEDGEMENTS

This work was supported by Jiangsu Province Association for Endemic Disease Control and Prevention (X200327). We thank Prof BH Chen (School of Public Health, Fudan University for her valuable suggestions.

### REFERENCES

Fuzun C, Xu C, Yonggui D, Yunxing C, Yawei H, Jing M, (2001). Effects of selenium at different concentrations on lipid peroxidation.the capacity of anti-oxidation and thyroid in rats with fluorosis. Chinese J. Control Endem. Dis. 16(4): 213-214.

Xiang QY, Liang LC, Wang CB, Chen XC (2003). Effect of fluoride in drinking water on children's intelligence. Fluoride 36(2): 84-94.

Xiang QY, Chen LS, Chen XD, Wang CS, Liang YX, Liao QL (2005). Serum fluoride and skeletal fluorosis in two villages in Jiangsu Province, China. Fluoride 38(3): 178-184.

Xiang QY, Liang YX, Chen BH, Wang CS, Zhen SQ, Chen XD (2004). Serum fluoride and dental fluorosis in two villages in China. Fluoride 37(1): 28-37.

Baum K, Borner W, Reiners C, Moll E (1981). Bone density and thyroid gland function in adolescents in relation to fluoride content of drinking water. Fortschr. Med. 99(36): 1470-1472.

Chinese Ministry of Health, Department of Endemic Diseases (1989). Manual of Prevention and Treatment of Endemic Iodine Deficiency.

Chinese Ministry of Health, Department of Endemic Diseases (1991). Manual of Prevention and Treatment of Endemic Fluorosis.

Chuanhua L, Xiaowei G, Jianchao B, Pin Y, Shumei Q, Yuan L (1998). An experimental study on effects of high fluoride and supplied selenium on thyroid in rats. Chinese J. Endemiol. 17(2): 105-107.

Desun X, Yanling W, Ye L (1994). Study the effect of fluoride on the 125I distribution in the thyroid in the rats. Chinese J. Control Endemic Dis. 9(4): 218-21.

Editorial (1976). Target organs in fluorosis. Fluoride 9(1): 1-4.

Eichner R, Borner W, Henschler D, Kohler W, Moll E (1981). Osteoporosis therapy and thyroid function. Influence of 6 months of sodium fluoride treatment on thyroid function and bone density. Fortschr. Med. 99(10): 342-348.

Guimin W, Zhiya M, Zhongjie L, Zhi C, Jiandong T, Ruilan Z (2001). Determination and analysis on multimark of test of th e patients with endemic fluorosis. Chinese J. Endemiol. [Chinese] 20(2): 137-139.

Wang H, Yang Z, Zhou B, Gao H, Yan1 X, Wang J (2009). Fluoride-induced thyroid dysfunction in rats: roles of dietary protein and calcium level. Toxicology and Industrial Health 25: 49-57.

Hu A, Liu Xiaoyang, Qin Yide (2007). Effect of fluorine on triiodothyronine and thyroxin in mice. J. Bengbu. Med. Coll. 32(4): 392-394.

Juvenal GJ, Kleiman de Pisarev DL, Crenovich L, Pisarev MA (1978). Role of neurotransmitters, prostaglandins and glucose on precursor incorporation into the RNA of thyroid slices." Eur. J. Endocrinol. 87(4): 776.

Liu G, Zhang W, Gu J, Chai C (2008). Effects of Fluoride on Morphological Structure and Function of Thyroids in Rats. J. Shanghai Jiaotong Univ. (Agric. Sci.) 26(6): 537-539.

Mingyin F, Enxiang G, Xioude Z, Yuting J (1994). The test and analysis of the thyroid function of the subjects in high fluoride areas. Shanghai J. Med. La. Sci. 9(3): 217.

National Standard of P. R. China (1999). Method for determination of fluoride in drinking water of endemic fluorosis areas (WS/T 106-1999).

Schuld A (2005). Is dental fluorosis caused by thyroid hormone disturbances?. Fluoride 38(2): 91-4.

Susheela AK, Bhatnagar M, Vig K, Mondal NK (2005). Excess fluoride ingestion and thyroid hormone derangements in children living in Delhi, India. Fluoride 38(2): 98-108.

Xiaoli L, Zhongxue F, Jili H, Qinlan W, Hongyin W (1999). The detection of children's T3, T4 and TSH contents in endemic fluorosis areas. Endemic disease bulletin 14(1): 16-17.

Xiaowei G, Xiaohong L (1994). The effects of high fluoride drinking water on the function of appendix cerebri–thyroid of the residents. Chinese J. Endemiol. 13(6): 378-9.

Xiuan Z, Jianxin L, Min W, Zirong X (2006). Effects of fluoride on growth and thyroid function in young pigs. Fluoride 39(2): 95-100.

Yaming G, Hongmei N, Shaolin W, Jundong W (2005). DNA damage in thyroid gland cells of rats exposed to long-term intake of high fluoride and low iodine. Fluoride 38(4): 318-23.

# 24

# Studies on the pollution potential of wastewater from textile processing factories in Kaduna, Nigeria

Asia Imohimi Ohioma[1]*, Ndubuisi Obejesi Luke[2] and Odia Amraibure[1]

[1]Department of Chemistry, Ambrose Alli University, Ekpoma Nigeria.
[2]Department of Civil Engineering, Ambrose Alli University, Ekpoma Nigeria.

Samples of effluents from 3 textile processing factories F(1), F(2) and F(3), were characterized for their pollution potential. The concentrations of solids were found to be 1020, 790 and 1,380 mg/l total solids for the factories 1, 2 and 3, respectively. The BOD's and COD's were 342.8 and 542.4 mg/l for F(1), 123.2 and 224.6mg/l for F(2) and 456 and 738.4mg/l for F(3). The pH for the effluents were 9.36, 8.98 and 9.44 for F(1), F(2) and F(3), respectively. This implies that the effluents were alkaline. The nitrogen and phosphorus concentrations were 56 and 2.13 mg/l for F(1), 51 and 1.14 mg/l for F(2) and 43 and 0.73 mg/l for F(3), respectively. The levels of copper (Cu), zinc (Zn), iron (Fe), manganese (Mn), lead (Pb) and chromium (Cr) were higher than the Federal environmental protection agency (FEPA) standards for effluent discharge. This shows that the textile effluents have severe pollution potentials since the parameters measured have values above the tolerable limits compared to the FEPA standards. The results also showed that the ratio of COD: BOD were 1.58, 1.82 and 1.62 for F(1), F(2) and F(3), respectively, indicating that the effluents may not be able to undergo up to 50% substrate biodegradation, thus biological processes may not be feasible for the treatment of these effluents. The high values obtained for the parameters assessed, especially those of the concentrations of the solid and of the oxygen demands, call for a pretreatment of the effluent before its discharge into water body. Also, the high conductivity observed shows that sufficient ions are present in the effluents, thus suggesting that the chemical method of coagulation and flocculation may be an ideal treatment method.

Keywords: Textile wastewater, factory, pollution, substrate, biodegradation, coagulation, flocculation.

## INTRODUCTION

The degradation of the environment due to the discharge of polluting wastewater from industrial sources is a real problem in several countries. This situation is even worse in developing countries like Nigeria where little or no treatment is carried out before the discharge. In spite of the many steps taken to maintain and improve the quality of surface waters, the quantities of wastewaters generated by these industries continue to increase, and municipalities and industries are confronted with an urgent need to develop safe and feasible alternative practices for wastewater management.

Some industrial wastewaters contain high concentra-

tion of nitrogen which may exist in the forms of ammonia, nitrate (v), nitrate (iii) and organic nitrogen (Priestly, 1991). It is widely acknowledged that nitrogen in wastewater has become one of the major pollutants for our water resources. Environmental legislation requires the removal of nitrogen from wastewater before being discharged (Zhiguo et al., 2000). Nitrogen can pose serious public health threat when present in drinking water above certain concentrations. Nitrogen is commonly found in oxic water as trioxonitrate (V), that is $NO_3^-$ The nitrate (v) ion, is not dangerous as such. It is reduced to the highly toxic dioxonitrate (iii), that is, $NO_2^-$ by certain bacteria at suboxic conditions commonly found in the intestinal tract. Nitrate (Iii) causes the disease known as methemoglo binemia in infants (Ademoroti, 1996a). Furthermore, ni-

*Corresponding author's E-mail: imoasia2000@yahoo.com.

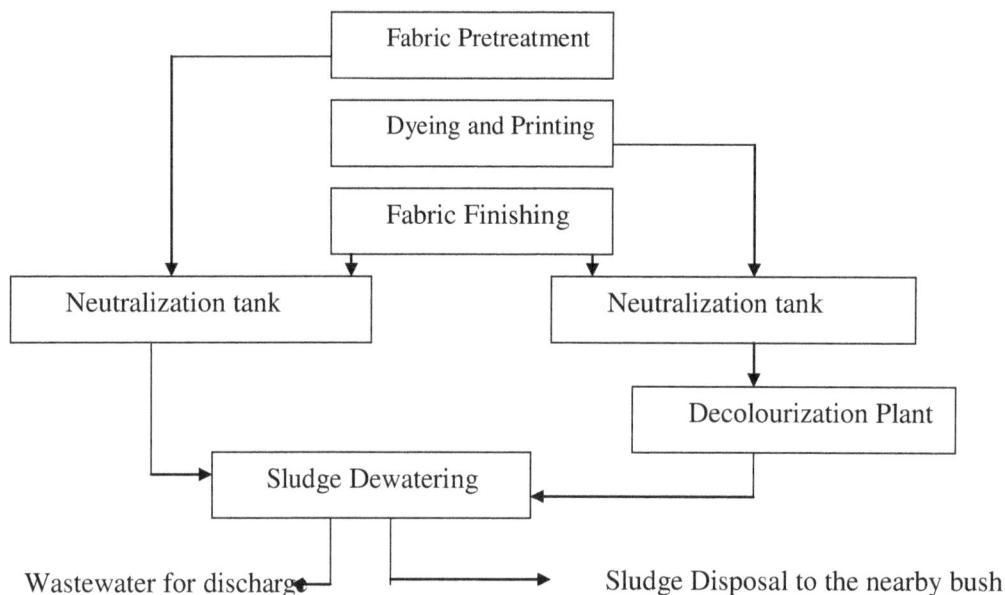

**Figure 1.** Wastewater flowchart of textile processing factory.

trates and phosphates, derived from wastewaters, are main nutrients, which promote growth of plants and algae. The amounts necessary to trigger algae blooms in water bodies are not well established but concentrations as low as 0.01 mg/l for phosphorus and 0.1 mg/l for nitrate may be sufficient for eutrophication when other elements are optimal (Henry and Heinke, 1989). In addition to having a detrimental aesthetic effect on lakes (odour and appearance), some algae are toxic to cattle, spoil the taste of the water, plug filtration units and increase the requirements for chemicals in the water treatment (Metcalf and Eddy, 1991)

The textile industry uses large volumes of waters in their operations and therefore discharges large volumes of wastewater into the environment, most of which is untreated. The wastewaters contain a variety of chemicals from the various stages of process operations which include desizing, scouring, bleaching and dyeing.

The textile industry is distinguished by the raw material used and this determines the volume of the water required for production as well as the wastewater generated. The production covers raw cotton, raw wool and synthetic materials. The industries studied in the present report are raw cotton based. In this type of production, slashing, bleaching, mercerizing and dyeing are the major sources of the wastewater generated. The main products of the factories are superprint, guarantee-superprint and minibrocade.

The factories consist of various departments, each of which carries out different operations and produces one type of specific wastewater. The wastewater contains acids used in desizing, dyeing bases like caustic soda used in scouring and mercerization. It also contains inorganic chlorine compounds and other oxidants, e.g. hypochlorite of

of sodium, hydrogen peroxide and peracetic acid for bleaching and other oxidative applications. Organic compounds are also present, e.g. dyestuff, optical bleachers, finishing chemicals, starch and related synthetic polymers for sizing and thickening, surface active chemicals are used as wetting and dispersing agents and enzymes for desizing and degumming. Salts of heavy metals are also present, e.g. of copper and zinc, and iron (iii) chloride used as printing ingredients. All these wastes are passed into an effluent tank and then drained into a drainage system.

**MATERIALS AND METHODS**

Textile wastewaters was collected from 3 textiles factories located in Kaduna. Kaduna lies at 10.52°N and 7.44°E. It is the capital city of Kaduna State and is situated in the central part of northern Nigeria (Kaduna, 2008). Concerning industry, it is one of the most developed cities in Northern Nigeria and textile industries are its dominating industries. Here the first textile industry in Nigeria was established (Jibrin, 2004; Yusuff and Sonibare, 2004). River Kaduna, a major river in the city receives the effluents from these industries, adding to the pollution load already present in the environment and a good reason for the importance of this research. This work is aimed at characterizing the wastewater obtained from 3 textile industries in Kaduna to assess their pollution potentials and to recommend appropriate treatment processes for the wastewaters. The wastewater flowchart of these industries is shown in Figure 1.

**Sampling of wastewaters**

Composite samples of the wastewaters were obtained from primary sedimentation tanks of the factory. Plastic bowls of 1 l capacity each were used to take samples manually over 12 h sampling period with 2 h interval starting at 7. 00a.m. and ending at 7.00p.m.

**Table 1.** Characterization of the effluents from three textile processing factories in Nigeria.

| Parameters | F(1) | F(2) | F(3) | FEPA standards |
|---|---|---|---|---|
| pH | 9.36± 0.02 | 8.98 ± 0.03 | 9.44 ± 0.04 | 6.9 |
| Temp °C | 31.8 ± 2.0 | 29.7 ± 3.0 | 29.1 ± 1.0 | NA |
| Conductivity µS | 740 ± 8.5 | 850 ± 4 | 940± 12 | NA |
| TSS mg/l | 370 ± 3 | 260 ± 1 | 180 ± 3 | 30 |
| TDS mg/ l | 650 ± 12 | 530 ± 13 | 1,200 ± 15 | 2000 |
| TS mg/l | 1020 ± 32 | 790 ± 27 | 1,380 ± 50 | 2000 |
| DO mg/ l | 4.49 ± 0.01 | 8.01 ± 0.03 | 3.44 ± 0.03 | NA |
| $BOD_5$ mg/l | 342.8 ± 0.6 | 123.2 ± 0.8 | 456.0 ± 1.6 | 50 |
| COD mg/l | 542.4 ± 1.3 | 224.6 ± 1.0 | 738.4 ± 10.3 | NA |
| Nitrate Nitrogen mg/l | 56 ± 3.2 | 51 ± 2.0 | 43 ± 2.4 | 10 |
| $PO_4$ mg/l | 2.13 ± 0.03 | 1.14 ± 0.01 | 0.73 ±0.01 | 5.0 |
| Ca mg/l | 67.32 ± 0.03 | 68.12± 0.02 | 74. 31 ± 0.04 | 75 |
| Mg mg/l | 43.11 ± 0.02 | 42.11 ± 0.02 | 31.03 ± 0.01 | 50 |
| Na mg/ l | 71.04 ± 0.11 | 41.23 ± 0.01 | 73.34 ± 0.01 | NA |
| K mg/l | 10.01 ± 0.03 | 24.21 ± 0.01 | 14.22 ± 0.13 | NA |
| Cu mg/l | 2.2 ± 0.02 | 4.5 ± 0.02 | 3.6 ± 0.4 | 1.00 |
| Zn mg/l | 8.7 ± 0.01 | 7.3 ± 0.2 | 6.1 ± 0.3 | 5.0 |
| Fe mg/l | 27.21 ± 0.13 | 37.00 ± 0.31 | 28.0 ± 0.01 | 0.3 |
| Mn mg/l | 3.80 ± 0.12 | 13.20 ± 0.01 | 0.90 ± 0.02 | 0.05 |
| Pb mg/l | 1.90 ± 0.02 | 1.70 ± 0.01 | 1.12 ± 0.03 | 0.05 |
| Cr mg/l | 0.70 ± 0.02 | 0.50 ± 0.08 | 0.24± 0.01 | 0.05 |

| | | |
|---|---|---|
| NA | = | Not Available |
| F(1) | = | Effluent from factory 1 |
| F(2) | = | Effluent from factory 2 |
| F(3) | = | Effluent from factory 3 |

This coincided with the working period in the factory and sampling was most convenient during this period. Composite samples were collected from each industry once a week for 7 weeks and analyzed. Where analysis could not be carried out immediately, samples were preserved in a refrigerator maintain at 4°C. At this temperature, biodegradation is minimal.

Every week the weekday for the sample collection was changed to account for cyclic and intermittent variations occurring at the work site.

### Methods of samples analysis

All samples were analyzed as described in the standard methods for the examination of water and wastewater (APHA, 1995) and standard methods for water and effluents analysis (Ademoroti, 1996b). The pH was determined with a pH meter, temperature was measured with a thermometer, conductivity and TDS were measured using a TDS/conductivity/salinity meter. TS and TSS were determined by a gravimetric method, DO was determined by the sodium azide modification of the Winkler method, as well as the BOD after appropriate dilutions. The COD was determined by the dichromate digestion method, nitrate nitrogen by using the indophenol colorimetric method. Phosphate was determined by the use of a spectrophotometer, calcium and magnesium were determined by gravimetric method and potassium by a flame photometric method, heavy metals, Cu, Zn, Fe, Mn, Pb and Cr were determined using an atomic absorption spectrophotometer (AAS). Where the analysis was not immediately possible, samples were preserved to inhibit biodegradation (Manual of Practice No. OM - 1, 1980). All the

reagents used for the analyses were of analytical grade and obtained from BDH chemicals limited poole England.

### RESULTS AND DISCUSSION

Table 1 show results of the detailed analyses carried out on the effluents obtained from the 3 textile processing factory F(1), F(2) and F(3). The pH values were slightly alkaline being 9.36, 8.98 and 9.44 for F(1), F(2) and F(3), respectively, typical for textile processing factories.

The effluent has high levels of solids. The total solids (TS), total suspended solids (TSS) and total dissolved solids (TDS) were 1020, 370 and 650 mg/l for F(1), 790, 260 and 530 mg/l for F(2) and 1380, 180 and 1200 mg/l for F(3), respectively. The high values of solids in F(3) may be caused of other domestic activities going on at that point.

The concentration of dissolved oxygen (DO) in the effluents of F(1), F(2) and F(3) were 4.49, 8.01 and 3.44 mg/l, respectively. The BOD and COD were 342.8 and 542.4 mg/l for F(1), 123.2 and 224.6 mg/l for F(2), and 456 and 738.4 mg/l for F(3), respectively. From these results a ratio of COD: BOD was calculated, resulting in 1.58, 1.82 and 1.62 for F(1), F(2) and F(3), respectively. This indicates that these effluents are high in recalcitrant and hardly degradable compounds and may not undergo more than 50% substrate biodegradation, as it is known that organic matter with 50 - 90% substrate biodegradation has a COD: BOD ratio between 2 and 3.5 (Quano et al., 1978)

The concentrations of solids and the oxygen demand were quite high when compared to the effluent discharge standard set by Federal environmental protection agency (FEPA, 1988). The water of all 3 effluents also show a high conductivity, indicating that sufficient ions are present in the effluent. The high conductivity suggests that the effluent could be treated by physicochemical method of coagulation and flocculation (Asia and Ademoroti, 2002).

The levels of nitrogen, 56, 51 and 43 mg/l for F(1), F(2) and F(3) respectively and phosphate, 2.13, 1.14 and 0.73 mg/l for F(1), F(2) and F(3), respectively) were also higher than FEPA and WHO standard.

The amount of calcium (67.32, 68.12 and 74.31 mg/l for F(1), F(2) and F(3) respectively) and magnesium, 43.11, 42.11 and 31.03 mg/l for F(1), F(2) and F(3) respectively in the effluent were within the federal environmental protection agency (FEPA) and world health organisation (WHO) standards of 75 and 50 mg/l for calcium and magnesium respectively.

The levels of heavy metals present in the effluent were quite high. The amount of copper (Cu), zinc (Zn), iron (Fe), manganese (Mn), lead (Pb) and chromium (Cr) obtained for F(1) were 2.2, 8.7, 27.21, 3.80, 1.9 and 0.70 mg/l as against the effluent discharge standard of 1.0, 5.0, 0.3, 0.05, 0.05 and 0.05 mg/l respectively for these parameters set by federal environmental protection agency (FEPA). The values obtained for F(2) were 4.5, 7.3, 37.0, 13.20, 1.7 and 0.5 mg/l and for F(3), the values were 3.6, 6.1, 28.0, 0.9, 1.12, and 0.24 mg/l for Cu, Zn, Fe, Mn, Pb and Cr respectively. These values are higher than the FEPA and WHO standards (FEPA, 1988; WHO, 1971).

Heavy metals if present even in low concentrations are toxic to living organisms, including humans as well as the microbial population present in the effluent treatment processes. Furthermore, heavy metals may limit the use of the effluent for irrigation in agriculture due to its toxicity (Page et al., 1987).

## Conclusion

The results obtained from this study showed that the effluents from the textile processing factories was alkaline and had a high salt concentration.

The amount of nitrogen, phosphorus and heavy metals present in the effluent were significantly higher than the standards given by the federal environmental protection agency (FEPA). The results also showed that the values of the concentrations of solids and of the oxygen demands were quite high. The results of this study indicate that biological treatment may not be feasible for these

effluents and that physicochemical method may be a better alternative.

Based on this study, we call for treatment of all effluents generated by the textile factories in Nigeria before its discharge into a natural water body. These effluents can be treated by chemical methods of coagulation and flocculation.

## REFERENCES

Ademoroti CMA (1996a). "Standard methods for water and effluents Analysis" foludex press Ltd., Ibadan.

Ademoroti CMA (1996b). Environmental Chemistry and Toxicology. Foludex Press Ltd., Ibadan.

APHA (1995). "Standard Method for Examination of Water and Wastewater". 19th Edition. American Public Health Association, Washington D.C.

Asia IO, Ademoroti CMA (2002). "The Application of physiochemical methods in the treatment of Aluminum extrusion sludge" Afr. J. Sci. 3(2): 609- 623.

FEPA (1988). Effluent Discharge Standard, Federal Environmental Protection Agency, Lagos

Henry JG, Heinke G (1989). "Environmental Science and Engineering" Prince Hall, Eaglewood Cliffs, N.J. 07632.Imohimi Ohionia Asia.

Jibrin W (2004). Dilemma of Textile Industries in Nigeria. In 1st Economic Submit, Arewa House Kaduna.

Kaduna (2008). The official website of Kaduna State. Nigeria. www.kaduna-state.com

Manual of Practice No. OM – 1 (1980). "Waste water sampling for process and quality control" Water pollution – control Fed. Alexandria, Va . pp. 895-911.

Metcalf and Eddy Inc. (1991). wastewater engineering treatment, disposal and re-use 3rd edition McGraw Hill New York.

Page AL, Logan TJ, Ryan JA (1987). Land Application of Sludge Lewis Publishers Inc. pp. 5-20.

Priestly AT (1991). "Report on Sewage Sludge Treatment and Disposal-Environmental Programs and Research Needs from an Australian Perspective". CSIRO, Division of chemicals and Polymers pp. 1 – 44.

Quano EAR, Lohani BN, Thanh NC (1978). "Water Pollution Control in Developing Countries" Asian Institute of Technology. p. 567.

Yusuff RO, Sonibare JA (2004). Characterization of Textile Industries Effluents in Kaduna, Nigeria and Pollution Implications. Global Nest: the Int. J. 6 (3): 212-221.

WHO (1971). "International Standards for Drinking Water". 3rd ed., Geneva.

Zhiguo Y, Herwig B, James L, Willy V (2000). 'Reducing the Size of Nitrogen Removal Activated Sludge Plant by Shorting the Retention Time of Inert Solid via Sludge Storage'. Wat. Res. 34(2): 611- 619.

# 25

# Influence of starch addition on properties of methylolated urea/starch copolymer blends for application as a binder in the coating industry

**S. A. Osemeahon[1], O. N. Maitera[1], A. J. Hotton[2] and B. J. Dimas[1]**

[1]Department of chemistry, Modibbo Adama University of Technology, Yola, Nigeria.
[2]National Agency for Food and Drug Administration and Control (NAFDAC) Jalingo, Nigeria.

Urea formaldehyde resin was reactively blended with various concentrations (10 to 70%) of Cassava starch in order to formulate a paint binder for emulsion paint formulation. Some physical properties and formaldehyde emission of the blended resin were investigated. Viscosity initially decreased before a gradual increase was noted with increase in starch concentration. Refractive index and elongation at break increased initially but gradually decreased with Cassava starch content in the blend. Gel time, density, melting point, moisture uptake and formaldehyde emission decreased with increase in starch inclusion. The interaction between the two different polymers shows that 50% starch was the optimal loading inclusion. This new system has advantages of low brittleness, low formaldehyde emission and water reduction characteristics. Therefore the polymer blend can be recommended as binder for coating industry.

**Key words:** Urea formaldehyde, starch, copolymer, binder.

## INTRODUCTION

The development of reliable high performance coating materials with excellent thermal and mechanical properties is the focus of modern technology. Waterborne coatings are finding more application due to increased legislative restrictions on the emission of volatile organic materials to the atmosphere (Motawie et al., 2010). Depending on the type of binder, paints are classified into 2 main categories i.e. oil paint which is oil based and emulsion paint which has synthetic resin as the binder, and is water soluble as- against oil paint which is solvent (organic solvent) soluble. Although oil-based paints display a lot of advantages such as water resistance, durability and flexibility, its major drawback is its use of organic compound as solvent, which is threatened by growing proliferation of VOC regulations imposed worldwide. Water-borne resins are polymeric materials whose composition enables them either to dissolve or to swell in water. Most buildings are protected and decorated using water-borne paints due to their ease of applications, fast drying, non-odour, good wash-ability and finish. Although most household paints are water based, this is not true of industrial paints because of the special requirement of the industrial coatings. Hence satisfactory water based polymers with the required properties have not yet been developed thus leading to emphasis on sustainable commercial production of water-borne paints (Hasmukh and Sumeet, 2010; Motawie et al., 2010; Osemeahon, 2011). Developing nations such as ours will need to put more effort into the development of local technologies for the purpose of achieving the above target.Polymer blends have grown tremendously in leaps and bonds over the past few decades (Oluranti et al., 2011). Blending could be a simple process for developing new composites for coating systems. Polymer blend is one of the most useful approaches to prepare new materials with specially tailored and improved properties that are often absent in a single polymer. The performance of a polymeric material can be improved by selection of suitable ingredients and their ratios, leading to the formation of new materials with enhanced physical, chemical and mechanical properties (Kaniappan and Latha, 2011; Hwang et al., 2012). The synthesis of a new class of urea formaldehyde (UF) resin through a one step process was reported by Osemeahon and Barminas (2007). Although this new class of resin showed much improvement in terms of formaldehyde emission and moisture uptake (water resistant), compared with the traditional urea formaldehyde (UF) resin, its brittleness and hardness remains a source of concern (Osemeahon and Archibong ., 2011 ). Hence the need to modify this new class of urea formaldehyde (UF ) resin in order to address the problem of brittleness and further reduce both formaldehyde emission and moisture uptake. This work seeks solution to the problem, by the copolymerization reaction between urea formaldehyde/Cassava starch (UF/CS). Starch is one of the most abundant natural polymers, used in a wide range of products including binders, sizing materials, glues and pastes. Inexpensive materials such as starch are biodegradable additives, which are appropriate for blending with synthetic polymers (Amine et al., 2010; Borghei et al., 2010). High paste viscosity, clarity and freeze-thaw stability are some of the excellent properties of starch, which are of advantage to many industries. Cassava is a renewable and one of the most abundant substances in nature used for consumption and also as raw materials for industries (Akpa, 2012).

## MATERIALS AND METHODS

Urea formaldehyde, sodium dihydrogen phosphate, sulphuric acid, sodium hydroxide pellets, and sucrose, were reagent grade products from British Drug house (BDH). The materials were used as received. Cassava starch was collected from a farm in Yola, Nigeria.

### Treatment/preparation of starch

The Cassava starch was washed, dried and grounded into powder. It was treated by dispensing it in cold water and filtering. The filtrate was heated using hot plate with occasional stirring (5 min intervals) until colloidal suspension was observed which formed a gel on cooling.

### Method

The method used for the synthesis and determination of the film properties was according to Osemeahon and Archibong (2011).

### Resin synthesis

Urea formaldehyde (UF) resin was prepared by reacting one mole (6.0 g) of urea with three moles (24.3 ml) of 37% (w/v) formaldehyde using 0.2 g of sodium di-hydrogen phosphate as catalyst. The pH of the solution was adjusted to 6 using 0.5 m $H_2SO_4$ and 1.0 m NaOH solution. The reaction was then allowed to proceed for 2 h in a thermostatically control water bath at 70°C, after which the sample was removed and kept at room temperature (30°C).

### Blend/film preparation

Blending of urea formaldehyde resin with Cassava starch was carried out by preparing 10% Cassava starch in urea formaldehyde at room temperature (30°C). The solution was mixed thoroughly using magnetic stirrer. The above procedure was repeated at different Cassava starch concentrations (10, 20, 30, 40, 50 and 70%) and the resulting blends analyzed. Copolymer of the different resins obtained with various Cassava starch concentrations were introduced into a glass petri dish for casting. The resins were then allowed to cure and set for seven days at 30°C and the physical properties of these copolymers were carefully investigated.

### Determination of viscosity and gel time

The viscosity of the UF/CS resin was evaluated in relation to that of the standard sucrose solution at 30°C, using capillary viscometer. Five different readings were taken for each sample and the average value calculated. The gel point of the resin was determined by monitoring the viscosity of the resin with time until a constant viscosity profile was obtained.

### Determination of density, melting point and refractive index

The density of the different resins was determined by taking the weight of a known volume of resin inside a density bottle using metller (Model, AT 400) weighing balance. Five different readings were made for each sample and the average value calculated. The melting point of the different copolymer samples were determined by using Galenkamp melting point apparatus (model MF B 600 to 010F). The different copolymer samples were ground into powder and some quantity of each sample was introduced into different capillary tubes. The melting point was then taken one after the other for all the samples. The refractive indices of the resins samples were determined with Abbe refracttometer. Five readings were taken for each of the parameters and the average value calculated. The properties of the blends were also determined according to standard methods (AOAC, 2000).

### Determination of moisture uptake

The moisture uptake of the resin film was determined. Gravimetrically known weights of the sample were introduced into a desiccator containing a saturated solution of calcium chloride. The

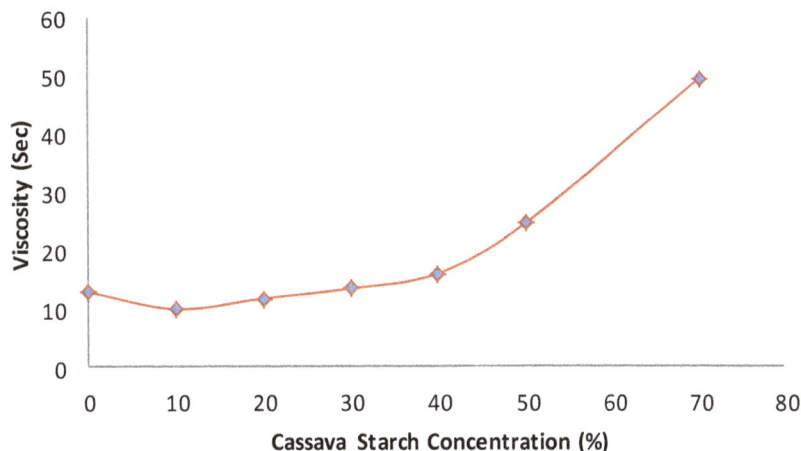

**Figure 1.** Effect of Cassava starch concentration on viscosity of methylol urea resin.

increase in weight (wet weight) of each sample was monitored until a constant weight was obtained. The difference between the wet weight and dry weight of each sample was then recorded as the moisture intake by resin. Triplicate determinations were made for each sample and the average value recorded.

### Determination of formaldehyde emission

Formaldehyde emission test was performed by using the standard 2 h desiccator method. The evaluation of the absorbed formaldehyde by the 25.0 ml water was obtained from standard calibration curves derived from refractometric technique using Abbe refractometer. In brief, the prepared resin was aged for 2 days. At the end of this period, the resin was poured into a mold made from aluminium foil with a dimension of 69.6 × 126.5 mm and thickness of 1.2 mm. The mold and its content was then allowed to equilibrate for 24 h in the laboratory after which it was then placed inside a desiccator along with 25 ml of water, which absorbed the formaldehyde emitted. The set up was allowed to stay for 2 h after which the 25 ml water was removed and analyzed for formaldehyde Content. Triplicate determinations were made for each sample and mean value recorded.

### Elongation at break

Elongation at break was measured, using Inston testing machine (model 1026). Resin films of known dimension 50 mm long, 10 mm wide and 0.15 mm thick were brought to rapture at a clamp rate of 20 mm/min and a full load of 20 kg. A number of five runs were done for the sample and the average elongation evaluated and expressed as the percentage increase in length.

## RESULT AND DISCUSSION

### Viscosity and gel point

The viscosity of a substance (liquid, gas or fluid) is its resistance to flow. Studying rheological properties of

fluids and gels are very important, since operation processes design depends on the way the product flows through a pipe, stirring in a mixer and packaging into containers. Emulsion lattices has many sensory attributes which are related to their rheological properties, example; creaminess, thickness, smoothness, spread ability, flow ability, brittleness and hardness (Hussain and Nasr, 2010; Akpa, 2012). Due to the presence of functional groups in the polymeric backbone, inter-polymeric specific interactions have long been known to result in unusual behavior and material properties that are dramatically different from those of the nonfunctional polymers. These interactions include ion-ion coulombic interaction, hydrogen bonding and transition metal complexation of the component polymer chains, resulting in solution viscosity variation. Rheological properties such as the viscosity can be directly correlated to the evolving physical and mechanical properties during resin cure (Derkyi et al., 2008; Osemeahon, 2011). Figure 1 shows the effect of Cassava starch on the viscosity of urea formaldehyde resin. It can be observed that at low concentration of 10% Cassava, the viscosity decreased slightly and then increased with increase in Cassava starch concentration. This phenomenon can be explained in terms of specific interactions between urea formaldehyde and Cassava starch. In a dilute system, there are strong specific interactions and the complexes are isolated from each other with the formation of compact structure, which reduced the viscosity of the blend solution. However, as the blend concentration increases, the isolated complexes combined and lead to the formation of a gel-like intermolecular complex structure leading to increase in viscosity of the copolymer blend system. Since the variation is linear, it indicates the miscibility of polymer blend (Osemeahon and Barminas, 2007; Reddy et al., 2008). Viscosity also increases with increase in solid content as a consequence of higher

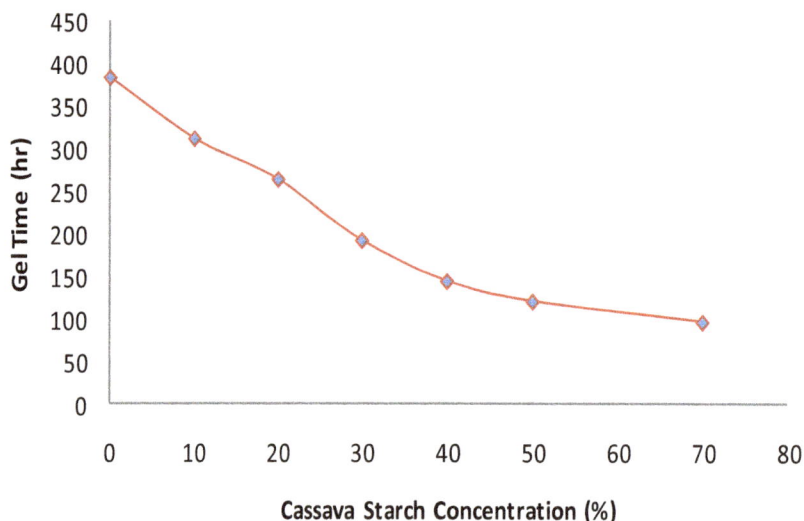

**Figure 2.** Effect of Cassava starch concentration on gel time of methylol urea resin

surface interactions among the particles. Concentrated polymer shows a great deal of interaction between the macromolecule. The higher the concentration, the higher the viscosity observed (Palma, 2007; Taghizadeh and Toroutan, 2005).

The gel time or pot life is the maximum length of time the system remains in sufficiently fluid condition to be applied to a substrate. The dry time of any paint is a function of its binders gel time among other factors. On the technical front, gel time enable paint formulator to ascertain the optimum storage period of a binder before its utilization for paint formulation. The gel time is an important kinetic characteristics of curing because it describes the attainment of certain critical conversion responsible for the transition from liquid to solid state of the curing process .The gel point is characterized by the appearance in the reactive system of macro molecule with an infinitely large molecular weight (Desai et al., 2003 ; Derkyi et al., 2008). Figure 2 shows the effect of Cassava starch concentration on gel time. It can be observed that the gel time decreases with increase in Cassava starch concentration. The reaction between monomers leads to the formation of network, hence gelation. Both molecular weight and poly-dispersity increased until one single macro molecule is formed. At this point, the behavior of the system changes from liquid-like to rubber-like thus the reactive system becomes a gel (Gonzalez et al., 2012).

## Density

Density is a physical property of matter that expresses a ratio of mass to volume. It is very useful for identification and characterization of substances. The density of a

paint binder in the coating industry had a profound influence on factors such as pigment dispersion, brushability of paint, flow, leveling and sagging (Kazys and Rekuviene , 2011 ; Osemeahon and Archibong, 2011). Figure 3 shows the effect of Cassava starch on the density of urea formaldehyde resin. The gradual decrease observed in density with increase in Cassava starch concentration can be as a result of differences in the molecular features and morphology which influenced the packing nature of resin molecules as the concentration of starch increases (Osemeahon and Barminas, 2007). This result is similar to the findings of Barminas and Osemeahon, (2007) when natural rubber was blended with methylol urea resin.

## Melting point

The melting point of a polymer has a direct bearing to its thermal property. Melting point of polymer varies depending on molar mass, intermolecular van der waal interactions and intrinsic structures that affects the rigidity. In the case of coating industry, the melting point of a binder is related to its thermal resistance as well as its brittleness. Urea formaldehyde resin is known to compose of molecules that crosslink into clear hard plastics (Afsoon et al., 2011; Osemeahon et al., 2010). Figure 4 exhibits the effect of Cassava starch on the melting point of methylol urea resin where the melting point decreased slightly at the beginning up to 10% Cassava starch. Thereafter, a sharp decrease in melting point was observed. The melting points of the two monomers are very different which probably resulted from different contributing factors. At a certain domain of concentrations of the two copolymers, their melting point

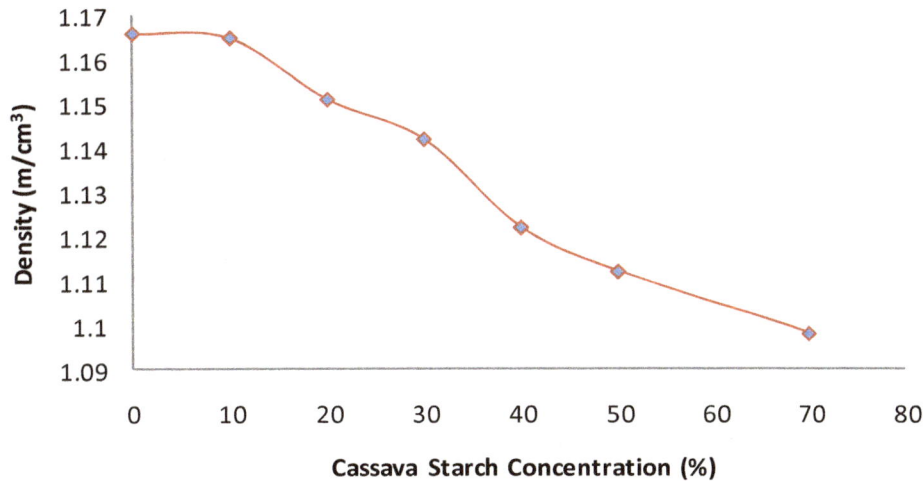

**Figure 3.** Effect of Cassava starch concentration on the density of methylol urea resin.

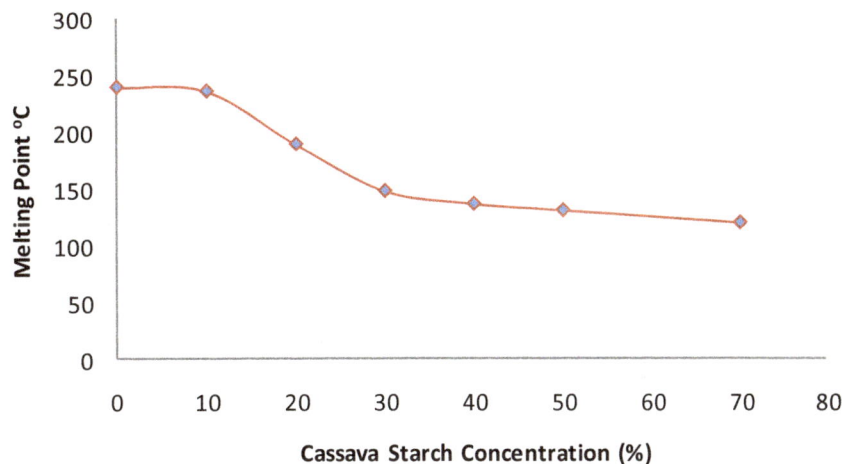

**Figure 4.** Effect of Cassava starch on the melting point of methylol urea resin.

decreased even below the melting point of each monomer. The melting point of the blended films shifted due to interaction of the two biopolymers as influence by the lower melting temperature characteristic of Cassava starch (Afsoon et al., 2011; Wirongong et al., 2011).

### Refractive index

Gloss is a measure of the ability of the coated surface to reflect light. Reflection of light from surfaces can be classified according to the diffuse component or the specular component, which is expressed as a function of the incidence angle and refractive index of the material, the surface roughness and a geometrical shadowing function. Gloss is a necessary coating property when the purpose is for surface aesthetic or decoration (Kaygin and Akgun, 2009: Yumiko et al., 2010). Figure 5 presents

the effect of Cassava starch on the refractive index of urea formaldehyde resin. Initially, the refractive index increased from 0 to 20% Cassava starch inclusion after which a gradual decrease in refraction was observed with increase in Cassava starch concentration. This result is due to differences in the level of specific interaction between the two polymers resulting in molecular weight, molecular features and molecular orientations depending on morphology and crosslink density (Qi et al., 2002). This suggests that from 0 to 20% Cassava starch inclusion, the gloss property especially on a smooth surface increases after which a decrease is observed.

### Moisture uptake

Polymeric binders play a major role in moisture transport properties of paint because it is one of the major

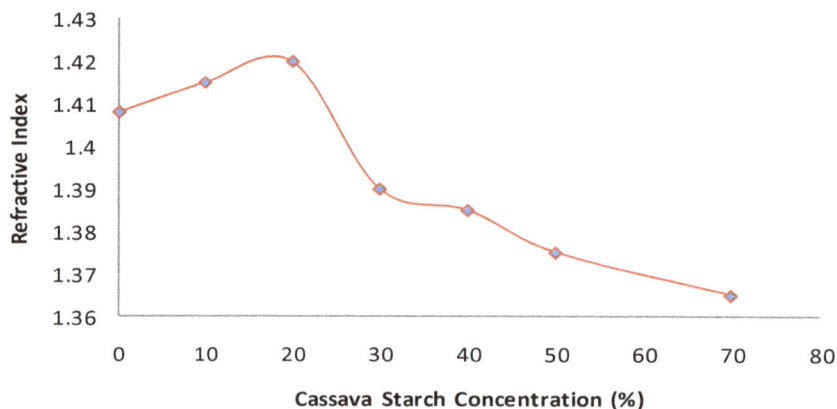

**Figure 5.** Effect on Cassava starch concentration on the refractive index of methylol urea.

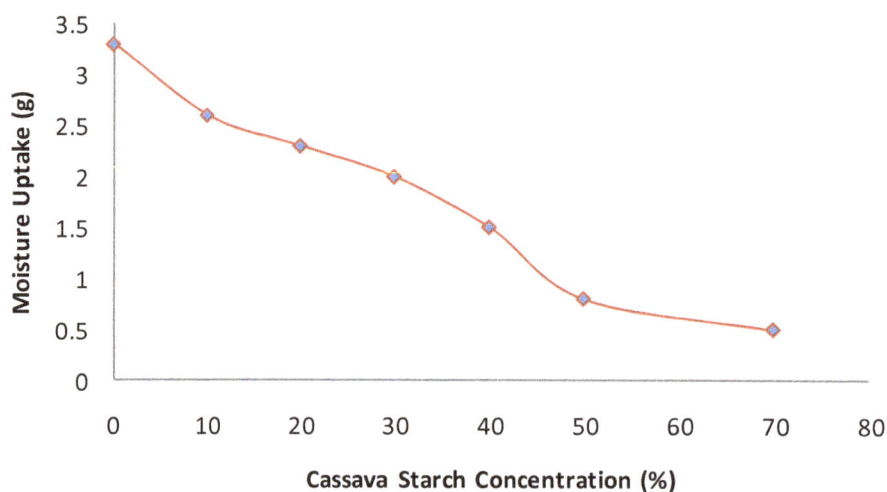

**Figure 6.** Effect of Cassava starch on the moisture uptake of methylol urea resin.

constituent. Susceptibility to durability issues pertaining poor water resistance is one of the drawback of water borne resin. The functional groups on polymers that are used can undergo hydrogen bonding or ionic. A detailed knowledge of moisture transport is essential for understanding the resistance of a material against attacks from its environment (Emile, 2003; Bharath and Swamy, 2009). Figure 6 shows the effect of Cassava starch on the moisture uptake of methylol urea resin. It can be observed that moisture uptake drastically dropped at the beginning and decreased steadily after 10% blend until after 40% blend. The different levels of interactions gave rise to polymers with different morphology and crosslink density. From 0 to 10% blend, the molecular size holes in the copolymer structure were rapidly reduced. After this period, the size of the molecular size holes might have slowly decreased with increase in Cassava starch loading; hence the steady reduction in

moisture uptake. After the steady decrease in moisture uptake between 10 to 40% blend, a sharp decrease in moisture uptake was noted. This might be as a result of the drastic decrease in molecular size hole with increase in Cassava starch inclusion. Blending improves the resistance to moisture susceptibility of mixtures as also seen in waste plastic coating of aggregates (Bindu and Beena, 2010).

## Formaldehyde emission

A serious drawback of urea formaldehyde resin is the emission of the hazardous formaldehyde during cure. The issue of formaldehyde exposure in homes is long standing and has been studied overtime. Hydrolysis of cured urea resins has been known to be responsible for

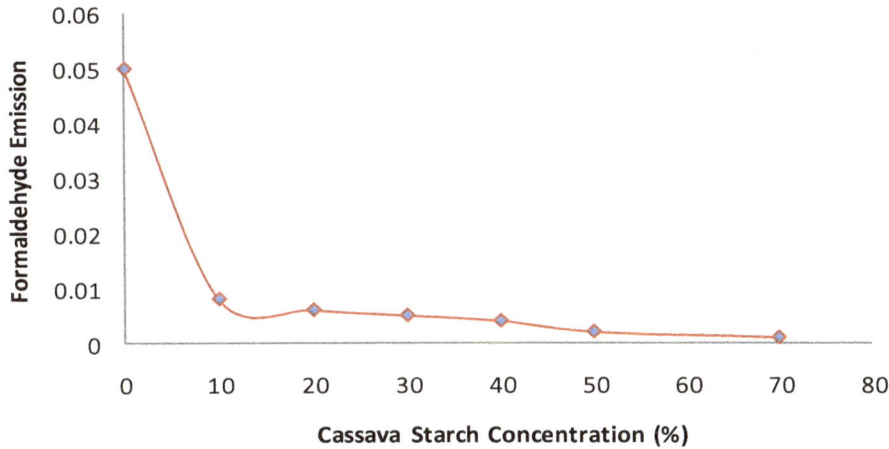

**Figure 7.** Effect of Cassava starch on the formaldehyde emission of methylol urea resin.

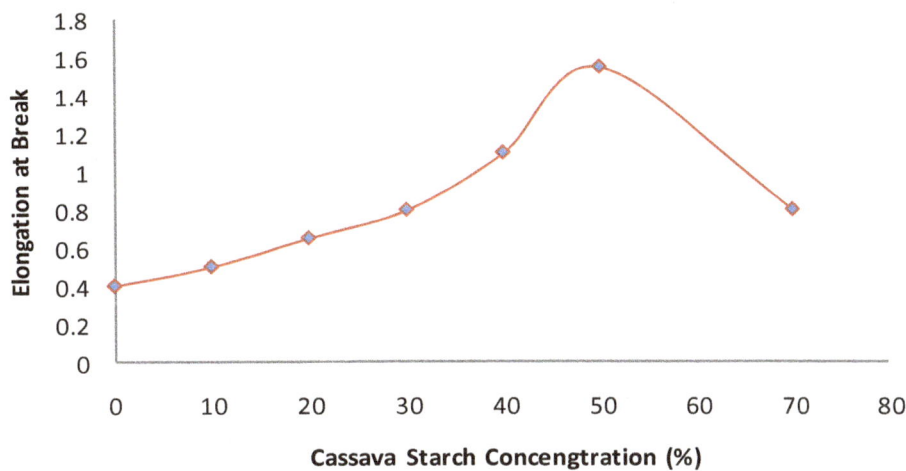

**Figure 8.** Effect of Cassava starch on the elongation at break of methylol urea resin.

formaldehyde emission leading to sick building syndrome. It is therefore important to determine the formaldehyde emission from synthesized urea formaldehyde resin (Derkyi et al., 2008; Park et al., 2010). It is observed in figure 7 that the level of emission decreased with increase in the concentration of Cassava starch in the blend. This important promising trend is due to the gradual decrease in the methylol urea content with increase in the Cassava starch content in the blend. The blending process has succeeded in reducing the emission lower than the permissible level of 0.1 ppm as stated by Pizza et al., (2002). This agrees with the findings of the copolymerization of urea formaldehyde with acrylamide (Abdullah et al., 2010).

## Elongation at break

The structure of thermosetting resins also leads to some unwanted mechanical properties that limit their structural applications. They are relatively brittle. Elongation at break can be a propel tool to determine the adhesion between phases because of its sensitivity for load transfer between phases (Cakir et al., 2012; Hwang et al., 2012). Figure 8 shows the effect of Cassava starch on the elongation at break of methylol urea resin. A steady increase in elongation at break was observed from 0 to 40% Cassava starch and an abrupt increase at 50% due to compatibilization effect and improved interfacial adhesion, hence flexibility is improved and brittleness reduced. Then a decrease was observed after 50%. This non-linear behavior in terms of the mechanical property is caused by the difference in intimate physical interlocking, and the extent of phase separation (Hwang et al., 2012) The result is in consonance with the report of Cardona and Moscou (2010) where resole phenolic resins were modified by forming copolymer with cardanol (main component of cashew nut shell liquid). The presence of cardinol decreases the crosslink density and makes the resole phenolic resins less brittle.

## Conclusion

The analyzed results obtained from the sample analysis showed that the interaction between methylol urea and Cassava starch has remarkable synergistic properties with 50% of Cassava Starch being the optimal inclusion value. The values of moisture uptake, formaldehyde emission and melting point obtained from this experiment are within the acceptable levels required in the coating industry. These results present Cassava starch modified methylol urea as a resin with relatively very low moisture uptake and formaldehyde emission against the traditional hard and brittle resin. The increase in elongation at break and reduction in density is also a positive value to the coating industry. Therefore, this work has shown that biodegradable starch films could be used to produce copolymer composite binder for the coating industry especially in the formulation of emulsion paints.

## Conflict of Interest

The authors have not declared any conflict of interest.

### REFERENCES

Abdullah ZA, Park BD (2010). Influence of acrylamide copolymerization of urea-formaldehyde resin adhesives to their chemical structure and performance. J. Appl. Polym. Sci. 117(6):3181-3186.

Afsoon F, Laleh R, Faramarz A (2011). DSC analysis of thermosetting polyimide based on three bismaleimide resin eutectic mixtures. Iranian Polym. J. 20(2):161-171.

Akpa JG (2012). Production of Cassava Starch-based Adhesives. Research J. Eng. Appl. Sci. 1 (4):219-214.

Amine M, Ahmed A, Antonio P, Fatima C, Bertrand C (2010). Preparation and Mechanical Characterization of Particle board made from Maritime Pine and glued with Bio-adhesive based on Cornstarch and Tannins. Moderas Ciencia Y Technol.12 (3):189-197.

AOAC (2000). Official method of analysis international, (Horwitz W. Editor). Gaithersburg Mongland, USA, 17th Edition, 1(41):1-68.

Barminas JT, Osemeahon SA (2007). Study of a composite from reactive blending of methylol urea resin with natural rubber. Afr. J. Biotechnol. 6(6):810-817.

Bharath NK, Swamy PR (2009). Adhesive tensile and moisture absorption characteristics of natural fibers' reinforced urea formaldehyde composites. Int. J. recent Trends Eng. 1(5):60-62.

Bindu CS, KS Beena (2010). Waste plastic as a stabilizing additive in stone mastic asphalt. Inter. J. Eng. Tech. 2(6):379-387.

Borghei M, Abdolreza K, Sharhrzad K, Abdolrasoul O, Amir H (2010). Microbial biodegradable potato starch base low density polyethylene. Afr. J. Biotechnol. 9(26):4075-4080.

Cakir M, Kartal I, Demirer H, Samur R (2012). Effect of water absorption on the wear behaviour of sol-gel processed epoxy/silica hybrids. Sci. Res. Essays. 7(7):805-812.

Derkyi NS, Darkwa NA, Yartey JG (2008). Effect of Cassava flour as urea formaldehyde adhesive extender on the bonding strength of plywood. Ghana J. For. 23(24):25-34.

Desai DS, Patel VJ, Sinha KV (2003). Polyurethane adhesive system from biomaterial-based polyol for bonding wood. Int. J. Adhesion Adhes. 23:393-399.

Emile G (2003). Moisture transfer properties of coated gypsum. Eindhoven university press, Eindhoven, Netherlands, pp. 2-6.

Gonzalez GM, Cabanelas JC, Baselga J (2013). Application of FTIR on epoxy resins-Identification, monitoring the curing process, phase separation and water uptake.Infrared Specroscopy-Material Science, Eng. Technol. pp. 261-283.

Hasmukh SP, Sumeet JP (2010). Novel surface coating system based on maleated shellac. E–J. Chem. 7(S1):S55-S60.

Hussain AI, Nasr HE (2010). The role of carboxylic acid on the characterization and evaluation seed emulsion of styrene/butyl acrylate copolymers lattices as paint. Nat. Sci. 8(8):94-103.

Hwang Y, Sangmook L, Youngjae Y, Kwangho J, Lee W (2012). Reactive Extrusion of Polypropylene/Polystyrene Blends with Supercritical CarbonDioxide. Macromol. Res.20 (6):559-567.

Kaniappan K, Latha S (2011). Certain Investigations on the formulation and characterization of polystyrene/poly (methyl methacrylate) blends. Int. J. Chem. Tech. Res. 3(2):708-717.

Kaygin B, Akgun E (2009). A nano-technological product: An innovative varnish type for wooden surfaces. Sci. Res. Essay. 4(1):001-007.

Motawie AM Sherif MH, Badr MM, Amer AA, Shehat AS (2010). Synthesis and characterization of waterborne epoxy resins for coating application. Austr. J. Basic Appl. Sci. 4(6):1376-1382.

Oluranti SA, Emmauel RS, Adesola TA, Olusesan FB (2011). Rheological properties of polymers: structure and morphology of molten polymer blends. Mater. Sci. Appl. 2:30-41.

Osemeahon SA, Barminas JT (2007). Development of amino resin for emulsion paint formulation: Reactive blending of methylol urea with soybean oil. Afr. J. Biotechnol. 6(6):803-809.

Osemeahon SA, Nkafamiya II, Milam C, Modibbo UU (2010). Utilization of amino resin for emulsion paint formulation; Effect of urea formaldehyde viscosity on urea and soybean oil copolymer composite. Afr. J. Pure Appl. Chem. 4(1):001-006.

Osemeahon SA, Archibong CA (2011). Development of urea formaldehyde and polyethylene waste as a copolymer binder for emulsion paint formulation. J. Toxicol. Environ. Health Sci. 3(4):101-108.

Osemeahon SA (2011). Copolymerization of methylol urea with ethylol urea resin for emulsion paint formulation. Afri. J. Pure Appl. Chem. 5(7)204-211.

Osemeahon SA, Ilesamin JO, Aliyu BA, Mishelia I (2011). Copolymerization of methylol urea with vegetable oil; Effect of using different types of vegetable oil on some physical properties of the copolymer composite. Int. J. Phy. Sci. 6(15):3629-3635.

Palma MSA (2007). Effect of monomer feed rate on the properties of the copolymer butyl acrylate/vinyl acetate in semi-batch emulsion polymerization. Indian J. Chem. Tech. 14:1-7.

Park B-D, Jeong H-W, Lee S-M (2010). Morphology and chemical elements detection of cured urea formaldehyde resins. J. Appl. Polym. Sci., 120 (3):1475-1482.

Pizza A, Beaujean M, Zhao C, Properzi M, Huang Z (2002). Acetal-induced strength increases and lower resin content of MUF and other polycondensates adhesives. J. Appl. Sci. 84:2561-2571.

Qi GR, Wang YH, Li XX, Peng HY, Yang SL (2002). Viscometric study on the specific interaction between proton-donating polymers and proton-accepting polymers. J. Appl. Polym. Sci. 85:415-421.

Reddy MM, John K, Naidu VS (2008). Study of viscosity & refractive index of nylon 6, 6 and Poly (methyl methacrylate) in formic acid. Indian J. Pure Appl. Phys. 46:209-211.

Rekuviene R, Kazys R (2011). Viscosity and density measurement methods for polymer melts. Ultragarsas (ultra sound), 66(4):20-25.

Taghizadeh MT, Foroutan M (2005). Hydrophobically Associated Polymer, Viscosity, R I, Critical Concentration, Vinylpyrrolidone-Vinyl acetate Copolymer. Iranian Polym. J. 14(1):47-54.

Wirongong T, Lisa M, Sasitorn W, Pensiri S, Pornchai R (2011). Effect of carboxymethyl cellulose concentration on physical properties of biodegradable Cassava starch based films. Chem. Cent. J. 5:6.

Yumiko H, Takanobu S, Tetsuo O, Toshiaki O, Masashi M , John M (2010). Effects of specular component and polishing on color resin composites. J. Oral Sci. 52(4):599-607.

# 26

# Identification and quantification of heavy metals in local drinks in Northern Zone of Nigeria

**Bakare-Odunola M. T.**[1] **and Mustapha K. B.**[2]

[1]Department of Pharmaceutical and Medicinal Chemistry, Faculty of Pharmaceutical Sciences, University of Ilorin, Ilorin, Kwara State, Nigeria.
[2]Department of Medicinal Chemistry and Quality Control, National Institute for Pharmaceutical Research and Development, Abuja, Nigeria.

**Nine heavy metals were studied in locally prepared drinks, namely, "Zobo" and "Kunnu Zaki". The samples were prepared from the outer covering (calyx) of the fruits of roselle *Hibiscus sabdariffa* and cereals (millet or guinea corn), respectively. Twenty samples of "Zobo" coded $Zb_1$-$Zb_{20}$ and twenty samples of "Kunnu Zaki" coded $Kz_1$-$Kz_{20}$ were bought from different parts of Samaru-Zaria, Nigeria. The qualitative analysis of the samples was by official methods and were quantitatively analyzed using Atomic Absorption Spectrophotometer (AAS). Iron (Fe), Copper (Cu), Zinc (Zn) and Lead (Pb) were detected in most Zb and Kz samples, while Chromium (Cr), Manganese (Mn), Silver (Ag), Mercury (Hg) and Bismuth (Bi) were absent in all the samples. The Fe values ranged from 3.13 to 5.48 mg/L; Cu ranged from 0.12 to 0.62 mg/L, Zn ranged from 0.02 to 0.22 mg/L and Pb ranged from 0.54 to 1.28 mg/L in Zb samples. The Fe values for Kz samples ranged from 18.63 to 31.25 mg/L, Cu ranged from 0.03 to 0.11 mg/L, Zn ranged from 0.08 to 0.39 mg/L and Pb ranged from 0.80 to 1.55 mg/L. The higher values detected in Fe, Zn and Pb for Kz samples compared with Zb samples could be due to the different materials used in their preparation. The implication of the results in public health is discussed.**

**Key words:** Cereals, drinks, heavy metals, samples, *Hibiscus sabdariffa*.

## INTRODUCTION

Metals are elements that cannot be decomposed to simpler units by chemical means. They are shinny, ductile, malleable and usually good conductor of heat, and electricity. Metals have high densities, high melting points, high molar heat of fusion and evaporation (Graham and John, 1978).

Local drinks are non-alcoholic drinks produced and consumed within a locality. Usually, they contain sweetening, flavouring and other naturally occurring or locally obtained ingredients. In Nigeria, the common local drinks are "Zobo" (Zb), "Kunnu Zaki" (Kz). Zb is a brick red non-alcoholic drink prepared by boiling the dried calyx of *Hibiscus sabdariffa* in water. The extract is filtered and more water can be added. The extract can then be sweetened with sugar after adding the flavouring agent. *H. sabdariffa* (*Roselle*) is an annual herb native to tropical Africa reaching up to 2 m. The dried calyces contain the flavonoids gossypetine, hibiscetine and sab-daretine. Small amount of delphinidin 3-monoglucoside, cyanidin 3-monoglucoside (chrysanthenin), and delphinidin are also present (Bernd and Franz, 1990). Juice made by cooking a quantity of calyces with water is used as a cold drink in the West Indies, tropical America, Jamaica, Mexico, Egypt and Nigeria as "Zobo" (Kolawole and Maduenyi, 2004). Kz is prepared from cereals (millet or guinea corn). Other ingredients used in its preparation are ginger, black pepper and sweet potatoes or sugar. It can be prepared by soaking the cereal in water for about twelve hours to get soft. The ginger and the black pepper together with the sweet potatoes are ground together, while the soaked cereal is ground separately using a grinding machine. Hot boiling water would be used to make a thick paste with the ground cereals. The ground ginger, pepper and potatoes mixture would be diluted very well with cold water and mixed with the thick paste with vigorous stirring. The mixture would be allowed for about six hours in which the chaffs from the cereals and the ground ingredient would have settled down. The mixture is filtered with fine sieve. The resulting mixture is colloidal in nature, made up of fine starch granules suspended in a sugar solution that settles on standing. The sweet potatoes can serve as a sweetening agent, but granulated sugar can be added to the desired taste. The ginger and black pepper can be used as flavouring agents.

The massive consumption of these drinks could be due to the poor economic state of the country, the nutritive and medicinal values. The production of these drinks in most cases goes through non-hygienic conditions. It lacks uniformity, specificity of source of water, purification of the ingredient and there is no specification of pack-aging materials and place of production; these are the sources of impurities in the finished product resulting in hazardous effect on the health and total well being of the consumers.

Heavy metals present in the body are of great danger particularly when present at a concentration above the tolerance limit (Nriagn, 1988). Most heavy metals are regarded as toxic to living organisms, because of their tendency to accumulate in selected tissues. More over their presence is a causative agent of various sorts of disorder including neuro-, nepro-, carcino, terato- and immunological (Zukowska and Biziuk, 2005). Accumulation of non essential investigated heavy metals (e.g. Pb, Cd, and Cr) in the environment could be useful indicators of the possible toxic effect for the consumers (Liu, 2003; Tasi, 2005).

Acute toxicity was observed in patient with renal failure following hemodialysis with water stored in Zinc galva-nized tank. The patient suffered nausea, vomiting, fever and severe anemia (Galley et al., 1972). Lead is toxic to

such as the nervous, gastrointestinal and genital system (Abou-Arab, 2001) and also a possible human carcinogen (Yakasai et al., 2004) and the accumulation of Mn may cause hepatic encephalopathy (Layrangues et al., 1998). Outbreak of "Minamata" disease caused by Mercury poisoning has been reported in Iraq and Canada. In Iraq, 7.2% of 6,350 people hospitalized died (Masazum and Smith, 1975). Péter et al. (2012) reported the accumulation of some heavy metals in milk of grazing sheep in North-East Hungary. Due to the health hazards caused by these toxic metals which might be present in the raw materials or water used in the preparation of drinks. The present study was carried out to investigate the presence of metallic impurities and to determine the quantity in these local drinks.

## MATERIALS AND METHODS

Measurement were made with a Buck Model 210 Variant Giant Pulses Correction (VGP) system Atomic Absorption Spectrophotometer (AAS) equipped with the corresponding hollow cathode lamp (Lead, Copper, Iron and Zinc) at the time of analysis. Lamp current 10 mA, wavelength 217.0 nm, band pass 0.5 nm with flame type consisting of air/acetylene and stoichiometric fuel flow at 0.9 to $1.21 \text{ min}^{-1}$.

### Samples

The samples were bought from Samaru-Zaria, Nigeria and were coded $Zb_1$-$Zb_{20}$, (Zobo), $Kz_1$-$Kz_{20}$ (Kunnu Zaki); they were refrigerated until the time of use.

### Reagent

All the reagents were analytical grade from British Drug House (BDH).

Official methods were used for the identification of the metals (USPXX, 1990). Stock solution of each metal was prepared as follows: Iron (Fe), 1.00 g of iron powder was dissolved in 40 ml of 2 M hydrochloric acid and 10 ml of 2 M nitric acid; this was made up to 1 L in volumetric flask with deionized water to give 1000 mg/L Fe solution; Lead (Pb), 1.598 g lead nitrate was dissolved in 50 ml nitric acid and the solution was made to volume in 1 L volumetric flask with deionized water to make 1000 mg/L Pb solution; copper (Cu), 1.00 g of copper metal was dissolved in 50 ml nitric acid and was diluted to mark in 1 L volumetric flask with deionized water to 1000 mg/L Cu solution; Zinc (Zn), 1.245 g of zinc oxide was dissolved in 50 ml of 2 M hydrochloric acid. This was diluted to mark in 1 L volumetric flask with deionized water to 1000 mg/L Zn solution.

### Preparation of calibration curve

Standard solutions were prepared from each metal stock solution of 1000 mg/L by further dilution using deionized water. Working standard solutions of Fe ranged from 2.00 to 10.00 mg/L for Zb samples and from 10.00 to 50.00 mg/L for Kz samples. Cu ranged from 0.05 to 0.80 mg/L for Zb samples and from 0.02 to 0.10 for Kz samples. Zn ranged from 0.02 to 0.40 mg/L for both samples. Pb ranged from 0.50 to 2.00 mg/L for both samples. 100 ml of the

**Table 1.** Qualitative analysis of heavy metals impurities in twenty "Zobo" (Zb1-Zb20) samples.

| Sample code | Zinc (Zn) | Copper (Cu) | Iron (Fe) | Lead (Pb) |
|---|---|---|---|---|
| Zb1 | A | A | P | P |
| Zb2 | P | P | P | P |
| Zb3 | P | P | P | P |
| Zb4 | P | P | P | P |
| Zb5 | P | P | P | P |
| Zb6 | P | P | P | P |
| Zb7 | P | P | P | P |
| Zb8 | P | P | P | P |
| Zb9 | P | P | P | P |
| Zb10 | P | P | P | P |
| Zb11 | A | A | P | P |
| Zb12 | P | P | A | P |
| Zb13 | P | P | A | P |
| Zb14 | P | P | P | A |
| Zb15 | P | P | P | P |
| Zb16 | P | P | P | P |
| Zb17 | P | P | A | P |
| Zb18 | P | P | P | P |
| Zb19 | A | P | A | P |
| Zb20 | P | A | A | P |

P: Present; A: Absent.

standard solution of each metal was adjusted to pH of 2.5 by adding 1 M trioxonitrate (v) acid. Each standard solution and blank was transferred into an individual 250 ml separating funnel. One milliliter ammonium pyrrolidine dithriocarbamate was added followed by the addition of 10 ml methyl isobutyl ketone and the solution was shaken vigorously for 2 min and allowed to settle. The aqueous layer was drained off and discarded, while the organic layer was then aspirated directly into the flame (zeroing the instrument on methyl isobutyl ketone) and the absorbance was recorded. The nebulizer, atomizer and burner were flushed each time with distilled water after each sample solution was aspirated before the next. The stability of the instrument was checked at intervals by introducing the highest working standard solution and the blank.

### Pre-treatment of the samples

100 ml of each of the samples were measured into series of weighed platinum crucibles and labeled accordingly. The platinum crucibles were then placed on series of hot plates for about three hours to evaporate to dryness with low heat. The dried crucibles were then cooled in a desiccator and the weight recorded. The differences in the weight were recorded and the residue removed from the crucible. Method digestion in mixture acids was employed using nitric acid, perchloric acid and hydrofluoric acid mixture. 0.2 g of each pre-treated sample was treated with 5 ml of deionized water to dampen the sample; 6 ml of concentrated nitric acid were then added, followed by 1 ml of perchloric acid and heated on a water bath to the appearance of white fumes. 5 ml of hydrofluoric acid was added after cooling and the resulting mixture boiled for 10 min. This was filtered and made up to mark with deionized water in a 100 ml volumetric flask. The sample solutions were then analyzed as described under preparation for calibration curve. The

concentration of each metal from sample was determined from the calibration curve.

## RESULTS AND DISCUSSION

The result of qualitative tests using official methods showed that Iron (Fe), Copper (Cu) and Zinc (Zn) and Lead (Pb) were present in most Zb and Kz (Tables 1 and 2), respectively. Chromium (Cr), Manganese (Mn), Silver (Ag), Mercury (Hg) and Bismuth (Bi) were absent in all the samples.

The result of the quantitative tests for Zb and KZ in Tables 3 and 4 show that higher values of Fe, Zn and Pb were detected in Kz compared with Zb. However, the value of Cu detected in Zb was more than that of Kz. This could be due to the different material used in their preparation. The values of Fe and Pb were both high and not within the tolerance limits of metals set by World Health Organization (WHO, 1996) (Table 5). Although, Iron performs important roles in the body but when in excess, especially the ferric salt, produces irritation of the gastro-intestinal tract which is characterized by abdominal pain and diarrhea most especially when on empty stomach. Lead can be described as an element that is purely toxic. Some elements, although toxic at high levels, are actually required nutrients at lower levels. This is clearly not the case for lead. No nutritional value or positive biological effect has been shown to result from lead exposure. Also,

**Table 2.** Qualitative analysis of heavy metals impurities in twenty "Kunnu Zaki" (KZ1-KZ20) samples.

| Sample code | Zinc (Zn) | Copper (Cu) | Iron (Fe) | Lead (Pb) |
|:---:|:---:|:---:|:---:|:---:|
| KZ1 | P | P | P | P |
| KZ2 | A | P | P | P |
| KZ3 | P | A | P | P |
| KZ4 | P | P | P | P |
| KZ5 | P | P | P | P |
| KZ6 | P | P | P | P |
| KZ7 | P | P | P | P |
| KZ8 | P | P | P | P |
| KZ9 | P | P | P | P |
| KZ10 | A | A | P | P |
| KZ11 | A | P | P | P |
| KZ12 | P | P | A | P |
| KZ13 | P | P | A | P |
| KZ14 | P | P | P | A |
| KZ15 | P | A | P | P |
| KZ16 | P | A | P | P |
| KZ17 | A | P | A | P |
| KZ18 | P | A | P | P |
| KZ19 | P | P | A | P |
| KZ20 | P | A | A | P |

P: Present; A: Absent.

**Table 3.** The quantitative analysis of 20 samples Zobo (Zb1 – Zb20) concentration (mg/L).

| Sample code | Iron (Fe) | Lead (Pb) | Zinc (Zn) | Copper (Cu) |
|:---:|:---:|:---:|:---:|:---:|
| Zb1 | 5.48 | 0.54 | N/D | N/D |
| Zb2 | 3.90 | 0.74 | 0.08 | 0.18 |
| Zb3 | 5.48 | 0.80 | 0.06 | 0.22 |
| Zb4 | 4.68 | 0.90 | 0.03 | 0.18 |
| Zb5 | 5.47 | 0.74 | 0.19 | 0.15 |
| Zb6 | 4.68 | 0.80 | 0.20 | 0.18 |
| Zb7 | 3.90 | 1.06 | 0.22 | 0.25 |
| Zb8 | 3.90 | 0.54 | 0.05 | 0.22 |
| Zb9 | 5.48 | 0.96 | 0.08 | 0.12 |
| Zb10 | 4.68 | 1.12 | 0.09 | 0.15 |
| Zb11 | 5.48 | 1.20 | N/D | N/D |
| Zb12 | N/D | 0.96 | 0.06 | 0.15 |
| Zb13 | N/D | 1.12 | 0.03 | 0.46 |
| Zb14 | 4.68 | N/D | 0.02 | 0.56 |
| Zb15 | 5.48 | 0.80 | 0.06 | 0.62 |
| Zb16 | 3.13 | 0.84 | 0.08 | 0.15 |
| Zb17 | N/D | N/D | 0.09 | 0.25 |
| Zb18 | 3.90 | 0.80 | 0.11 | 0.18 |
| Zb19 | N/D | 0.74 | N/D | 0.22 |
| Zb20 | N/D | 1.28 | 0.06 | N/D |

N/D: Not detected.

**Table 4.** The quantitative analysis of 20 samples Kunnu Zaki (Kz1-Kz20) concentration (mg/L).

| Sample code | Iron (Fe) | Lead (Pb) | Zinc (Zn) | Copper (Cu) |
|---|---|---|---|---|
| Kz1 | 19.40 | 0.96 | 0.12 | 0.09 |
| Kz2 | 21.75 | 27 | N/D | 0.06 |
| Kz3 | 19.40 | 1.01 | 0.14 | N/D |
| Kz4 | 19.40 | 1.12 | 0.11 | 0.03 |
| Kz5 | 20.20 | 1.33 | 0.15 | 0.06 |
| Kz6 | 23.50 | 0.96 | 0.19 | 0.03 |
| Kz7 | 25.05 | 0.90 | 0.12 | 0.11 |
| Kz8 | N/D | N/D | 0.14 | 0.09 |
| Kz9 | N/D | 0.85 | 0.31 | 0.09 |
| Kz10 | 24.25 | 0.80 | N/D | N/D |
| Kz11 | 21.75 | 1.01 | N/D | 0.11 |
| Kz12 | 25.83 | 1.07 | 0.39 | 0.11 |
| Kz13 | N/D | 0.96 | 0.34 | 0.08 |
| Kz14 | 18.63 | N/D | 0.22 | 0.10 |
| Kz15 | 27.40 | 1.50 | 0.14 | N/D |
| Kz16 | 31.25 | 1.55 | 0.09 | N/D |
| Kz17 | N/D | 1.12 | N/D | 0.06 |
| Kz18 | N/D | N/D | 0.08 | N/D |
| Kz19 | 28.98 | 0.96 | 0.19 | 0.09 |
| Kz20 | N/D | 0.90 | 0.39 | N/D |

N/D: Not detected.

**Table 5.** The mean ± standard deviation concentration of heavy metals in mg/L.

| Sample | Fe | Pb | Zn | Cu |
|---|---|---|---|---|
| Zb | 3.54±2.19 | 0.84±0.29 | 0.08±0.06 | 0.22±0.16 |
| Kz | 23.34±3.3 | 2.54±6.29 | 0. 20±0.11 | 0.08±0.03 |
| WHO limits | 0.100 | 0.01 | 0.01–0.075 | 2.00 |
| RDA | 0.01-0.06 | 3mg/week | 0.10 | 7.45 |

WHO limit: World Health Organization Limit; RDA: Recommended Daily Allowance.

no case of lead deficiency has ever been noted in the medical literature; for lead therefore any exposure is of potential concern. The metallic impurities detected can be traced to the water used in the production, equipment, ingredients added, containers, packaging materials and environmental pollutants. Zn and Cu are within acceptable limit (WHO, 1996). The study recommends that the health authorities should think how to control the quality of these local drinks; specified amount of each ingredient should be used in the production. This might reduce the concentration of Pb in both Zobo and Kunnu Zaki. For now, the producer should reduce the ingredients and used purified water for the production of the drinks.

## Conclusion

Since one means of exposure route of human to heavy metals is through ingestion of contaminated foods, drinks and beverages, efforts should be focused on the estimation of dietary intakes of potential toxic agents by consumers.

## Conflicts of interest

No competing interests exist.

### REFERENCES

Abou-Arab AAK (2001). Heavy Metal Content in Egyptian Meat and the Role of detergent washing on their levels. In Food and Chemical Toxicology. (39)593-599.

Bernd M, Franz MJ (1990). Hibiscus Flowers a Mucilage Drug? J. Agric Food Chem. 1,116D.

Galley ED, Blomfield J, Dixon SR (1972). Acute Zinc Toxicity in Hemodialysis. Br. Med. J. 4(5836): 331-333.

Graham C, John S (1978). Properties of Metals. In: Chemistry in Context. Thompson Nelson and Sons Ltd. Lincoln Way Windmill

Road, Sunburg Middlesex T.W. 16 7HP. Singapore p.157.

Kolawole JA, Maduenyi A (2004). Effect of Zobo Drink (*Hibiscus sabdariffa* water extract) on the Pharmacokinetics of acetaminophen in Human Volunteers. Eur. J. Drug Metab. Pharm. 29:25-29.

Layrangues GP, Rose C, Sphahr, L, Zayed, J, Normandin L, Butterworth RF (1998). Metabolism 13:311-318.

Liu ZP (2003). Lead poisoning combined with cadmium in sheep and horses in vicinity of ferrous metal smelters. Sci. Total Environ. (309):117-126.

Masazum H, Smith AU (1975). Minamata disease. Medical Report, Canada pp. 180-192.

Nriagn JO (1988). A Silent Epidemic of Environment Poisoning Environ. Pollut. 50:139-161.

Péter P, Ference P, Akos B, Laszlo B (2012). Accumulation of Some Heavy metals (Pb, Cd and Cr) in Milk of grazing sheep in North-East Hungary. J. Microbiol. Biotechnol. Food Sci. 2(1)389-394.

Tasi J (2005). Heavy metal, Macro and micro element content of grass species and dicotyledons. Acta Agron. Hung. (53)349-352.

United States Pharmacopoeia (1980). Identification Test for Metals The USPXX. Convention IPC 12601 905-907.

World Health Organization (WHO) (1996). Guidelines for Drinking Water Quality. 2nd Edition Geneva. (2)152-279.

Yakasai IA, Salawu F, Musa H (2004). The Concentrations of Lead in Ahmadu Bello University Dam, Raw, Treated (Tap) and ABUCONS Pure Waters. Chem. Class. J. 14:86-90.

Zukowska J. Biziuk M (2005). Methodological Evaluation of method for dietary heavy metal intake. J. Food Sci. 73(2):R1-R9.

# 27

# Antibiogram of food-borne pathogens isolated from ready-to-eat foods and Zobo Drinks Sold Within and Around PRESCO Campus of Ebonyi State University (EBSU), Abakaliki, Ebonyi State, Nigeria

Iroha Ifeanyichukwu[1], Afiukwa Ngozi[1], Nwakaeze Emmanuel[1], Ejikeugwu Chika[2], Oji Anthonia[1] and ILang Donathus[3]

[1]Department of Applied Microbiology, Ebonyi State University, P.M.B 053, Abakaliki, Nigeria.
[2]Department of Pharmaceutical Microbiology and Biotechnology, Nnamdi Azikiwe University, P.M.B 5025, Awka, Nigeria.
[3]Department of Microbiology, Federal University Ndufu-Alike, Ikwo, Nigeria.

**Food poisoning (food-borne disease) is an infection that occurs after consuming food contaminated by sufficient numbers of viable pathogens and their toxins. It is a common and costly preventable infection that is of public health concern, and which is treated with available antibiotics. Jellof-rice, abacha, moi-moi and zobo drinks are some ready-to-eat foods sold within the PRESCO campus of Ebonyi State University (EBSU), Abakaliki, Nigeria. These foods are commonly patronized by students and other unsuspecting visitors in this region, and they have been implicated in a handful of bacterial related infections in recent times. Random samples of the food items were collected from shops selling them, and these were analyzed microbiologically to determine the most prevalent organisms. Suspect isolates were identified and tested for antibiotic susceptibility profiles. *Escherichia coli, Klebsiella pneumoniae* and *Pseudomonas aeruginosa* were the commonest microbes isolated, and these showed varying rates of resistance and susceptibility to the tested drugs. Clindamycin, ampicillin and ofloxacin were less effective against the test organisms while gentamicin, erythromycin and ciprofloxacin showed substantial activity. The findings in this study showed that some ready-to-eat foods and zobo drinks sold within PRESCO campus of EBSU, Abakaliki, Nigeria were considerably contaminated with resistant pathogenic bacteria, hence, the need for constant monitoring of ready-to-eat foods in order to prevent the outbreak of food-borne illnesses in this region.**

**Key words:** Zobo drinks, ready-to-eat foods, bacteria, antibiotic resistance.

## INTRODUCTION

According to the New South Wales Food Authority (NSWFA), ready-to-eat foods are foods that are originally consumed in the same state as that in which it is sold and does not include nuts in the shell and whole, raw fruits and vegetables that are intended for hulling, peeling or these foods are usually hazardous in that they support the growth of pathogens when not properly handled, prepared or stored; and they can serve as route for the onward transmission of food-borne pathogens in human population. Therefore, in as much as food supports life, it washing by the consumer (NSWFA, 2009). Some of has been described as a vehicle for the transmission of microbial diseases, and among which are those caused by *E. coli* and other medically important bacteria (Ifediora et al., 2006; Kornacki et al., 2004; Muinde et al., 2005). According to the US Department of Health and Human Services (USDHHS) website, food-borne illnesses (which are commonly referred to as food poisoning) are diseases that results from eating contaminated food (USDHHS, 2013). Food poisoning can ensue after eating food contaminated by considerable number of viable pathogens, and this commonly occurs after eating at picnics, restaurants or fast food joint. Poor handling of these foods play critical role in the onward transmission of food-borne pathogens including *Escherichia coli* and *Klebsiella pneumoniae* to unsuspecting patronisers who eat them. Bacterial pathogens have been implicated in a handful of food-borne diseases in recent times, and these microbes are resistant to some available antimicrobial agents (Kornacki et al., 2001; Ifediora et al., 2006; Elkholy et al., 2003). In addition, infections can also occur from toxin production by the organisms. Zobo drink is sourced from the water extract of dried calyx of *Hibiscus sabdariffa* plant (Haji-Faraji et al., 1999). It is an indigenous drink consumed in Africa, Asia and some parts of South America owing to its perceived medicinal benefits which include antioxidant effect, anti-diabetic effect, and anti-hypertensive effects (Kolawole et al., 2004; Lin et al., 2011; Fullerton et al., 2011). Some of the organisms that contaminate these foods are regarded as indicator organisms (for example, *E. coli*) due to their feacal origin (Kornacki et al., 2001). The growing resistance of pathogens (including *E. coli* and *Klebsiella* species) isolated from locally prepared ready-to-eat foods is a public health concern in both developed and developing countries (Elkholy et al., 2003). Bacterial-related resistant infections have posed a serious problem in the treatment of infectious diseases in health care delivery systems due to limited therapeutic options.

In view of this, this work is aimed at detecting the presence of some enteric pathogens from some ready-to-eat foods and zobo drinks sold within the PRESCO campus of Ebonyi State University, Abakaliki, Nigeria.

### MATERIALS AND METHODS

#### Sample collection

Different types of ready to eat foods (how many samples collected) including rice (n = 50), abacha (n = 50), moi-moi (n = 50) and zobo drinks (n = 50) were aseptically and randomly collected from 20 food vendors within and around the PRESCO campus of Ebonyi State University (EBSU), Abakaliki, Nigeria. The samples were transported to the Microbiology Laboratory of EBSU, Abakaliki in transport media where they were analyzed following standard microbiology techniques.

#### Analysis of samples

Each food sample was macerated using a sterile marble mortar. One gram (1 g) of each food sample was homogenized in sterile

water and the volume of the homogenate was made up to 10 ml to obtain a 1:10 suspension. 0.1 ml of the suspension was inoculated on Trypton Soy broth and incubated at 37°C for 18 to 24 h. A loopful of the culture was then transferred to MacConkey agar plates and incubated for 18 to 48 h at 37°C. Suspect colonies of *E. coli* and *Klebsiella* species were transferred to eosin-methylene blue (EMB) agar for proper differentiation. All growth media were procured from Oxoid (Oxoid, UK). Also, 10 fold serial dilutions of zobo drink samples were performed using each sample of zobo drink, and these were inoculated into nutrient broth. They were incubated for 18 to 24 h at 37°C. Loopful of the culture were transferred to MacConkey agar plates and EMB agar plates and were also incubated for 18 to 24 h at 37°C. Suspected colonies of *E. coli* and *Klebsiella* species were transferred to nutrient agar slants from which they were subjected to Gram staining, Indole test, Methyl red test, Voges proskauer and Citrate tests for proper identification (Cheesbrough, 2000).

#### Antibiogram

Antimicrobial susceptibility test was performed on Mueller-Hinton (MH) agar (Oxoid, UK) plates by the Kirby-Bauer disk diffusion method as per the Clinical Laboratory Standard Institute (CLSI) criteria (CLSI, 2010). The tested antibiotics included erythromycin (15 µg), ciprofloxacin (5 µg), ofloxacin (10 µg), gentamicin (10 µg), clindamycin (10 µg) and ampicillin (10 µg). All antibiotic disks were procured from Oxoid, UK. Standard strains of *E. coli* ATCC 25922, *Pseudomonas aeruginosa* ATCC 27853, *K. pneumoniae* ATCC 700603 were used as controls. Plates were incubated at 37°C, and zones of inhibition were measured using meter rule as per the CLSI criteria.

### RESULT

*E. coli* were the most prevalent organism isolated from the ready-to-eat foods sold around the PRESCO campus of EBSU, Nigeria. A total of 82 *E. coli* isolates was isolated from the ready-to-eat foods and zobo drinks included in this study. On the other hand, 31 *K. pneumoniae* and 20 *P. aeruginosa* were also isolated from these food samples (Table 1). The antimicrobial susceptibility profile of the isolated bacteria from the ready-to-eat foods is shown in Table 2. Result of the antimicrobial sensitivity pattern of the isolates to different antibiotics showed that percentage susceptibility of *E. coli; K. pneumoniae* and *P. aeruginosa* isolates to the tested antibiotics were 83.33, 50 and 66.67%.

### DISCUSION

A handful of Nigerian students and other individuals depend mainly on food vendors, fast food centres and nearby restaurants that sell a variety of ready-to-eat foods (including jellof-rice, abacha, moi-moi, and zobo drinks) for their daily meal. A number of reasons abound for this rising development, but most importantly they patronize these fast food centres for want of time or sheer laziness in taking time out to cook the food themselves. As a result, they are at high risk of exposure to food-borne diseases due to poor handling and poor

Table 1. Distribution of isolated bacterial pathogens in the ready-to-eat food samples and zobo drinks.

| Food samples | *Escherichia coli* (n) | *Klebsiella pneumoniae* (n) | *Pseudomonas aeruginosa* (n) |
|---|---|---|---|
| Rice (n=50) | 20 | 9 | 4 |
| Abacha (n=50) | 15 | 10 | 6 |
| Moi-moi (n=50) | 25 | 5 | 3 |
| Zobo drink (n=50) | 22 | 7 | 7 |
| Total | 82 | 31 | 20 |

Table 2. Antibiotic susceptibility pattern of bacterial isolates from Zobo drink.

| Isolates | Zones of inhibition (mm) | | | | | | |
|---|---|---|---|---|---|---|---|
| | E | CIP | OFX | CN | AMP | DA | % Susceptibility |
| *Pseudomonas aeruginosa* | 20 | 23 | 18 | 5 | 9 | 16 | 4 (66.67) |
| *Escherichia coli* | 19 | 33 | 12 | 27 | 21 | 20 | 5 (83.33) |
| *Klebsiella pneumoniae* | 19 | 26 | 16 | 21 | 17 | 17 | 3 (50) |

AMP = ampicillin, CIP = ciprofloxacin, OFX = ofloxacin, DA = clindamycin, E = erythromycin, CN = gentamicin.

preparation of these foods, a practice that allows pathogenic microorganisms to thrive in them and cause infection upon consumption. *E. coli*, *K. pneumoniae* and *P. aeruginosa* were the organisms isolated from ready-to-eat foods including zobo drinks sold around the PRESCO campus of EBSU, Abakaliki, Nigeria, but *E. coli* (a uropathogen that indicates feacal contamination) was the most prevalent bacteria isolated (Table 1). This was followed by *K. pneumoniae* and *P. aeruginosa*.

Studies both within and outside Nigeria have shown that *E. coli* and other enteric pathogens including *K. pneumoniae* and the non-enteric organism *P. aeruginosa* are responsible for many of the global cases of food poisoning (Ifediora et al., 2006; Muinde et al., 2005; Marwa et al., 2012). This is not far from the truth owing to the variety of bacteria isolated from the food samples in this study (Table 2). Lack of access to portable water and poor handling of foods in this area may have contributed to the worrisome frequency of pathogenic microbes in ready-to-eat foods and zobo drinks at the PRESCO campus of EBSU, Abakaliki, Nigeria. The antimicrobial susceptibility studies of the recovered bacterial isolates from ready-to-eat foods in this work to some selected antibiotics showed that the *E. coli* isolates were completely resistant to ampicillin, ofloxacin and clindamycin (Table 2). However, the isolate was susceptible to ciprofloxacin, erythromycin and gentamicin.

According to a recent report, multidrug resistance in *E. coli* strains from food origin was significantly higher than those from clinical origin, and this has been associated to the feacal source of the pathogen (Ochman et al., 2000). Feacal contamination of food portends danger to the health of those consuming them, owing to the notoriety of *E. coli* in multidrug resistant diseases. The percentage

susceptibility of the bacterial isolates from the ready-to-eat foods and zobo drink revealed percentage susceptibilities of 66.7% (*P. aeruginosa*), 83.33% (*E. coli*) and 50% (*K. pneumoniae*) to the tested antibiotics. Frequency of *K. pneumoniae*, *E. coli*, and *P. aeruginosa* has also been reported from fermented zobo drinks in southwest Nigeria, and these are responsible for some of the food-borne illnesses in that region (Ojoko et al., 2002).

The presence of bacterial pathogens in the marketed zobo drinks is probably related to the source or quality of water used for their processing. The hawking of ready-to-eat foods and zobo drinks also predisposes them to dust particles which may harbour pathogens that lead to food poisoning upon consumption. The frequency of bacterial pathogens from food samples and their resistance to some available antibiotics as obtained in this study (Tables 2) is worrisome due to the notorious nature of the isolated pathogens (*E. coli*, *K. pneumoniae* and *P. aeruginosa*) in terms of drug resistance. Routine microbiological analysis of ready-to-eat foods including zobo drink around this region and other parts of Nigeria is paramount to curtail any disease outbreak in form of food poisoning due to these pathogens. Such practice if dutifully followed will ensure that quality foods are sold to unsuspecting customers, and the emergence and spread of resistant microbes through them will also be contained Finally, this study revealed the presence of enteric and non-enteric organisms including *E. coli*, *K. pneumoniae* and *P. Aeruginosa* in the ready-to-eat food and zobo drink sold around the PRESCO campus of EBSU, Abakaliki, Nigeria, and the microbes are resistant to some available drugs. Regular monitoring of the quality of foods and drinks sold to students and other unsuspecting members of the public in this region is required to forestall any imminent health danger. Food handlers should also be educated and be observant of current public health guidelines in their profession so as to minimize food-borne related illnesses.

## REFERENCES

Cheesbrough M (2000). District Laboratory Practice in Tropical Countries. Part II. 2nd edition. Cambridge University Press, UK. pp. 178-187.

Clinical Laboratory Standard Institute, CLSI (2010). Performance Standards for Antimicrobial Susceptibility Testing. CLSI Approved Standards CLSI M100-S20, Wayne, PA. USA.

Elkholy A, Baseem H, Hall C, Longworth DL (2003). Antimicrobial resistance in Cairo Egypt 1999-2000: A survey of five hospitals. J Antimicrob. Chemother. 51:625-630.

Fullerton M, Khatiwada J, Johnson JU, Davis S, Williams LL (2011). Determination of Antimicrobial Activity of Sorrel (*Hibiscus sabdariffa*) on *Escherichia coli* O157:H7 Isolated from Food, Veterinary, and Clinical Samples. J. Med. Food 14(9):950-956.

Haji-Faraji M, Haji-Tarkhani A (1999). The effect of sour tea (*Hibiscus sabdariffa*) on the essential hypertension. J. Ethno Pharmacol. 65:231-236.

Ifediora AC, Nkere CK, Iroegbu CU (2006). Weaning food preparations consumed in Umuahia, Nigeria: Evaluation of the bacteriological quality. J. Food Technol. 4:101-105.

Kornnacki J. Johnson L (2001). *Enterobacteriaceae*, Coliforms, and *Escherichia coli* as quality and safety indicators. *In*: Downees FP (eds), Compendium of methods for the microbiological examination of foods, 4th ed. American Public Health Association Washington D.C pp 69-82.

Lin HH, Chen JH, Wang CJ (2011). Chemopreventive properties and molecular mechanisms of the bioactive compounds in *Hibiscus sabdariffa* Linne. Curr. Med. Chem. 18(8):1245-54.

Marwa EA Aly, Tamer M, Essam, Magdy A, Amin (2012). Antibiotic resistance profile of *E. coli* strains isolated from clinical specimens and food samples in Egypt. Int. J. Microbiol. Res. 3(3):176-182.

Muinde OK, Kuria E (2005). Hygienic and sanitary practices of vendors of street foods in Nairobi, Kenya. J. Food Agric. Nutri. Dev. 5:1-15.

New South Wales Food Authority, NSWFA (2009). Microbiological quality guide for ready-to-eat foods: A guide to interpreting microbiological results. Available online at http://www.foodauthority.nsw.gov.au Accessed on 5th September, 2013-09-14

Ochman H, Lawrence JG, Groisman EA (2000). Lateral gene transfer and the nature of bacterial innovation. *Nature*, 405:299-304.

Ojoko AO, Adetuyi FC, Akinyosoye FA, Oyetayo VO (2002). Fermentation studies on rosellle (*Hibiscus sabdariffa*) calyces neutralized with trona. J.Food Technol. Afr. 7(3):75-78.

U.S. Department of Health and human Services, USDHHS (2013). Food Poisoning. Available online at: www.foodsafety.gov/foodpoisoning Accessed on 13th September, 2013.

World Health Organization WHO (2003). Participants Manual. Module A: Decentralization Policies and Practices: Case Study Ghana. Geneva: World Health Organization P 10.

# Permissions

# List of Contributors

**Trupti D. Chaudhari**
Research Scholar, Environmental Biotechnology Laboratory, Department of Life Sciences, University of Mumbai, Santacruz (E), Mumbai-400 098, India

**Susan Eapen**
Nuclear Agriculture Biotechnology Division, Bhabha Atomic Research centre, Trombay-400 085 Mumbai, India

**M. H. Fulekar**
Department of Life Sciences, University of Mumbai, Santacruz (E), Mumbai-400 098, India

**Edith .M. Williams**
Institute for Partnerships to Eliminate Health Disparities, University of South Carolina, 220 Stoneridge Drive, Suite 208 Columbia, SC 29210, Columbia

**Robert Watkins**
Department of Family Medicine, State University of New York at Buffalo 173 CC, ECMC, 462 Grider Street, Buffalo, NY 14215, USA

**Judith Anderso**
Jericho Road Ministries, 318 Breckenridge Buffalo, NY 14213, USA

**Laurene Tumiel-Berhalter**
Department of Family Medicine, State University of New York at Buffalo 173 CC, ECMC, 462 Grider Street, Buffalo, NY 14215, USA

**A. O. Ajayi**
Department of Microbiology, Adekunle Ajasin University, P.M.B. 01, Akungba-Akoko, Ondo state, Nigeria

**T. O. Adejumo**
Department of Microbiology, Adekunle Ajasin University, P.M.B. 01, Akungba-Akoko, Ondo state, Nigeria

**Avnish K. Verma**
Limnology Research Unit, Aquatic Biology Laboratory, SOS in Zoology, Jiwaji University, Gwalior-474011 (M. P.), Madhya Pradesh, India

**D. N. Saksena**
Limnology Research Unit, Aquatic Biology Laboratory, SOS in Zoology, Jiwaji University, Gwalior-474011 (M. P.), Madhya Pradesh, India

**Albert SALAKO**
Obafemi Awolowo College of Health Sciences, Olabisi Onabanjo University, Sagamu, Nigeria

**Oluwafolahan SHOLEYE**
Department of Community Medicine and Primary Care, Olabisi Onabanjo University Teaching Hospital, Sagamu, Nigeria

**Sunkanmi AYANKOYA**
Obafemi Awolowo College of Health Sciences, Olabisi Onabanjo University, Sagamu, Nigeria

**Patricia VÁZQUEZ-ALVARADO**
Dentistry Academic Area, Science Health Institute, Autonomous University of the State of Hidalgo, Mexico

**Arcelia MELÉNDEZ-OCAMPO**
Faculty of Dentistry, National Autonomus University of México, Mexico

**Rosa María ORTIZESPINOSA**
Medicine Academic Area, Science Health Institute, Autonomous University of the State of Hidalgo, Mexico

**Sergio MUÑOZ-JUÁREZ**
Medicine Academic Area, Science Health Institute, Autonomous University of the State of Hidalgo, Mexico

**Alejandra HERNANDEZ-CERUELOS**
Medicine Academic Area, Science Health Institute, Autonomous University of the State of Hidalgo, Mexico

**Pierre MANDA**
Laboratoire de Toxicologie et Hygiène Agro-industrielle, UFR des Sciences Pharmaceutiques et Biologiques, Université de Cocody, BPV 34 Abidjan, Côte d'Ivoire

**Djédjé Sébastien DANO**
Laboratoire de Toxicologie et Hygiène Agro-industrielle, UFR des Sciences Pharmaceutiques et Biologiques, Université de Cocody, BPV 34 Abidjan, Côte d'Ivoire

**Ehouan Stephane-Joel EHILE**
Laboratoire de Toxicologie et Hygiène Agro-industrielle, UFR des Sciences Pharmaceutiques et Biologiques, Université de Cocody, BPV 34 Abidjan, Côte d'Ivoire UFR des Sciences et Techniques des aliments, Université d'Abobo Adjamé, Abidjan Côte d'Ivoire

**Mathias KOFFI**
Laboratoire Central pour l'Hygiène Alimentaire et l'agro-industrie, Laboratoire National pour le Développement Agricole, Abidjan Côte d'Ivoire

**Ngeussan AMANI**
UFR des Sciences et Techniques des aliments, Université d'Abobo Adjamé, Abidjan Côte d'Ivoire

**Yolande Aké ASSI**
Laboratoire Central pour l'Hygiène Alimentaire et l'agro-industrie, Laboratoire National pour le Développement Agricole, Abidjan Côte d'Ivoire

**Khusnul Yaqin**
Department of Fisheries, Faculty of Marine science and Fisheries, Hasanuddin University, Jalan Perintis Kemerdekaan Km 10, Makassar 90245, Indonesia

**Bibiana Widiati Lay**
Environmental Science Study Programme, Bogor Agricultural University, Darmaga Campus, Bogor 16680, Indonesia

**Etty Riani**
Environmental Science Study Programme, Bogor Agricultural University, Darmaga Campus, Bogor 16680, Indonesia

**Zainal Alim Masud**
Environmental Science Study Programme, Bogor Agricultural University, Darmaga Campus, Bogor 16680, Indonesia

**Peter-Diedrich Hansen**
Department of Ecotoxicology, Technische Universitaet, Faculty VI, Franklin Strasse 29 (OE4), D-10587 Berlin, Germany

**Oriakpono Obemeata**
Department of Animal and Environmental Biology, Faculty of Science, University of Port Harcourt, P. M. B. 5323 Rivers State, Nigeria

**Hart Aduabobo**
Department of Animal and Environmental Biology, Faculty of Science, University of Port Harcourt, P. M. B. 5323 Rivers State, Nigeria

**Ekanem Wokoma**
Department of Crop and Soil Science, Faculty of Agriculture, University of Port Harcourt, P. M. B. 5323 Rivers State, Nigeria

**M. A. Adeleke**
Public Health Entomology and Parasitology Unit, Department of Biological Sciences, Osun State University, P. M. B 4429, Osogbo, Nigeria

**A. O. Hassan**
Microbiology Unit, Ladoke Akintola University Teaching Hospital, Osogbo, Nigeria

**T. T. Ayepola**
Public Health Entomology and Parasitology Unit, Department of Biological Sciences, Osun State University, P. M. B 4429, Osogbo, Nigeria

**T. M. Famodimu**
Public Health Entomology and Parasitology Unit, Department of Biological Sciences, Osun State University, P. M. B 4429, Osogbo, Nigeria

**W. O. Adebimpe**
Department of Community Medicine, College of Health Sciences, Osun State University, Osogbo, Nigeria

**G. O. Olatunde**
Public Health Entomology and Parasitology Unit, Department of Biological Sciences, Osun State University, P. M. B 4429, Osogbo, Nigeria

**Shakeel Ahmed Ibne Mahmood**
Bangladesh Arsenic Control Society, BACS, C/o, 2/24 Babar Road, Mohammadpur, Dhaka 1207, Bangladesh

**Amal Krishna Halder**
Bangladesh Arsenic Control Society, BACS, C/o, 2/24 Babar Road, Mohammadpur, Dhaka 1207, Bangladesh

**Peter Musagala**
Department of Chemistry, Busitema University, P. O. Box 236, Tororo, Uganda

**Henry Ssekaalo**
Department of Chemistry, Makerere University, P. O. Box 7062, Kampala, Uganda

**Jolocam Mbabazi**
Department of Chemistry, Makerere University, P. O. Box 7062, Kampala, Uganda

**Muhammad Ntale**
Department of Chemistry, Makerere University, P. O. Box 7062, Kampala, Uganda

**Esmail Gharedaashi**
Department of Fishery, Gorgan University of Agricultural Sciences and Natural Resources, Gorgan, Iran

**Mohammad Reza Imanpour**
Department of Fishery, Gorgan University of Agricultural Sciences and Natural Resources, Gorgan, Iran

**Vahid Taghizadeh**
Department of Fishery, Gorgan University of Agricultural Sciences and Natural Resources, Gorgan, Iran

**Tobias I. Ndubuisi Ezejiofor**
Department of Biotechnology, Federal University of Technology, Owerri, Imo State Nigeria

**A. N. Ezejiofor**
Toxicology Unit, Department of Clinical Pharmacy University of Port Harcourt, Rivers State, Nigeria

**A. C. Udebuani**
Department of Biotechnology, Federal University of Technology, Owerri, Imo State Nigeria

**E. U. Ezeji**
Department of Biotechnology, Federal University of Technology, Owerri, Imo State Nigeria

**E. A. Ayalogbu**
Department of Biotechnology, Federal University of Technology, Owerri, Imo State Nigeria

**C. O. Azuwuike**
Department of Biotechnology, Federal University of Technology, Owerri, Imo State Nigeria

**L. A. Adjero**
Toxicology Unit, Department of Clinical Pharmacy University of Port Harcourt, Rivers State, Nigeria

**C. E. Ihejirika**
Department of Environmental Technology, Federal University of Technology, Owerri, Imo State, Nigeria

**C. O. Ujowundu**
Department of Industrial Biochemistry, Federal University of Technology, Owerri, Imo State, Nigeria

**L. A. Nwaogu**
Department of Industrial Biochemistry, Federal University of Technology, Owerri, Imo State, Nigeria

**K. O. Ngwogu**
Department of Chemical Pathology, College of Medicine and Health Sciences, Abia State University Teaching Hospital, Aba, Nigeria

**Doris A. Fiasorgbor**
Faculty of Development Studies, Presbyterian University College, Ghana, Akuapem Campus, Ghana

**O. M. Agbolade**
Department of Plant Science and Applied Zoology, Parasitology and Medical entomology laboratory, Olabisi Onabanjo University, P. M. B. 2002, Ago-Iwoye, Ogun State, Nigeria

**O. O. Adesanya**
Department of Microbiology, Olabisi Onabanjo University, P. M. B. 2002, Ago-Iwoye, Ogun State, Nigeria

**T. O. Olayiwola**
Department of Chemical Sciences, Olabisi Onabanjo University, P. M. B. 2002, Ago-Iwoye, Ogun State, Nigeria

**G. C. Agu**
Department of Microbiology, Olabisi Onabanjo University, P. M. B. 2002, Ago-Iwoye, Ogun State, Nigeria

**E. E. Ezenwaji**
Department of Geography and Meteorology, Nnamdi Azikiwe University, Awka, Nigeria

**A. C. Okoye**
Department of Environmental Management, Nnamdi Azikiwe University, Awka, Nigeria

**V. I. Otti**
Department of Civil Engineering, Federal Polytechnic, Oko, Nigeria

**Aras Mohammed Khudhur**
Soil and Water Department, College of Agriculture, University of Salahaddin, Erbil, Iraq
School of Biomedical and Biological Sciences, University of Plymouth, Plymouth, PL4 8AA, United Kingdom

**Kasim Abass Askar**
School of Biomedical and Biological Sciences, University of Plymouth, Plymouth, PL4 8AA, United Kingdom

**Khallef Messaouda**
Department of Biology, Research laboratory, Biology, Water and Ecology, University of 8 Mai 1945, Guelma, Algeria

**Merabet Rym**
Department of Biology, Research laboratory, Biology, Water and Ecology, University of 8 Mai 1945, Guelma, Algeria

**Benouareth Djamel Eddine**
Department of Biology, Research laboratory, Biology, Water and Ecology, University of 8 Mai 1945, Guelma, Algeria

**Adamu Uzairu Sani Uba**
Department of Chemistry, Ahmadu Bello University, Zaria, Nigeria

**Muhammad Sani Sallau**
Department of Chemistry, Ahmadu Bello University, Zaria, Nigeria

**Hamza Abba**
Department of Chemistry, Ahmadu Bello University, Zaria, Nigeria

**Okunola Oluwole**
Department of Chemistry, Ahmadu Bello University, Zaria, Nigeria

**Joshua**
National Research Institute for Chemical Technology, Zaria, Nigeria

**Ayaz A. Naik**
Biological Oceanography Division, National Institute of Oceanography, Dona Paula, Goa-403004, India

**A. Wanganeo**
Department of Environmental Science and Limnology B.U Bhopal-462026, M.P India

**Jia-Ren Liu**
Public Health College, Harbin Medical University, 157 BaoJian Road, NanGang District, Harbin, P. R. China 150086
Harvard Medical School, 300 Longwood Avenue, Boston, MA, 02115-5737, USA

**Hong-Wei Dong**
Public Health College, Harbin Medical University, 157 BaoJian Road, NanGang District, Harbin, P. R. China 150086

**Xuan-Le Tang**
Public Health College, Harbin Medical University, 157 BaoJian Road, NanGang District, Harbin, P. R. China 150086

**Jia Yu**
Public Health College, Harbin Medical University, 157 BaoJian Road, NanGang District, Harbin, P. R. China 150086

**Xiao-Hui Han**
Public Health College, Harbin Medical University, 157 BaoJian Road, NanGang District, Harbin, P. R. China 150086

**Bing-Qing Chen**
Public Health College, Harbin Medical University, 157 BaoJian Road, NanGang District, Harbin, P. R. China 150086

**Chang-Hao Sun**
Public Health College, Harbin Medical University, 157 BaoJian Road, NanGang District, Harbin, P. R. China 150086

**Bao-Feng Yang**
Department of Pharmacology, Harbin Medical University, Harbin, Heilongjiang, P. R. China 150081

**Quanyong Xiang**
Jiangsu Province Center for Disease Control and Prevention, 172 Jiangsu Road, Nanjing 21009, P. R. China

**Liansheng Chen**
Jiangsu Province Center for Disease Control and Prevention, 172 Jiangsu Road, Nanjing 21009, P. R. China

**Youxin Liang**
School of Public Health, Fudan University (formerly Shanghai Medical University), 138 Yixueyuan Rd., Shanghai (200032), China

**Ming Wu**
Jiangsu Province Center for Disease Control and Prevention, 172 Jiangsu Road, Nanjing 21009, P. R. China

**Bingheng Chen**
School of Public Health, Fudan University (formerly Shanghai Medical University), 138 Yixueyuan Rd., Shanghai (200032), China

**Asia Imohimi Ohioma**
Department of Chemistry, Ambrose Alli University, Ekpoma Nigeria

**Ndubuisi Obejesi Luke**
Department of Civil Engineering, Ambrose Alli University, Ekpoma Nigeria

**Odia Amraibure**
Department of Chemistry, Ambrose Alli University, Ekpoma Nigeria

**S. A. Osemeahon**
Department of chemistry, Modibbo Adama University of Technology, Yola, Nigeria

**O. N. Maitera**
Department of chemistry, Modibbo Adama University of Technology, Yola, Nigeria

**A. J. Hotton**
National Agency for Food and Drug Administration and Control (NAFDAC) Jalingo, Nigeria

**B. J. Dimas**
Department of chemistry, Modibbo Adama University of Technology, Yola, Nigeria

**M. T. Bakare-Odunola**
Department of Pharmaceutical and Medicinal Chemistry, Faculty of Pharmaceutical Sciences, University of Ilorin, Ilorin, Kwara State, Nigeria

**K. B. Mustapha**
Department of Medicinal Chemistry and Quality Control, National Institute for Pharmaceutical Research and Development, Abuja, Nigeria

**Iroha Ifeanyichukwu**
Department of Applied Microbiology, Ebonyi State University, P.M.B 053, Abakaliki, Nigeria

**Afiukwa Ngozi**
Department of Applied Microbiology, Ebonyi State University, P.M.B 053, Abakaliki, Nigeria

**Nwakaeze Emmanuel**
Department of Applied Microbiology, Ebonyi State University, P.M.B 053, Abakaliki, Nigeria

**Ejikeugwu Chika**
Department of Pharmaceutical Microbiology and Biotechnology, Nnamdi Azikiwe University, P.M.B 5025, Awka, Nigeria

**Oji Anthonia**
Department of Applied Microbiology, Ebonyi State University, P.M.B 053, Abakaliki, Nigeria

**ILang Donathus**
Department of Microbiology, Federal University Ndufu-Alike, Ikwo, Nigeria